Individuals Across the Sciences

Individuals Across the Sciences

Edited by ALEXANDRE GUAY
AND THOMAS PRADEU

UNIVERSITY PRESS

Oxford University Press is a department of the University of
Oxford. It furthers the University's objective of excellence in research,
scholarship, and education by publishing worldwide.

Oxford   New York
Auckland   Cape Town   Dar es Salaam   Hong Kong   Karachi
Kuala Lumpur   Madrid   Melbourne   Mexico City   Nairobi
New Delhi   Shanghai   Taipei   Toronto

With offices in
Argentina   Austria   Brazil   Chile   Czech Republic   France   Greece
Guatemala   Hungary   Italy   Japan   Poland   Portugal   Singapore
South Korea   Switzerland   Thailand   Turkey   Ukraine   Vietnam

Oxford is a registered trademark of Oxford University Press
in the UK and certain other countries.

Published in the United States of America by
Oxford University Press
198 Madison Avenue, New York, NY 10016

© Oxford University Press 2016

All rights reserved. No part of this publication may be reproduced, stored in
a retrieval system, or transmitted, in any form or by any means, without the prior
permission in writing of Oxford University Press, or as expressly permitted by law,
by license, or under terms agreed with the appropriate reproduction rights organization.
Inquiries concerning reproduction outside the scope of the above should be sent to the
Rights Department, Oxford University Press, at the address above.

You must not circulate this work in any other form
and you must impose this same condition on any acquirer.

Library of Congress Cataloging-in-Publication Data
Individuals across the sciences / [edited by] Alexandre Guay [and] Thomas Pradeu.
pages cm
Includes bibliographical references and index.
ISBN 978–0–19–938251–4 (hardcover : alk. paper)   1. Individuality.   2. Individual
differences.   I. Guay, Alexandre, joint editor.
BF697.I637 2015
111—dc23
2015003229

9 8 7 6 5 4 3 2 1
Printed in the United States of America
on acid-free paper

CONTENTS

Acknowledgments vii
List of Contributors ix

CHAPTER 1 Introduction: Progressive Steps toward a Unified Conception of Individuality Across the Sciences 1
Alexandre Guay and Thomas Pradeu

PART I | Metaphysical and Logical Foundations to Individuality

CHAPTER 2 Why Individuality Matters 25
Stéphane Chauvier

A Foreword to E. J. Lowe's CHAPTER 46
Alexander Bird

CHAPTER 3 Non-individuals 49
E. J. Lowe

CHAPTER 4 Individuality, Quantum Physics, and a Metaphysics of Nonindividuals: The Role of the Formal 61
Décio Krause and Jonas R. Becker Arenhart

PART II | Puzzles about Individuals in Biology and Physics

CHAPTER 5 Individuality and Life Cycles 85
Peter Godfrey-Smith

CHAPTER 6 What Biofilms Can Teach Us about Individuality 103
Marc Ereshefsky and Makmiller Pedroso

CHAPTER 7   Cell and Body: Individuals in Stem Cell
Biology   122
Melinda Bonnie Fagan

CHAPTER 8   Collective Individuals: Parallels Between Joint Action
and Biological Individuality   144
Cédric Paternotte

CHAPTER 9   On the Emergence of Individuals in Physics   165
Simon Saunders

CHAPTER 10   Are There Individuals in Physics, and If So, What
Are They?   193
James Ladyman

CHAPTER 11   Minimal Structural Essentialism: Why Physics
Doesn't Care Which Is Which   207
David Glick

CHAPTER 12   Bohm's Approach and Individuality   226
Paavo Pylkkänen, Basil J. Hiley, and Ilkka Pättiniemi

CHAPTER 13   Branch-Relative Identity   250
Christina Conroy

PART III | Beyond Disciplinary Borders

CHAPTER 14   The Metaphysics of Individuality and
the Sciences   273
Matteo Morganti

CHAPTER 15   The Biological and the Mereological: Metaphysical
Implications of the Individuality Thesis   295
Matthew H. Haber

CHAPTER 16   To Be Continued: The Genidentity of Physical
and Biological Processes   317
Alexandre Guay and Thomas Pradeu

CHAPTER 17   Experimental Realization of Individuality   348
Ruey-Lin Chen

CHAPTER 18   Eliminating Objects Across the Sciences   371
Steven French

Name Index   395
General Index   399

ACKNOWLEDGMENTS

We would like to thank all the contributors to this volume, who, in addition to writing their own chapter in time, commented (often very extensively) on chapters written by others. We also thank the many colleagues who, without contributing directly to this volume, reviewed one or several chapters, contributing decisively to the overall quality of this collective work: Guido Bacciagaluppi, Giovanni Boniolo, Adam Caulton, Ellen Clarke, David Crawford, John Dupré, Michael Esfeld, Adam Ferner, James Griesemer, Evelyn Houliston, Philippe Huneman, Maximilian Kistler, Lucie Laplane, Tim Lewens, Cynthia Macdonald, Maureen O'Malley, Kerry McKenzie, Jean-Pierre Marquis, Julian Reiss, Stéphanie Ruphy, Claudine Tiercelin, David Wallace.

Special thanks to Alexander Bird, who so kindly agreed to write a very nice and touching presentation of Jonathan Lowe—undoubtedly one of the most brilliant and innovative metaphysicians of our time.

Finally, our gratitude goes to Peter Ohlin and Lucy Randall, of Oxford University Press, who, as always, have been extremely efficient in dealing with our questions and doubts.

LIST OF CONTRIBUTORS

Jonas R. Becker Arenhart is Professor of Philosophy at the Federal University of Santa Catarina. His work concerns mainly logical and metaphysical aspects of quantum mechanics.

Alexander Bird is Professor of Philosophy at the University of Bristol. His work deals in particular with metaphysics and epistemology of science and medicine. He is the author of many papers and books, including *Nature's Metaphysics: Laws and Properties* (Oxford University Press, 2007), where he argued for a dispositional essentialist account of natural properties and necessitarianism about the laws of nature.

Stéphane Chauvier is Professor of Philosophy at Paris-Sorbonne University. His works in metaphysics deals mainly with the relation between language, thought, and reality. He is the author of various papers on the transition from linguistic reference to existence. His latest book in the field (*Le Sens du Possible*, Vrin, 2010) won the Montyon Prize 2011 of the French Academy.

Ruey-Lin Chen is a Philosophy of Science Professor at National Chung Cheng University, Taiwan. His current research interest is in the philosophy of science across physical, biological, and experimental cases. He is the author of four books in history and philosophy of science in Chinese. He also published some articles in English.

Christina Conroy is an Assistant Professor of Philosophy at Morehead State University in Morehead, Kentucky. She is currently working on a single-world interpretation of Everettian quantum mechanics. She is the author of "The Relative Facts Interpretation and Everett's Note Added in Proof" in *Studies in the History and Philosophy of Modern Physics* 43: 112–120, 2012.

Marc Ereshefsky is Professor of Philosophy at the University of Calgary. His research focus includes the nature of species, scientific classification, biological taxonomy, natural kinds, individuality, historicity, and homology.

Melinda Bonnie Fagan is Sterling McMurrin Associate Professor of Philosophy, Department of Philosophy, University of Utah. Her research focuses on experimental practice in biology, with particular emphasis on explanation and modeling. She is the author of *Philosophy of Stem Cell Biology* (Palgrave Macmillan, 2013) as well as over twenty-five articles and book chapters on topics in philosophy of science and biology.

Steven French is Professor of Philosophy of Science at the University of Leeds and co-editor in chief of the *British Journal for the Philosophy of Science*. He is coauthor, with Décio Krause of *Identity in Physics* (Oxford University Press, 2006) and recently published *The Structure of the World* (Oxford University Press, 2014).

David Glick is Visiting Assistant Professor of Philosophy at the University of Rochester, United States. His work focuses on structuralism and holism in metaphysics and physics. He received his PhD in 2014 from the University of Arizona. His dissertation, "Structures and Objects: A Defense of Structural Realism," was supervised by Richard Healey.

Peter Godfrey-Smith is Distinguished Professor of Philosophy at the CUNY Graduate Center in New York City. He works in the philosophy of biology and philosophy of mind, and is the author of four books, including *Theory and Reality* (2003), *Darwinian Populations and Natural Selection* (2009), and *Philosophy of Biology* (2014).

Alexandre Guay is Professor of Philosophy of Natural Sciences and Analytical Philosophy at the Université catholique de Louvain, Belgium. Most of his research focuses on ontological puzzles in physics. He is the editor in chief of *Lato Sensu: Revue de la Société de Philosophie des Sciences*.

Matthew H. Haber is an Associate Professor in the Department of Philosophy at the University of Utah, with an adjunct appointment in the Center for Quantitative Biology. His work primarily deals with topics in biological systematics, taxonomy, classification, and phylogenetics.

Basil J. Hiley is Emeritus Professor of Physics at Birkbeck, London University and Honorary Research Associate of University College, London. His research involves a mathematical study of symplectic and orthogonal Clifford algebras and their applications to foundational questions in quantum phenomena. At present, he is part of a team investigating experimentally the properties of weak measurements on atoms. He is coauthor with David Bohm of *The Undivided Universe: An Ontological Interpretation of Quantum Theory* and coeditor with David Peat of *Quantum Implications: Essays in Honour of David Bohm.*

Décio Krause is Associate Professor of the Department of Philosophy of the Federal University of Santa Catarina, Brazil. His works deal mainly with the philosophy of logic, analytic ontology, and the foundations of quantum physics. He is author, with Steven French, of *Identity in Physics: A Historical, Philosophical, and Formal Analysis* (Oxford University Press, 2006), and of several papers dealing with the question of the individuality of quantum objects.

James Ladyman is Professor of Philosophy at the University of Bristol. He is coeditor of *Studies in History and Philosophy of Modern Physics* and was previously coeditor of the *British Journal for the Philosophy of Science.* He is author with Don Ross of *Every Thing Must Go: Metaphysics Naturalised* (Oxford University Press, 2007).

E. J. Lowe was Professor of Philosophy and Director of Postgraduate Studies at the Department of Philosophy at Durham University, United Kingdom. He made critical contributions to metaphysics, philosophy of mind, philosophical logic, and the history of early modern philosophy. He died prematurely on January 5, 2014, shortly after finishing his contribution to this volume.

Matteo Morganti works at the "Roma TRE" University of Rome. He teaches and does research in the philosophy of science, and is particularly interested in the interplay between the natural sciences (mainly physics) and analytic metaphysics. He is the author of *Combining Science and Metaphysics* (Palgrave Macmillan, 2013), and of several papers in international journals.

Cédric Paternotte is an Assistant Professor of Philosophy at Paris-Sorbonne University, France. His research concerns the rational, psychological, epistemic, biological, and evolutionary aspects of cooperation. His current topics of interest include social norms, scientific groups, organisms, and biological adaptations.

Ilkka Pättiniemi studies theoretical philosophy at the University of Helsinki. He is currently writing a master's thesis on the philosophy of quantum mechanics.

Makmiller Pedroso is an Assistant Professor in the Department of Philosophy and Religious Studies at Towson University, United States. His work deals with conceptual issues in evolutionary biology and biological systematics. His articles have appeared in journals such as *Synthese,*

Philosophical Quarterly, Biology and Philosophy, and *Studies in History and Philosophy of Biology and Biomedical Sciences*.

Thomas Pradeu is a full-time permanent researcher in Philosophy of Science at the CNRS in Bordeaux, France, and an associated member of IHPST, Panthéon-Sorbonne University. From 2008 to 2014, he was an Associate Professor in the Philosophy department at Paris-Sorbonne University. His work deals mainly with the topic of biological individuality, particularly in immunology and developmental biology. He has published several papers in both science and philosophy journals, and he is the author of *The Limits of the Self: Immunology and Biological Identity* (Oxford University Press, 2012).

Paavo Pylkkänen is Senior Lecturer in theoretical philosophy at the University of Helsinki, Finland, and the University of Skövde, Sweden. His research areas are philosophy of mind, philosophy of science, and the foundations of quantum theory. In particular, he has studied the relevance of Bohm's interpretation of quantum theory to problems in scientific metaphysics. He is the author of *Mind, Matter and the Implicate Order* (Springer, 2007).

Simon Saunders is Professor of Philosophy of Physics at Oxford University, and fellow of Merton College. His work has focused on the interpretation of quantum mechanics and quantum field theory, on structuralism in philosophy of science, and most recently on identity, particle-indistinguishability, and space-time symmetries. He is coeditor of *Many Worlds? Everett, Quantum Theory, and Reality* (Oxford University Press, 2010).

Individuals Across the Sciences

CHAPTER 1 | Introduction
Progressive Steps toward a Unified Conception of Individuality Across the Sciences

ALEXANDRE GUAY AND THOMAS PRADEU

1.1 Connecting Metaphysics and the Sciences

How to define and identify individuals has been a recurring issue throughout the history of philosophy. It was, for example, pointedly studied by Aristotle in his *Metaphysics* and *Categories*, by Locke in his *Essay*, and by Leibniz in his *New Essays*. Most contemporary philosophers consider the problem of individuality from a general, metaphysical, point of view, as is the case, in particular, of Peter Strawson in his landmark book *Individuals: An Essay in Descriptive Metaphysics* (Strawson 1959) and David Wiggins in *Sameness and Substance* (1980) and subsequently in *Sameness and Substance Renewed* (2001). In sharp contrast, the preferred approach in philosophy of science has been to define, in a very focused and circumscribed way, the ontological status of certain individuals, most often within a specific scientific domain, typically physics (e.g., Saunders 2006, French and Krause 2006, Ladyman and Ross 2007, Muller and Seevinck 2009, Caulton and Butterfield 2012, Dorato and Morganti 2013, Morganti 2013, French 2014), or biology (e.g., Hull 1978, 1980, 1992, Buss 1987, Maynard-Smith and Szathmáry 1995, Michod 1999, Wilson 1999, Gould 2002, Wilson 2005, Dupré and O'Malley 2007, Godfrey-Smith 2009, Clarke 2010, Pradeu 2012, Bouchard and Huneman 2013, Wilson and Barker 2013). Today, many consider that the approach used in philosophy of science has been much more precise and globally much more fruitful than the purely metaphysical approach, often criticized for being excessively general and at odds with "real" science (this view is defended, in particular, in Redhead 1995, Maudlin 2007, Ladyman and Ross 2007, Ross, Ladyman, and Kincaid 2013; see also French 2014). It seems crucial to spell out in detail this conflict between the highly general approaches favored by metaphysicians and the much more focused approaches favored by philosophers of science, and to determine to what extent this conflict could be overcome.

This volume stemmed from two observations, which can be seen as two reactions to the conflicting landscape just described. First, the recent developments in the notion of individuals in physics or in biology, fascinating as they are, have remained in most cases "regional," in the sense that very few transversal reflections, offering a meticulous comparison between various conceptions of individuality and/or various scientific individuation criteria across different sciences, have been done (a recent and stimulating exception is French 2014). In an attempt to at least partially fill in this gap, the present volume offers a confrontation of philosophers of physics and philosophers of biology on their definitions of individuation criteria. Several chapters of the volume are intrinsically interdisciplinary, while others rather belong to either philosophy of physics or philosophy of biology but at the same time take into account lessons that can be learnt from the other field. Second, the advantage of the regionalism that characterizes scientific approaches to individuality is, naturally, that it is well focused and precise, insofar as it isolates a relatively small field of study, but the difficulty is that too often it lacks ambition compared to the aim of metaphysical approaches, namely to uncover the general conditions for individuating any object (e.g., Strawson 1959, Wiggins 2001, Macdonald 2005, Lowe 1989, 2006). To address this difficulty, the present volume attempts to articulate the perspective of general metaphysics with that of philosophy of physics and philosophy of biology on individuals. This confrontation between metaphysics and science should not be a "battlefield" (to use Kant's famous image), in which each side would misunderstand the basic methodology of the other; instead, progress will be made only thanks to the work of metaphysicians informed by today's sciences, and philosophers of science who keep in mind metaphysics' goal to develop a unifying and operative notion of the individual. The present volume, in other words, aims at clarifying and overcoming the difficulties that hold back the construction of a general conception of the individual that would be adequate for both physics and biology, and perhaps beyond.

The kind of convergence between metaphysics and science advocated here can be partly related to "metaphysics of science," a label that has recently aroused much enthusiasm in the philosophical community (e.g., Mumford and Tugby 2013). Logical empiricists famously looked at metaphysics with suspicion, and even suggested that most metaphysical statements were meaningless (because they were neither analytic nor synthetic) (see, in particular, Carnap 1935). The critique of logical empiricism in the second half of the twentieth century led, among other things, to a resurgence of metaphysics in Anglophone philosophy, and to the view that there was perhaps no strict and clear-cut boundary between metaphysics and science. The 2000s witnessed a burst in "metaphysics of science" with, notably, fascinating work about laws, causation, dispositions, natural properties, and natural kinds (see, e.g., Kistler 2006, Bird 2007, Chakravartty 2007, and, for an overview, Mumford and Tugby 2013). We believe that metaphysics of science as it has been done in the last 15 years or so is right in its most fundamental project, which is both very ambitious and legitimate: *aiming at building a general worldview on the basis of several sciences* and, if possible, a worldview that would hold for *all* the sciences (for an explicit claim that such is the fundamental goal of metaphysics of science see, again, Mumford and Tugby 2013, 6). The problem is that some claims made in metaphysics of science are, according at least to many practitioners of science and/or to philosophers who are very close to science itself, excessively general and far from "real" science; it is certainly not clear, for instance, that all sciences offer laws, pace a majority of metaphysicians of science (it is likely, in particular, that a majority of biologists and philosophers of biology would reject the idea that there are biological "laws").

In contrast with some (though certainly not all) "metaphysicians of science," when we talk about the articulation of metaphysics and science, we have in mind a metaphysics that is well informed by actual and present-day science, in a precise and therefore limited way: metaphysicians interested in building a scientific metaphysics must put a lot of effort in knowing in detail one scientific field, without presupposing the existence of structural features of an allegedly general way of doing science that would hold for all the sciences (Maudlin 2007, Ladyman and Ross 2007, Ross, Ladyman, and Kincaid 2013, French 2014). This investment into science itself is necessary for the construction of a solid scientific metaphysics, but it is by no means sufficient: indeed, the sciences do not by themselves offer a pertinent and well-articulated metaphysics, which is to say that metaphysics cannot directly be read off the sciences, as argued convincingly by Morganti (2013). Philosophers interested in science and scientists interested in reflecting on metaphysics need to patiently *build* this scientific metaphysics, using scientific theories and practices, as well as metaphysical concepts and approaches, as their main tools. French and McKenzie (2012) have suggested that metaphysics should be used as a "toolbox" for the construction of a pertinent and scientifically sound worldview; what we suggest here is that both the sciences and general metaphysics should be used as "toolboxes" for the construction of an adequate scientific metaphysics.

In accordance, therefore, with the recent "metaphysics of science" trend, we think that *generality* is a very important aim for science and metaphysics. But we are also convinced that *general claims of a science-based metaphysics have to be patiently constructed*, rather than being assumed in the first place. For instance, because so many physicists and philosophers of physics have cast doubt on the idea of causality, their arguments have to be carefully examined before making any claim about the supposedly self-evident view that all sciences need the concept of causality and, even more tentatively, a concept of causality that would be common to all scientific domains. The strategy followed in this volume is then more modest and more progressive than that of many proponents of recent "metaphysics of science": instead of offering from the start a concept of an individual that would hold for all the sciences, we try to determine *whether or not such a concept can emerge from the actual confrontation of two different scientific fields*, namely physics and biology. In our view, this modest strategy is in fact the only way to reach the aforementioned ambitious aim, that is, the construction of a general and scientifically informed worldview. What we need to do, then, is to explore the resemblances and differences between the physical and the biological senses of the term "individual." Let us now justify this choice of limiting our inquiry to physics and biology, and then explain how the concept of individual is used in physics and in biology, respectively.

1.2 The Many Faces of Individuality in the Sciences

The notion of individuality is used extensively, and differently, in many sciences. This includes, naturally, physics and biology, as this volume will show in detail. But the notion of individuality also plays a critical role, for instance, in economics (e.g., Davis 2010), sociology (e.g., Elias 1991), anthropology (e.g., Dumont 1986), and the social sciences in general (Rosenberg 1988), as well as in chemistry—where a major aim is to understand how individual entities *aggregate* (for an overview, see Weisberg, Needham, and Hendry 2011). Despite this rich diversity, we have decided, in this volume, to focus almost exclusively on the physical and the biological sciences. At least two reasons explain this choice.

First, as this volume will make clear, the articulation of the physical and the biological discourses on individuality is already extremely complex; by no means would we want to suggest by this choice that the physical and the biological approaches are the *most* complex, or the *most* interesting; we are simply convinced that, to address an extremely complex problem, one has to start somewhere, and if possible at a place where it seems clear that progress is likely to be made.

Second, the incomparably long and rich traditions in philosophy of physics and philosophy of biology have given rise to many debates about the notion of individuality in these fields, and it seemed a reasonable strategy to start with, and build on, these previous debates. We hope that the present volume will stimulate others to explore other scientific domains, and that future work will indeed address the problem of how to understand individuality in those scientific fields and, even more importantly, of how to articulate the different conceptions of individuality conveyed by these various fields. Opening up some possibilities, this volume, here and there, touches upon other scientific fields (Paternotte, for instance, uses the evolutionary notion of a collective entity to shed light on how the social sciences conceptualize what a collective being is and how joint action should be understood).

So, what is an individual, and how is this notion used in the sciences? Though there is much disagreement among metaphysicians about how to define an individual, most would probably agree that an individual is, minimally, an entity that can be singled out and counted, possesses acceptably clear-cut boundaries, and shows some identity through time ("persistence"). Every aspect of this apparently simple definition deserves to be analyzed—and perhaps put into question, as will appear clearly in the contributions of Chauvier and Lowe, in particular. In addition, one must address the question of whether or not such an understanding of what an individual is could be applicable and useful in today's sciences. In what follows, we tentatively lay the foundations of possible ways to define individuals in physics and in biology. Needless to say, the different chapters of this volume will explore in much more detail what physics and biology have to say (or don't have to say) about the notion of an individual.

1.3 Individuality in Biology

Not only have philosophers traditionally been interested in determining what a biological individual is, but many of them (e.g., Aristotle and Locke) also have used biological individuals as *paradigmatic* individuals, that is, as models to better understand what an individual in general is (see Hull 1978, 1992). One important difficulty, though, is that the individuation of biological entities is much more complex and surprising than might be expected by those having in mind familiar and relatively "big" animals such as mammals (Hull 1992), and therefore it is perhaps not entirely certain that biological entities should still be viewed as models for the understanding of individuals more generally. In the last decades, both biologists and philosophers of biology have suggested finding *biological* criteria (as opposed to common-sense criteria, which lead to what can be called *phenomenal individuation*) for the individuation of biological entities, and this has led to one of the most active debates within philosophy of biology (e.g., Hull 1978, 1980, 1992, Buss 1987, Maynard-Smith and Szathmáry 1995, Michod 1999, Wilson 1999, Gould 2002, Wilson 2005, Okasha 2006, Dupré and O'Malley 2007, 2009, Godfrey-Smith 2009, Clarke 2010, Pradeu 2012, Bouchard and Huneman 2013, Wilson and Barker 2013). In fact, many individuality criteria can be used to delineate living things, including physiological, embryological, genetic, evolutionary, immunological, and neurological criteria. For the sake of simplicity, two main sets of approaches can be distinguished among individuations based on the biological sciences:

1. *Physiological individuation*: this approach grounds biological individuality mainly in the numerous biological fields that are interested in the explanation of actual working of present-day living entities (those fields include, for instance, much of molecular and cellular biology, neurobiology, immunology, etc.). Very often (but certainly not always), the main focus in this approach is the organism, seen as a strongly cohesive and unified metabolic entity, with mutually dependent components (e.g., Perelman 2000, Pradeu 2010).
2. *Evolutionary individuation*: proponents of this approach consider, behind Hull (1978, 1992), that the theory of evolution by natural selection is our best tool to say what a biological individual is, because evolution is in the background of all biological processes, and because the theory of evolution by natural selection is arguably the most powerful and comprehensive theory of biology. (On this distinction between physiological and evolutionary individuation, see Pradeu 2012; see also Wilson and Barker 2013.)

Recent debates over individuality within the domain of philosophy of biology have been dominated by evolutionary approaches to individuality. Important contributions have concerned biological individuals as units of selection (e.g., Hull 1978, 1980, 1992, Gould 2002), as results of a "transition in individuality" (for instance from unicellularity to multicellularity: e.g., Buss 1987, Maynard-Smith and Szathmáry 1995, Michod 1999, Okasha 2006), or as "Darwinian individuals," that is, members of a "Darwinian population," itself defined as a set of entities characterized by variation, fitness differences, and heritability (Godfrey-Smith 2009, 2013). Other contributions have paid more attention to physiological individuation (e.g., Sober 1991, Mossio and Moreno 2010, Pradeu 2010, 2012; see also Dupré and O'Malley 2009, Godfrey-Smith 2013), but they have tended to be less influential than evolution-based approaches.

Overall, most philosophers of biology have considered that biological individuality

- Cannot be grounded in our everyday notion of what an individual is (e.g., Hull 1992)
- Has to be grounded in biological *theories* (an idea that is reminiscent of Quine (1948, 1960), but that can naturally be questioned, as does Chen in this volume, for instance)
- Is dependent on the question being asked, and, often, on the domain of study
- Can appear at several levels (it is a "hierarchical" or "multilevel" perspective on individuality: e.g., Hull 1980, Gould and Lloyd 1999, Gould 2002)
- Comes in degrees (e.g., Santelices 1999, Godfrey-Smith 2009)

1.4 Individuality in Physics

In physics, individuals are not an obvious given. There is no straightforward equivalent of the organism, for instance. After the rise of relativistic and quantum physics, the belief that the world is populated by objects with a sufficiently strong identity to be called "individuals" cannot be admitted without demonstration. This question has led to a real debate in physics and philosophy of physics, a debate that has focused on exactly the question of whether or not it was still possible to talk about physical "individuals." During this debate, new distinctions and definitions were proposed, and new approaches were explored. This brought the discussion farther and farther from its original sources, which lay in the metaphysical and biological traditions. Indeed, one could even wonder if the current discussion on physical individuals is about individuals at all. For example, the kind of individuals discussed in the quantum context would probably not be considered as "real individuals" by many metaphysicians. But this conclusion, of course, is disputable, in view of the suggestion made above that the a priori conceptions often favored by metaphysicians should not necessarily be the final word in metaphysics.

In the following, we will mention only a few of the avenues recently discussed in the literature. In most researches, a basic distinction is made between problems relating to the synchronic and the diachronic identity of individuals. The belief that it is legitimate to separate these classes is often founded on the belief that intemporal individuals, this is, individuals that do not persist, could exist. And since many, if not most, metaphysical conceptions of an individual include, as a necessary condition, persistence in time, it seems that philosophers of physics became interested in a new metaphysical concept, something different from the concept of individual as such.

Importantly, it is not a general reflection about the temporal dependence of individuals that originally motivated philosophers of physics to explore the possibility of intemporal individuals, but mainly permutation arguments involving possibilities equivalence, in other words an equivalence between models potentially describing different worlds. For example, if we rearrange space-time points in such a way that no empirical consequence follows, do we face a new physical situation or just another description of the same situation? The latter possibility makes it difficult to sustain that space-time points are individuals, while the former leaves this possibility open. But of course space-time points do not persist. If they are individuals, they are not temporally extended. One could argue that the individuality of space-time points explains why the situations after permutation are distinct. In this position, their individuality involves some kind of modal aspect that guarantees transworld identity. Many similar permutation arguments could be found in the literature, for example, about identical particles in quantum and classical physics (Castellani 1998, Brading and Castellani 2003).

It seems legitimate to discuss the possibility that individuals exist only at a certain time, an idea that has provided incentives for the bundle theory of physical individuality. In the context of this theory, an individual is defined by its properties at a certain time. Usually, a contrapositive version of the principle of identity of indiscernibles is invoked: no two distinct individuals share all the same individuating properties. "What is an acceptable individuating property?" is a difficult question. Should we include relational properties and/or space-time locations? If we do, the ontological status of these properties has to be clarified (French and Krause 2006), but this task is far from being simple (Earman 1989, Stachel 2002, 2004). An empiricist and more modest strategy is also possible (Saunders 2006). Instead of metaphysical individuating properties, this strategy consists in promoting the notions of absolute and weak discernibility as fundamental identity criteria (see Quine 1976, and Saunders's chapter in this volume). This makes the individuating properties more clearly dependent on our linguistic framework.

The bundle theory is not the only philosophical approach framing the debate. The particular kind of empiricist motivation (only properties indexed by space-time points are metaphysically acceptable), implicit to the bundle theory, is not shared by all. Moreover, one could believe that temporal extension is required to be an individual. Fortunately, other options are available. One could expand the bundle theory to include properties at different space-time locations, for example using genidentitical relations among events to define individuals (see Reichenbach 1956, as well as Guay and Pradeu's chapter in this volume). This approach is potentially efficient to sustain a strong diachronic identity criterion but seems less useful if ephemeral individuals are considered. Moreover, in this approach, quantum particles are generally considered as non-individuals. This position is the "received view" in physics' community. Within physics, the discussion generally stops there. But philosophers, for their part, ask for a more precise characterization of the non-individuality that is typically the result in quantum physics. Quantum particles are countable. They also are weakly discernable. So if they are not individuals, what are they?

In reaction to the above view, a more metaphysically inclined position would be to consider individuality as something beyond and above any individual's properties. Many variants are defendable, from haecceitism to bare individuals to systemic properties (Adams 1979, Lewis 1986, Morganti 2013). The absence of empirical access to the individuating features could be seen as a real problem for these positions. But if they manage to shed some light on certain ontological puzzles, like the origin of quantum statistics, they could be taken as viable metaphysical options (Morganti 2009, 2013).

Overall, current philosophers of physics consider physical individuality as at least one of the following:

- Nonexistent
- A kind of property profile
- A kind of space-time process
- Something primitive, beyond any properties

1.5 Building Bridges

Even though spelling out in detail what the notion of an individual means in physics and in biology is interesting by itself, the most stimulating task from the point of view defended here is to determine how to compare the conceptions of individuality found in physics and in biology, and if possible to articulate them into a more integrative framework. Yet this is far from being an easy task. At first sight, at least, physics and biology appear to differ very significantly in the way they conceptualize individuals and address the problem of individual identity through time (a more detailed analysis of these differences can be found in the chapter of Guay and Pradeu in this volume):

(1) Parts-whole questions seem crucial in biology. This probably has to do with the fact that most, if not all, biological entities appear to be constituted of smaller biological entities (as, for example, when one asks to what extent the cells constituting a multicellular organism are themselves "individuals"). Parts-whole questions play a less important role in physics, where many discussions concern particles

situated at a fundamental level, and/or "entities" that cannot be easily individuated (as an electromagnetic field, for instance).
(2) A crucial issue in physics is to determine how to distinguish one particle from several other, supposedly "identical," particles. In contrast, in biology, even individuals that are said to be "identical" express, most of the time, some significant differences and, at the very least, can usually be distinguished one from the other from a spatial point of view.
(3) In discussions about synchronic identity in physics, the principle of indiscernibles is critical. The same is not true in biology: in everyday practice, at least, biologists often say that two living things are "identical" even when they do not share all their properties, in particular their position in space. A nice example is that of "clones," which are often described as "identical"—even though many data suggest that clones always express differences. (Interestingly, the combination of (2) and (3) suggests that discernibility is easier but at the same time considered as less crucial in biology than in physics.)
(4) Discussions over structuralism are extremely important in physics. Many physicists aim at determining what remains invariant under transformations. By contrast, structuralism plays a limited role in biology, if any (exceptions include French 2011, 2014, as well as French's and Ladyman's chapters in this volume).

But do these differences between physics and biology undermine the project of articulating these two fields in the hope of building a more unified perspective on individuality? Several contributions to this volume suggest that the answer to this question is negative; in other words, they show that a fruitful dialogue can be established between physics and biology on the individuality issue. In some cases, they defend a general worldview initially grounded in physics, and then show how it can be extended to biology (e.g., French explains how ontic structural realism and an eliminative attitude toward individual objects can be applied not only to physics but also to biology); in other cases, they analyze two fundamental examples, one taken from physics and the other from biology, to defend a general approach to scientific objects (for instance, Chen uses the case of Bose-Einstein condensates in physics and that of genetic engineering in biology to defend the idea that scientific individuation should be grounded in experimental *practice* rather than in scientific theories); in still other cases, they make a point that clearly belongs to one scientific field, but show nonetheless how it can be related to a theory or a view belonging to the other field (for instance, Fagan defends an "uncertainty principle" with regard to the definition of stem cells, and then, building on Saunders and others, she explores the analogy between this principle and the famous uncertainty principle of quantum mechanics); finally, some chapters try to demonstrate that a common principle of individuation can be applied to both physical and biological cases, and that each illuminates the other (for instance, Guay and Pradeu argue that the so-called genidentity view sheds light on the individuation through time of both physical and biological entities).

But the network of disciplinary interactions to be found in this volume is even more complex than that, as it includes, in addition to a dialogue between physics and biology, a dialogue between metaphysics and those sciences. Indeed, several chapters show how metaphysical, physical, and biological views on individuality have been interacting in intricate and fascinating ways. Our metaphysical notions often come from "folk science"—in particular, as has been emphasized above, from "folk biology" (e.g., Aristotle, in his *Categories*, explains his metaphysical concept of *primary substance* by giving the example of "an individual horse"). In turn, science often cast doubt on our intuitive concepts of what an individual is, and this can probably lead to a transformation of these intuitive concepts, which are, therefore, probably much less eternal, immutable, and "a priori" than many metaphysicians have thought (a nice example of this complex interplay is the way the notion of an immunological individual has been inspired by the "self" concept found in the psychological and social sciences, and then has in turn modified our conception of what it means to be a human "self" in present-day societies: see, e.g., Tauber 1991, Cohen 2009). Several chapters of this volume, among others those of Morganti, Saunders, Haber, and Ereshefsky and Pedroso, demonstrate that general metaphysics and the sciences can and should talk to each other, and that much gain is to be expected from this dialogue.

1.6 Book Structure

Laying aside chapter 1 (which is the present introduction), this volume is composed of 17 chapters, organized into three parts. The first part concerns metaphysical and logical foundations to individuality. The second part explores puzzles about individuals in a specific science, most often biology or physics. Finally, the third part focuses on transversal problems, that is, problems that arise at the interface between physics and biology, or at the interface between metaphysics on the one hand, and biology and/or physics on the other hand.

1.6.1 Part I: Metaphysical and Logical Foundations to Individuality

Part I provides a general and wide-ranging perspective about the notion of individual, as the contributions gathered in this part explore what an individual is from a metaphysical and logical point of view. Therefore, they are situated at the most fundamental, and potentially most unifying, level. One possibility would be to consider that approaches of this kind define every individual, whatever the domain of inquiry; in that case, they would impose delimitations and restrictions on how the sciences, and in particular physics and biology, may define the notion of individual. Alternatively, one could consider that these approaches are to be built, and constantly modified, on the basis of what current sciences say (and future sciences will say)—in which case their objective would be to reflect on current sciences in order to offer an as good as possible (though always precarious) view on what an individual in general is.

The aim of the chapters gathered in this part, therefore, is to constitute a wide and unifying picture of individuality, one that should be used as a foundation for the development of more specific conceptions of individuality, as, for examples, those that different experimental sciences could offer. However, it will be apparent in Parts II and III that many philosophers of experimental sciences do not in fact use the logical and metaphysical discourses as foundations for the ontological investigation of physics and biology: these discourses are considered top-down approaches, and therefore concurrent to the bottom-up methodology characteristic of philosophy of science.

Chapter 2, written by Stéphane Chauvier, can be seen as a conceptual basis for the rest of the volume. Chauvier distinguishes two concepts of an individual: the *logical-cognitive* concept of a discrete particular, and the *ontological* concept, which applies more stringently to entities that exhibit ontological autonomy, formal unity, and qualitative singularity. He then shows that ontological individuality matters, in two senses. First, not every particular object of thought, even of scientific thought, is a real individual, since what is individuated by us, by our way of conceptually dividing the world, is not necessarily individuated *in itself*. Second, not every real being is a real individual or is an individual to the same degree. Thus, according to Chauvier, the ontological concept of an individual can be used as a basis for a complete division of real beings, by distinguishing *individuals* and *non-individuals* (aggregates), but also by distinguishing various *degrees* of individuality and aggregativity.

In chapter 3, Jonathan Lowe (who so sadly died on January 5, 2014, only a few weeks after having submitted his contribution for the present volume) argues for the possibility of entities that are not individuals. Having claimed that an entity x is an *individual* just in case (1) x determinately counts as *one* entity, and (2) x has a determinate *identity*, Lowe shows that it is both logically and metaphysically possible for there to be *non*-individuals, that is, entities which fail to satisfy either clause (1) or clause (2)—or both. Lowe then explores the potential application of this distinction between *individuals* and *non-individuals* within and across the sciences, in a spirit of fruitful cooperation (rather than mutual hostility) between analytic metaphysics and theoretical science.

In chapter 4, Krause and Arenhart argue for the importance of logical and formal considerations in any discussion about individuality. They start with an intuitive definition of an individual (as a unity, having identity and being such that it is possible at least in principle to discern it from any other individual). But they then show that, when we leave the intuitive realm and attempt a logical analysis, we find a cluster of problems that are difficult to overcome within standard logico-mathematical apparatuses. They question the intuitive notion of an individual in view of recent discussions that arise in quantum theory, and they push the discussion to a "logical" view. They characterize individuals by means of invariance by automorphisms and, finally, they propose a metaphysics where the notion of identity is replaced, for some objects, by a weaker notion of indiscernibility.

1.6.2 Part II: Puzzles about Individuals in Biology and Physics

In Part II, each chapter investigates a particular conception or a particular puzzle about the application of the individuality concept to biology or physics. Even though each chapter is strongly focused on a specific science, many of them draw parallels between different sciences, or between metaphysics and science.

Let us first discuss the chapters grounded in biology. As emphasized above, Peter Godfrey-Smith, in recent publications, has suggested the concept of a "Darwinian individual," understood as a member of a "Darwinian population" (Godfrey-Smith 2009, 2013). In chapter 5, entitled "Individuality and Life Cycles," Godfrey-Smith shows that many recent discussions over biological individuality have focused on *spatial* aspects of individuality. But, he claims, *temporal* aspects of individuality are as important as spatial aspects. Many familiar organisms are "reproducing continuants": they come into being, persist, and then die. Yet some living entities put pressure on this familiar view, and on the traditional conception of heredity to which it is associated. Godfrey-Smith analyses several cases of complex life cycles, in particular some featuring "alternation of generations" (a process in which entities of a kind A make entities of a kind B, which in turn make entities of a kind A—like ferns, for instance), and eventually offers a renewed and extended conception of reproduction.

In chapter 6, Marc Ereshefsky and Makmiller Pedroso use the example of biofilms to examine Hull's and Godfrey-Smith's accounts of biological individuality and to explore the nature of individuality more generally. According to Ereshefsky and Pedroso, the case of biofilms shows that

Godfrey-Smith's account of biological individuality is too restrictive, while Hull's interactor account is appropriately inclusive. The chapter then augments Hull's account in three ways. First, Hull's notion of interactor is embedded in a general theory of individuality that applies to individuals both in and outside of biology. Second, the sort of interaction required of the parts of an individual is explored and elaborated. Third, Hull's commitment to replicator theory is dropped and Griesemer's account of reproducers is adopted.

In chapter 7, Melinda Fagan considers the biological individuality of stem cells (i.e., undifferentiated cells that self-renew and give rise to differentiated cells). She argues that stem cells are not biological individuals in the same way as *cells* of multicellular organisms, but at the same time she claims that some stem cells at least are biological individuals in the way of *multicellular organisms*. Her approach sheds light on central concepts and practices of stem cell biology, as well as the relation between cell and organismal individuality. The stem cell case also exhibits an unexpected parallel with physics, specifically Bohr's view of complementarity.

In chapter 8, Cédric Paternotte develops the analogy between a collective engaged in a joint action and biological individuals. He shows first that there exist many definitions of human joint action, or of what makes a group similar to an individual, but these definitions do not agree and are not directly reducible to each other. Paternotte argues that these definitions should at least meet an efficiency constraint: any account of joint action has to justify how it reliably leads agents to cooperation. The avenue suggested by Paternotte consists in exploring the analogy between definitions of joint action and of biological individuality, because the main components for biological individuality have been identified and their relations are much better understood than those between the components of human joint action. Paternotte concludes that we can import some insights of the biological literature to define what a joint action is, and when a group can and should be considered as an individual.

Let us now move to the chapters that focus mainly on physics. In chapter 9, Simon Saunders shows that, at a fundamental level, the world is built up from quantum fields, and then asks how we pass from this to the objects of the special sciences. In his view, there are two critical transitions: one is from indistinguishable to distinguishable things that persist over time—individuals—and the other is the transition from quantum to classical or semi-classical systems. He argues that the two are interlinked, and that the key method is to pass from a description in terms of particles whose only intrinsic attributes are mass, charge, and spin, to one in terms of individuals whose intrinsic attributes include stable dynamical properties—among them, spatial location, as provided by state-collapse, whether effective (as in many-worlds and pilot-wave theory) or fundamental. Saunders adds that there is also a connection with ontological relativity (or Putnam's paradox): the same method, applied to model theory, leads from permutable particulars that have no intrinsic attributes to distinguishable bundles of properties.

In chapter 10, James Ladyman discusses the debate about individuality, quantum particles, and the principle of the identity of indiscernibles. He argues that if the definability of identity is at stake, then the cut does not come between individuals and non-individuals as usually defined, but rather between weakly discernible and completely indiscernible entities. If what is at stake is the principle that identity must be grounded in qualitative properties, even things that are absolutely discernible, in the sense of their being at least one property that one has and not the other, may only be so in virtue of relational properties. If individuality can be grounded in non-qualitative features of the world, or if a logically thin notion of individuality is adopted, then quantum particles may be individuals after all. Ladyman concludes that the metaphysically significant notion of individuality apt for quantum particles is that of "real pattern."

In chapter 11, entitled "Minimal Structural Essentialism: Why Physics Doesn't Care Which is Which," David Glick argues for a trans-structural identity. He starts by noticing that the ways in which space-time points and elementary particles are modeled share a curious feature: neither seems to specify which basic object has which properties. The aim of Glick's chapter is to explain this. After reviewing several proposals, he argues that objects occupy their place in a given relational structure essentially. This view, called "Minimal Structural Essentialism," provides a metaphysical grounding for the physical equivalence of models related by permutation. According to this view, space-time points and elemental particles turn out to be individuals, albeit of a rather different sort than has traditionally been considered.

In chapter 12, Paavo Pylkkänen, Basil J. Hiley, and Ilkka Pättiniemi start with Ladyman and Ross's view that quantum objects are not individuals (or are at most weakly discernible individuals), which constitute the basis of Ladyman and Ross's defense of "ontic structural realism," according to which relational structures are primary to things. In response, Pylkkänen et al. draw attention to a version of quantum theory, namely the Bohm theory, according to which particles do have definite trajectories at all times. According to them, this view suggests that quantum particles are individuals after all, with position being the property in virtue of which particles are always different from one another. However, Pylkkänen et al. also admit that the individuals of the Bohm theory are very different from those of classical physics, and they resort to structuralist considerations to better understand their nature.

Finally, Christina Conroy, in chapter 13, goes a step further in proposing that, even if we had a robust concept of physical individual at a certain time, this individuality would be weakened if we took into account Everett's interpretation of quantum mechanics, for which the world is constantly branching. She shows that, when considering persons, one must consider the criteria for re-identifying a person over time. She then explains that in the context of the metaphysical picture implied by Everettian quantum mechanics—one that includes some type of branching structure to the world—problems of diachronic identity arise. This is in fact a Ship of Theseus-type problem. Conroy argues that an answer to the question "With whom will I be identical post-branching?" can be found in analogy with a solution proposed by Derek Parfit in his "Personal Identity" (1971). She proposes that we use a notion of branch-relative identity instead of the traditional equivalence relation of identity to solve this problem. This chapter thus concludes rich discussions, both in biology and in physics, on how individuals can be conceived.

1.6.3 Part III: Beyond Disciplinary Borders

The third part of the book is composed of chapters that explicitly transcend disciplinary borders, be they the borders between metaphysics and science, or those between physics and biology.

In chapter 14, Matteo Morganti argues for a pluralistic approach to the concept of individual. First, he looks at the debate about identity and individuality in non-relativistic quantum mechanics and offers a defense of the view according to which identity facts are primitive in that domain. Second, his chapter constitutes a contribution to the clarification of the relationship between science and metaphysics, in particular with respect to what a proper "naturalistic" methodology should and should not be taken to entail as far as the theme of individuality is concerned. His guiding idea is that taking identity and individuality facts as basic is not necessarily in conflict with naturalism. The overall picture that emerges from Morganti's chapter is that of a "pluralistic" approach, whereby different scientific domains and theories are likely to allow, and in fact to ask for, different forms of individuality.

Matt Haber, in chapter 15, raises the issue of the relationship between metaphysical mereology and biological debates over the idea that species would be individuals. Haber starts with Michael Ghiselin and David Hull's "individuality thesis," which famously states that biological species are individuals. Philosophers, Haber shows, have often interpreted this thesis in a mereological way, species being seen as mereological sums. Yet Haber argues that this is a mistake, since biological part/whole relations often violate the axioms of mereology. According to Haber, conflating these projects confuses the central issues at stake in both, and makes the job of evaluating them extremely difficult. His clarification of this issue helps identify the genuine metaphysical implications of the individuality thesis, which serves as an exemplar of scientifically informed metaphysics.

In chapter 16, Guay and Pradeu offer a defense of the "genidentity" thesis, first put forward by Kurt Lewin (1922), and then by Hans Reichenbach (1956). They show that the original notion of genidentity was often imprecise and not easily applicable to real scientific cases. In their view, however, a renewed notion of genidentity can be suggested, and this notion can shed important light on physical and biological cases. Guay and Pradeu draw lessons from physical examples of the genidentity view, and apply them to biological cases, in particular cases discussed by David Hull—one of the very few philosophers of science having supported explicitly, in recent times, genidentity. In conclusion, they suggest that genidentity could be an important argument in defense of a processual worldview.

In chapter 17, Ruey-Lin Chen discusses the *experimental realization of individuality*—the production of individuals posited by scientific theories. Experimental realization refers to the processes by which scientists produce new phenomena, properties, entities, or individuals by means of experimental techniques and instruments. On this basis, Chen addresses two main questions: (1) Is there a conception of individuality in and across experimental sciences? (2) Under what conditions can scientists be said to realize the individuality of an object? By examining the creation of Bose-Einstein condensates in experimental physics and the modification of genes in genetic engineering, Chen suggests a conception of *experimental individuality* in experimental sciences and identifies three realization conditions that apply to these cases, namely *manipulation, separation*, and *maintenance of structural unity*.

In the final chapter (chapter 18), Steven French pleads for the elimination of the notion of individual in our scientific ontology. He starts with the observation that an eliminativist view of objects in physics has recently been suggested in the context of "ontic structural realism" (defended, in particular, by Ladyman and French: for overviews, see Ladyman and Ross 2007, French 2014). In this chapter, French explores the extent to which eliminativism can be articulated and defended in the philosophy of biology. Though the motivations are very different in these two sciences, French argues that a range of issues can be identified that pull us away from an object-oriented stance. He then suggests that various metaphysical resources can be deployed to help assuage concerns regarding such a move, and explores some of its consequences for the biological sciences.

References

Adams, Robert Merrihew. 1979. Primitive Thisness and Primitive Identity. *Journal of Philosophy* 76: 5–25.

Bird, Alexander. 2007. *Nature's Metaphysics: Laws and Properties*. Oxford: Clarendon Press.

Bouchard, Frédéric, and Philippe Huneman, eds. 2013. *From Groups to Individuals: Evolution and Emerging Individuality*. Cambridge, MA: MIT Press.

Brading, Katherine, and Elena Castellani, eds. 2003. *Symmetries in Physics: Philosophical Reflections*. Oxford: Oxford University Press.

Buss, Leo. 1987. *The Evolution of Individuality*. Princeton, NJ: Princeton University Press.

Carnap, Rudolph. [1935] 1996. *Philosophy and Logical Syntax*. Bristol, UK: Thoemmes.

Castellani, Elena, ed. 1998. *Interpreting Bodies: Classical and Quantum Objects in Modern Physics*. Princeton, NJ: Princeton University Press.

Caulton, Adam, and Jeremy Butterfield. 2012. On Kinds of Indiscernibility in Logic and Metaphysics. *British Journal for the Philosophy of Science* 63: 27–84.

Chakravartty, Anjan. 2007. *A Metaphysics for Scientific Realism*. Cambridge: Cambridge University Press.

Clarke, Ellen. 2010. The Problem of Biological Individuality. *Biological Theory* 5: 312–325.

Cohen, Ed. 2009. *A Body Worth Defending: Immunity, Biopolitics and the Apotheosis of the Modern Body*. Durham, NC: Duke University Press.

Davis, John B. 2010. *Individuals and Identity in Economics*. Cambridge: Cambridge University Press.

Dorato, Mauro, and Matteo Morganti. 2013. Grades of Individuality. A Pluralistic View of Identity in Quantum Mechanics and in the Sciences. *Philosophical Studies* 163(3): 591–610.

Dumont, Louis. 1986. *Essays on Individualism: Modern Ideology in Anthropological Perspective*. Chicago: University of Chicago Press.

Dupré, John, and Maureen A. O'Malley. 2007. Metagenomics and Biological Ontology. *Studies in History and Philosophy of Biological and Biomedical Sciences* 38(4): 834–846.

Dupré, John, and Maureen A. O'Malley. 2009. Varieties of Living Things: Life at the Intersection of Lineages and Metabolism. *Philosophy and Theory in Biology* 1. http://quod.lib.umich.edu/p/ptb/6959004.0001.003/—varieties-of-living-things-life-at-the-intersection?rgn=main;view=fulltext.

Earman, John. 1989. *World Enough and Space-Time: Absolute versus Relational Theories of Space and Time*. Cambridge, MA: MIT Press.

Elias, Norbert. 1991. *The Society of Individuals*. Oxford: Blackwell.

French, Steven. 2011. Identity and Individuality in Quantum Theory. In *Stanford Encyclopedia of Philosophy*, Summer 2011 ed., ed. Edward N. Zalta. http://plato.stanford.edu/archives/sum2011/entries/qt-idind/.

French, Steven. 2014. *The Structure of the World: Metaphysics and Representation*. Oxford: Oxford University Press.

French, Steven, and Décio Krause. 2006. *Identity in Physics: A Historical, Philosophical, and Formal Analysis*. Oxford: Oxford University Press.

French, Steven, and Kerry McKenzie. 2012. Thinking outside the Toolbox: Towards a More Productive Engagement between Metaphysics and Philosophy of Physics. *European Journal of Analytic Philosophy* 8(1): 42–59.

Godfrey-Smith, Peter. 2009. *Darwinian Populations and Natural Selection*. Oxford: Oxford University Press.

Godfrey-Smith, Peter. 2013. Darwinian Individuals. In *From Groups to Individuals: Evolution and Emerging Individuality*, ed. Frédéric Bouchard and Philippe Huneman, 17–36. Cambridge, MA: MIT Press.

Gould, Stephen J. 2002. *The Structure of Evolutionary Theory*. Cambridge, MA: Harvard University Press.

Gould, Stephen Jay, and Elizabeth A. Lloyd. 1999. Individuality and Adaptation across Levels of Selection: How Shall We Name and Generalize the Unit of Darwinism? *Proceedings of the National Academy of Sciences* 96(21): 11904–11909.

Hull, David L. 1978. A Matter of Individuality. *Philosophy of Science* 45(3): 335–360.

Hull, David L. 1980. Individuality and Selection. *Annual Review of Ecology and Systematics* 11: 311–332.

Hull, David L. 1992. Individual. In *Keywords in Evolutionary Biology*, ed. E. Fox-Keller and E. Lloyd, 181–187. Cambridge, MA: Harvard University Press.

Ladyman, James, and Don Ross. 2007. *Every Thing Must Go: Metaphysics Naturalized*. Oxford: Oxford University Press.

Kistler, Max. 2006. *Causation and Laws of Nature*. London: Routledge.

Lewis, David. 1986. *On the Plurality of Worlds*. New York: Wiley-Blackwell.

Lowe, E. Jonathan. 1989. *Kinds of Being: A Study of Individuation, Identity and the Logic of Sortal Terms*. Oxford: Blackwell.

Lowe, E. Jonathan. 2006. *The Four-Category Ontology: A Metaphysical Foundation for Natural Science*. Oxford: Oxford University Press.

Macdonald, Cynthia. 2005. *Varieties of Things: Foundations of Contemporary Metaphysics*. Malden, MA: Blackwell.

Maudlin, Tim. 2007. *The Metaphysics within Physics*. Oxford: Oxford University Press.

Maynard-Smith, John, and Eörs Szathmáry. 1995. *The Major Transitions in Evolution*. Oxford: Oxford University Press.

Michod, Richard E. 1999. *Darwinian Dynamics: Evolutionary Transitions in Fitness and Individuality*. Princeton, NJ: Princeton University Press.

Morganti, Matteo. 2009. Inherent Properties and Statistics with Individual Particles in Quantum Mechanics. *Studies in the History and Philosophy of Modern Physics* 40: 223–231.

Morganti, Matteo. 2013. *Combining Science and Metaphysics: Contemporary Physics, Conceptual Revision and Common Sense*. Houndmills, Basingstoke, Hampshire: Palgrave Macmillan.

Mossio, Matteo, and Alvaro Moreno. 2010. Organisational Closure in Biological Organisms. *History and Philosophy of the Life Sciences* 32: 269–288.

Muller, F. A., and M. P. Seevinck. 2009. Discerning Elementary Particles. *Philosophy of Science* 76: 179–200.

Mumford, Stephen, and Matthew Tugby, eds. 2013. *Metaphysics and Science*. Oxford: Oxford University Press.

Okasha, Samir. 2006. *Evolution and the Levels of Selection*. Oxford: Oxford University Press.

Perelman, Robert L. 2000. The Concept of the Organism in Physiology. *Theory in Biosciences* 119(3–4): 174–186.

Pradeu, Thomas. 2010. What Is an Organism? An Immunological Answer. *History and Philosophy of the Life sciences* 32(2–3): 247–267.

Pradeu, Thomas. 2012. *The Limits of the Self: Immunology and Biological Identity*. New York: Oxford University Press.

Quine, Willard V. O. 1948. On What There Is. In *From a Logical Point of View* (1981). Cambridge, MA: Harvard University Press.

Quine, Willard V. O. 1960. *Word and Object*. Cambridge, MA: MIT Press.

Quine, Willard V. O. 1976. Grades of Discriminability. *Journal of Philosophy* 73: 113–116.

Redhead, Michael. 1995. *From Physics to Metaphysics*. Cambridge: Cambridge University Press.

Reichenbach, Hans. 1956. *The Direction of Time*. Berkeley: University of California Press.

Rosenberg, Alexander. 1988. *Philosophy of Social Science*. Boulder, CO: Westview Press.

Ross, Don, James Ladyman, and Harold Kincaid. 2013. *Scientific Metaphysics*. Oxford: Oxford University Press.

Santelices, Bernabé. 1999. How Many Kinds of Individual Are There? *Trends in Ecology and Evolution* 14(4): 152–155.

Saunders, Simon. 2006. Are Quantum Particles Objects? *Analysis* 66: 52–63.

Sober, Elliott. 1991. Organisms, Individuals and Units of Selection. In *Organism and the Origins of Self*, ed. A. I. Tauber, 275–296. Dordrecht: Kluwer.

Stachel, John. 2002. The Relations between Things versus the Things between Relations: The Deeper Meaning of the Hole Argument. In *Reading Natural Philosophy: Essays in the History and Philosophy of Science and Mathematics*, ed. D. B. Malament, 231–266. Chicago: Open Court.

Stachel, John. 2004. Structural Realism and Contextual Individuality. In *Hilary Putnam*, ed. Y. Ben-Menahem, 203–219. Cambridge: Cambridge University Press.

Strawson, Peter F. 1959. *Individuals: An Essay in Descriptive Metaphysics*. London: Methuen.

Tauber, Alfred I., ed. 1991. *Organism and the Origins of Self*. Dordrecht: Kluwer.

Weisberg, Michael, Paul Needham, and Robin Hendry. 2011. Philosophy of Chemistry. In *The Stanford Encyclopedia of Philosophy,* Winter 2011 ed., ed. Edward N. Zalta. http://plato.stanford.edu/archives/win2011/entries/chemistry/.

Wiggins, David. 1980. *Sameness and Substance*. Cambridge, MA: Harvard University Press.

Wiggins, David. 2001. *Sameness and Substance Renewed*. Cambridge: Cambridge University Press.

Wilson, Jack. 1999. *Biological Individuality: The Identity and Persistence of Living Entities*. Cambridge: Cambridge University Press.

Wilson, Robert A. 2005. *Genes and the Agents of Life: The Individual in the Fragile Sciences: Biology*. New York: Cambridge University Press.

Wilson, Robert A., and Matthew Barker. 2013. The Biological Notion of Individual (first published 2007). In *Stanford Encyclopedia of Philosophy*, Spring 2014 ed., ed. E. Zalta. http://plato.stanford.edu/entries/biology-individual.

PART I | Metaphysical and Logical Foundations to Individuality

CHAPTER 2 | Why Individuality Matters

STÉPHANE CHAUVIER

IF METAPHYSICS IS THAT part of philosophy whose main purpose is to determine what there is or what really exists, making sense of the metaphysical enterprise requires us to understand why we can spontaneously delude ourselves about what there is and why we consequently need to engage in investigations on what really exists.

One of the main reasons, which is also one of the main incentives of metaphysics, is undoubtedly the well-known and widely explored fact that everything that is an *object* of thought, including true thoughts, is not necessarily a real component of the world or a real *being*. We can form true thoughts about Madame Bovary, although she is not a real being. But we can also form true thoughts about numbers or about universal qualities such as courage or justice though, for some metaphysicians, there is no such thing as numbers or universal qualities.

Yet if this inclination of our mind to think about types of objects that turn out to be only logical-linguistic constructs is certainly one of the main sources of ontological illusion and thus one of the main spurs to do metaphysics, there is another source, less systematically explored, that might be called the propensity to individuate what is not an individual or to treat as an individual object what is not a "real" individual.

Let's take a coffee cup and break it in two halves. Give to one of the two resulting pieces, the one that contains the handle, a nickname, for example *Bambinette*. Bambinette is a true being: it is by no means a fiction of our mind. Bambinette is also a being that we can perfectly individuate: we can single it out from all other half-cups with a handle, we can think of it in its absence, we can reidentify it every time we meet it again. Bambinette can also verify necessary (a posteriori) propositions: it necessarily has a handle, it is necessarily earthenware, it is necessarily "born" at a given instant of time from the breaking of a particular coffee cup. Suppose further that a small group of scientists, working in different parts of the world, but forming a close-knit network of research, focuses on the *kind* of individual objects of which Bambinette is an example. They are interested in how the half-cups-with-a-handle were "born," what variations of shapes and sizes they can support, and so on. They index all the half-cups and develop a complex taxonomy, forging scholarly names derived from a now extinct Caucasian language. Bambinette and its fellows form, for our scholars, a class of individual objects whose behavior can be the material for a domain of nomological knowledge. But for us, who are observing these scholars, it will probably be difficult to find much of interest in their results. And the reason is that we will probably be pretty hesitant to consider Bambinette and its fellow half-cups as *true individuals*. While the science developed by the fans of the half-cups-with-a-handle contains genuine truths and while these truths also have a *fundamentum in re*, their science seemingly failed to *carve the world at its real joints*. It has extracted from the frame of the world a class of individual objects, which certainly are not indebted to our mind for *existing*, but which are indebted to it for existing *apart* from all the rest.

The case of Bambinette is, of course, merely a pleasant fiction, but it can serve to highlight what I believe to be the main issue of a metaphysical investigation on individuality. If, as Quine (1969, 1) has it, "we are prone to talk and think of objects," it is surely not sufficient to think of an object to be entitled to conclude that there exists an object that we are thinking about. But it is also not sufficient to think of a really existing individual object, which our mind sets apart from the rest of the whole world in order to think of it, to be entitled to view that object as being really or in itself apart from the rest of the world. If we are to avoid taking the fictional for the real, we must also avoid dividing the world in our own way.[1] The concept of an individual, as a being that has a real or intrinsic individuality, can then be used to capture the difference between what is individuated by us and what is individuated in itself or what has a real individuality.

The whole problem, as I shall try to show, is that if the concept of an individual that we mobilize is sufficiently discriminating, the class of the real individuals can become quite limited. Therefore, paying attention to what is really an individual requires in part that we be prepared to recognize that there may be *degrees* in individuality, but it also requires us to pay attention to the ontological status of what is not individual and to be prepared to recognize that to be and to be an individual can be two distinct things.

2.1 Individual versus Discrete Particulars

Assuming that the metaphysics of individuality finds its root and its stimulus in the fact that everything that is an individual object of thought is not necessarily a "real" individual is also assuming that we must be able to distinguish these two concepts: the concept of an individual object of thought and the concept of a real individual. However, this distinction is somewhat obscured by the fact that the unique word "individual" is used in the philosophical language to designate an *object that is logically and cognitively individuated* as well as to designate an *object that has an individuality or that has a certain degree of individuality*. This double meaning of the word "individual" is quite traditional. For example, in one of the main lexicons of metaphysical language written in the seventeenth century, that of Stephanus Chauvinus, we find, in the entry for "individuum," the following two definitions:

- "We call *individual* everything that ... may be designated by a proper name or a demonstrative" [Individuum *vocatur quicquid ... proprio nomine aut demonstrativo declarari queat*].
- "We call properly *individual*, not what cannot in any way be divided, but what cannot be divided into several individuals of the same kind as itself or into several individuals specifically similar to it" [Individuum *vero dicitur, non quod omnino dividi not possit, sed quod ita ut singulare sit ut dividi neutiquam queat in plura particularia, in plura talia quale est ipsum, in pluri sibi similia individua*]. (Chauvinus 1692, 315)

But despite this ambivalence in the philosophical use of the *word* "individual," it seems to me crucial to realize that we have here two *concepts* of an individual, two concepts that are certainly *partially coextensional* but that, from an intensional point of view, are two distinct concepts.

Consider first what we shall label the *logico-cognitive concept* of an individual. According to this concept, an individual is what we can think about via a demonstrative expression or concept of the form "that F," for example, "that man," "that sandbox," "that lightning," " that battle," and so forth. It is this logico-cognitive concept of an individual that is especially at work in Aristotle's *Categories* (1A20–1B10) when he characterizes the primary substances and the singular accidents, in contrast with the secondary substances and the universal attributes. The examples he gives are indeed "that man" *ho tis anthropos*, "that horse" *ho tis hippos*, but also "that grammatical science" *hē tis grammatikē*, "that white" *to ti leukon*. An individual, in this sense, is a *tode ti*, a "that something," that is to say any object of an *act* of *demonstrative identification*. An individual can therefore, on the one hand, be distinguished from a *universal*, such as courage or war, because there is a logical-cognitive difference between thinking about courage and thinking about this or that act of courage. But, on the other hand, an individual may also be distinguished from a *mass*, such as wine or water, because there is also a logico-cognitive difference between thinking about water and thinking about this or that puddle of water.[2] These two kinds of nonindividuals, the universals and the masses, certainly do not have the same ontological status. Universals are abstract entities: they have spatiotemporal instances, but are not themselves localized in time and space. Masses are, rather, concrete entities: they are localized in time and space, even if sparsely. However, despite this difference in the ontological status of universals and masses, these two kinds of nonindividuals can serve to define more precisely the concept of an individual, when the term "individual" is taken to designate a *tode ti*, a "that something." An individual is indeed a nonuniversal, that is, a *particular*. An individual is also a nonmassive entity, what may be called a *discrete* or *countable* entity. If both determinations are synthesized, we obtain a definition of an individual as a *discrete particular*.

The problem is that this logico-cognitive concept of an individual is precisely a *logico-cognitive* concept and not a purely *ontological* concept of an individual. It means that an individual, in the logico-cognitive sense of the word, is certainly a *real item*, but a real item that is defined not by internal or constitutive characteristics, but *by the specific way in which we think through to it or we have a cognitive access to it*. The logico-cognitive concept of an individual contains a description of the logico-linguistic format through which we apprehend the individual object and of the cognitive operations that we need to accomplish in order to access to it. An individual is what we apprehend via demonstrative expressions and concepts of the form "*that F.*" In the thought of an individual, understood as a discrete particular, we find a general component ("F") and an individuative component ("that"). The thought of that F, of that chip of wood or of that rabbit, can then be analyzed into two abilities: the ability to detect a woody presence or a rabbithoody presence and the ability to isolate an individual bearer of that specific or qualitative identity, an ability that can itself be analyzed into an individuating capacity (*one* chip of wood, *one* rabbit) and an individualizing or singularizing capacity (*that* chip of wood, *that* rabbit).

From a cognitive point of view, these capabilities seem prioritized. If we follow Quine (1959, § 19), or Strawson (1959, 202 ff.), we can imagine a thinker who is only capable of detecting woody features or rabbithoody features. Such a thinker would think by *feature-concepts*, in the words of Strawson, or by *mass-terms*, in the words of Quine; that is, he would think that there is wood or rabbit in the vicinity, as we think that there is blood or wind. But he would be blind to the individuality of the woody or rabbithoody features. This suggests that the ability to detect woody features or rabbithoody features can be implemented in a thinking creature, without the sensitivity to the individuality of these features being also implemented. However, it seems that we cannot conceive, conversely, a thinker who would be sensitive to the individuality of presences, but not to their specific or generic identity, a thinker who would detect bare individuals, without essence or nature. This asymmetry implies that the ability to think about a chip of wood or a rabbit is an enrichment of the ability to detect those specific features. Demonstrative reference to an individual, that is, demonstrative focus on an individual, seems to wrap two basic cognitive operations:

- A transition from "wood" to "a chip of wood" or from "rabbit" to "a rabbit" that may be called *individuation*
- A transition from "a chip of wood" to "that chip of wood" or from "a rabbit" to "that rabbit" that may be called *individualization*

However, it is crucial to understand that these operations of individuation and individualization are *ideal or cognitive operations rather than real operations*. We can compare them to the tightening of a photographic zoom that we focus, with an increasing resolution, over a portion of reality. Zooming obviously does not affect reality, and only a radical philosophical idealism could imagine that *cognitive* individuation and individualization are *real* individuation and individualization.

Still, we would perhaps not need to insist on the difference between the ideal and the real if the cognitive operations of individuation and of individualization always found their terminus in objects with undeniable individuality and if, conversely, no individual was thinkable without passing through these operations. Cognitive individuation and individualization would then be *reliable trackers* of ontological individuality: all that would be cognitively individualized would be an authentic individual, and every authentic individual would only be thinkable or cognitively accessible by means of cognitive individuation and individualization.

The problem is precisely that this correspondence between the cognitive and the ontological dimensions of individuality is far from systematic:

- On the one hand, it is clear that among the objects that have the logical-cognitive status of discrete particulars, some may be regarded as individuals only with a certain hesitation. Let's simply mention some of the very disparate objects that, in the context of an episode of thought or of speech, can be treated *as* individuals: that man or that rabbit of course, but also that chip of wood, that spark, that white lightning, that nostril, that hole in the Gruyère, that afternoon, that silence, and so on.
- On the other hand, it seems there are objects with an authentic individuality that we do not need to individuate or to individualize demonstratively in order to think to them, because the very concept of these objects is necessarily the concept of a unique existing object. This is, for example, the case of what Kant calls

the transcendental ideal of pure Reason[3] (A568/B596). This is the case of the World, at least when this concept is taken to mean, as in traditional metaphysics, the totality of everything that exists or the *universitas creaturarum*.[4] It is also the case of the One, of the Dyad, and of the Triad in the Platonic philosophy of numbers as exposed by Aristotle (1928, M, 6–10). And it is also the case of institutions such as the United Nations or the World Trade Organization. More generally, it is the case with all the individuals that are *necessarily unique in their species or in their gender* and for which therefore the problem of their individualization or of their demonstrative isolation never arises.[5]

This dual divergence suggests that the concept of an individual as discrete particular is in competition with *another concept of an individual*, a concept that we mobilize when we see a difference *from the point of view of individuality*, between a rabbit or the World on one side, a hole in a piece of Gruyère or a corner of a door on the other side. If we have a tendency to cognitively individuate and individualize, a tendency that must be linked with our penchant for predicative thought, the concept of a discrete particular to which we arrive when we explore the contents of the concept of an individual object of thought is not the only concept of an individual that we have, since we are perfectly able to perceive a difference, *from the point of view of individuality*, between a human being, a corner of a door, and a pile of sand, which however are all discrete particulars. It is very clear that many individuals are genuine discrete particulars. But it is equally clear that the concept of a discrete particular cannot be the only concept of an individual that we have, because we have the *logical space* to consider that everything that is a discrete particular is not an individual and that everything that is an individual is not a discrete particular.

2.2 Components of Individuality

What is this other concept of an individual that we are using when we ask whether the corner of a door, a gene, and a galaxy are real individuals, even though they are all discrete particulars in the same degree?

The highlight of this other concept of an individual, as the previous considerations suggest, is that it is, at least partially, a scalar concept.[6] Consider the following set of objects: a human person, a rabbit, a television, a pile of rice, a corner of a door. We deal every time with discrete particulars. But it seems obvious at the same time that there is a meaning or a use of the word "individual" that permits us to say that a corner of door has less individuality than a pile of rice, that a pile of rice has less individuality than a television, and that a television has less individuality than a rabbit and a rabbit than a human person.

That this is another concept of the individual is made manifest by these comparative judgments. The logical-cognitive concept of an individual does not permit gradations of this kind; it is not a *scalar concept*, but a *classifying concept*: one is not *more or less* a discrete particular. The problem is therefore to understand what an individual must be when we define an individual, not in terms of the cognitive operations that we need to accomplish to have access to it, but in terms of its internal or constitutive features, when we define an individual as something that has an individuality and that can have more or less individuality.

It seems to us that this other concept of an individual can be analyzed according to *three independent but complementary dimensions* that can be called the dimension of the (ontological) *autonomy*, the dimension of the (formal) *unity*, and the dimension of the (qualitative) *singularity* of an individual. In order to be an individual in the ontological sense or to have individuality, an object must be positively determinable in each of these three dimensions. But each of these three dimensions is actualized in the form of graduated determinations. This explains, in our opinion, the scalar nature of the concept of an individual: the individuality of an object is measured by its simultaneous position on each of these three scales of determinations, so that an individual object may have more or less individuality than another.

2.2.1 Individuality and Ontological Autonomy

Consider first the dimension that we call ontological autonomy, thereby trying to capture in a modern language the old distinction between *ens per se* and *ens in alio*. An object is an individual if it has its own being (*ens per se*) rather than having the being of another or that being inherent to another (*ens in alio*).

We can illustrate the traditional distinction between *ens per se* and *ens in alio* by the difference between Socrates and his nose. The nose of Socrates is undoubtedly a discrete particular of the general kind noses, but it is something that exists only in Socrates. If Socrates had two noses, as he has two ears, each of his noses would possess both the sortal identity of a nose and an individual identity: there would be, for example, the right nose and the left nose or the red nose and the blue nose. But these noses, although having a specific and an individual identity and therefore being objects of a cognitive individuation and individualization, would not be real individuals. We would treat them as real individuals only if they could show that they have not only an *identity* or *essence*, but their *own being*, for example by leaving the figure of Socrates to spend some time in a rest club for tired noses.

Suppose, to extend this idea, that we make the following discovery: human beings are really only appendages of a large organization, which has long remained invisible and unsuspected, like sores that appear and disappear on a face. We would probably have some trouble acclimating ourselves to such an humiliating discovery, but there seems little doubt that it would mean that we now have a hard time seeing every human being (including ourselves) as an individual, especially if we learned that everything that we have so far considered as the own actions and own thoughts of every human being is in fact the necessary effect of the circulation of animal spirits in the body of the Great Invisible Agent. By losing their ontological autonomy, humans would also lose their individuality, *although nothing would have changed in their specific identity and individual appearance.*[7]

Why this requisite of ontological autonomy? Why would individuality have to be refused to entities that have the being of another, and especially to the cognitively identifiable and individualizable parts of complete beings, like a nose or a door's angle?[8] The answer seems to be that a being that has an individuality is at least a being that is not only set apart from the rest of the world by the play of our cognitive attention, but that is apart by and in itself. To possess one's own being, an *ens per se*, is precisely to be *ontologically* separated from the rest of the things, in contrast with being simply *cognitively* separated from the rest of the world by an act of demonstrative focusing.

If, however, there is a difficulty with this requisite of ontological autonomy, it lies less in the combination of autonomy and individuality than in the development of a *criterion for* autonomy. For what does it mean to have one's own being, an *ens per se*? Surely it cannot mean being physically independent from the rest of the world or being causally encapsulated. Nor, to use another old concept, to be *causa sui*. Being an individual cannot mean being a world apart, being an *absolute* being without any structural or causal connection with the rest of the world. It is indeed clear that an absolute being is a being by itself, an *ens per se*. But the requisite of autonomy must be consistent with the fact that an individual can be spatially, causally, or genetically connected with other individuals or with nonindividuals in such a manner that it depends on them to exist or to continue to exist.

It is therefore essential to distinguish ontological autonomy from absoluteness or existential independence. An *ens absolutum* can be characterized as a being such that there is no other being that determines its existence or on whose existence it depends. In other words:[9]

Abs (a) $=_{Def}$: E!a $\land \neg \exists$x (E!a $\Rightarrow$ Ex).

This characterization can be applied to a transcendent being, such as God, or to the world, if it is considered as an uncreated absolute, but it does not apply to ordinary individuals.

By contrast, to capture the sense of the concept of ontological autonomy (*ens per se*), we need only pay attention to what it means not to have one's own being. It is to have the being of another, which can be expressed as

a exists in another $=_{Def}$: E!a $\Leftrightarrow \Box \exists$x (a << x).

We can therefore characterize ontological autonomy as not needing to be a proper part of something in order to exist:

a exists by itself $=_{Def}$: E!a $\Leftrightarrow \neg \Box \exists$x (a << x).[10]

The difficulty or, if one prefers, the element of gradation that is wrapped in the concept of ontological autonomy comes from the difference that can be made between two ways of having an *ens in alio*. On the one hand, if a being can exist only as a proper part of a determinate being, we may call that *token-inherence*. On the hand, a being might only be able to exist as part of a specific *type* of being, which can be called *type-inherence*. The difference can be expressed as follows:

Token-inherence: E!a $\Leftrightarrow \Box$(E!b $\land$ (a << b))
Type-inherence: E!a $\Leftrightarrow \Box \exists$x (Fx $\land$ (a << x))

Consider a living heart. A heart can only exist as a part of a living organism or of an artificial cardiac stimulation device.[11] A heart therefore has the being of the whole of which it is a member. But a heart is not necessarily a member of this or that determinate organism or artificial device: a heart can be transplanted or retain in vitro activity.

We can therefore assume that parts that are transplantable from a whole to another, that are not token-inherent, can be granted with a minimal form of ontological autonomy and therefore of individuality, provided that they also satisfy the second requisite of individuality that I detail below. Strictly speaking, a heart needs a concrete whole to exist as a living heart. But because it can migrate from one whole to another, it is as if the heart had a minimum ontological autonomy that we could call *potential*.[12]

This example suggests that, with regard to this determinable aspect of individuality, that is, the difference between *ens per se* and *ens in alio*, we can distinguish *levels of individuality*, going from the type-inherence of a heart, at one end, to, at the other extreme, the absoluteness of God or of the world. In the middle, we find ordinary individuality, that of beings that depend physically, causally, or genetically on other beings, but that do not need to be part of one of them to exist.

2.2.2 Individuality and Formal Unity

Does ontological autonomy suffice to constitute individuality? In fact, a great many entities that satisfy the criteria of *ens per se* do not have a genuine individuality: for instance, a pile of sand, a cloud, or a crowd. Eliminating beings that are inherent to others, such as a nose or the corner of a door, is clearly not sufficient to characterize the concept of an individual.

An additional and traditional feature of individuality is what may be called formal unity or cohesiveness, as opposed to material unity or cohesiveness. A being is an individual not only if it has its own being, but if, in addition, it has a constitutive or essential formal unity. This notion may seem vague, but we can actually associate more precise criteria to it.

Individual is the Latin decal of the Greek *atomos*, which literally means "indivisible." But, as the above quote of Stephanus Chauvinus recalls, an individual is more a formal than a material atom. A material atom would be a physically indivisible unit of matter. But many individuals can be cut into small pieces of material. The relevant indivisibility is not material but formal,[13] which can be illustrated this time by the difference between a pile of rice and a rabbit. The physical division of a pile of rice gives rise (until a certain point) to two piles of rice, while the division of an adult rabbit in two (by the middle) does not give rise to two rabbits, but to two pieces of rabbit. The difference between a pile of rice and a rabbit does not lie in their physical divisibility, but in the effect of the physical division on the "form" of the object, that is, on the kind or species to which it belongs. The "form" of a pile of rice does not disappear during the division, but divides itself or duplicates: we obtain two particulars of the kind "pile of rice." In contrast the "form" of a rabbit is indivisible: by physically dividing a rabbit, we remove its "form": what we obtain is only two pieces of rabbit, but not two rabbits.

This notion of formal atomicity, which explains the very word *in-dividual*, is however not sufficient to express the whole idea of formal unity. For what is true of the division is also true of the aggregation. If two piles of rice are aggregated, we get a big pile of rice. This time the form "pile of rice" is not divisible, but *miscible* with another. In contrast, if we aggregate two rabbits, even by trying to mix them, we do not obtain ultimately a larger rabbit, but simply an aggregation of rabbits (or of pieces of rabbit). The form "rabbit" is this time *immiscible*. An individual is then not only a formal atom: it is also a formal immiscible, so that the concept of formal unity, which may seem vague, can be more precisely analyzed into two more operational subconcepts: the concept of formal atomicity and the concept of formal immiscibility.

Why this double requisite of formal atomicity and immiscibility? Why should individuality be refused to formally divisible and formally miscible materials like a cloud, a crowd, a shoal of fish, or a pile of rice, despite the fact that they all have an *ens per se*? The explanation is probably that the double property of formal atomicity and immiscibility reveals the presence of a form proper to the individual or of an internal principle of *closing and completeness*. Consider three oranges and aggregate them by sticking them together irreversibly. Call *oranges-triad* such an agglomerate. Is this agglomerate an individual? The answer is yes, according to the previous criteria: a oranges-triad is formally indivisible and immiscible. The size of the three oranges can increase or decrease over time, their color may deteriorate, but the oranges-triad will persist, whereas it would disappear if we divided the agglomerate or if we mixed it with another.

This fanciful example suggests that individuality is not restricted to organizations or organized beings, whether natural or artificial. For the individuality of a being is less related to its organic nature than to the fact that it has its *own or proper boundaries* that separate it both from beings with which it coexists and from beings of lower levels (of which it can be composed) as well as from beings of higher levels (of which it may be a physical part). Any *item*, when it is cognitively or practically isolated by us, has boundaries. But there is a difference between internal boundaries and external boundaries. Let's just compare our oranges-triad with a state or with a cloud. All these beings have boundaries. But:

- It would be impossible to match the boundaries of two oranges-triads: the boundaries of two oranges-triads may be in contact, but cannot coincide. In contrast the boundaries of two clouds or of two states may coincide.[14]
- The boundaries of our oranges-triad are clear: one can say where exactly the triad ceases and where the outer world begins. But the boundaries of a cloud are fuzzy: is that droplet a part of it or is it outside? And if the boundaries of a state can be clear, it is because they are imposed by *fiat*.[15]

The connection between the notion of proper boundaries and the notion of individuality is quite easy to disclose. The idea of an individual without sharp boundaries or of a vague individual seems contradictory.[16] The beings that are individuals are precisely those that, because they have an internal form or an internal principle of unity and identity, also have their own boundaries. The concept of own or proper boundaries represents an internal or essential separation vis-à-vis the rest of the world. By contrast, an object whose boundaries are conventionally established or are vague does not have an essential separation from the rest of the world: two states, two cities, two clouds, two shoals of fish, two waves can become a single aggregate of the same kind.

Is formal unity as an essential dimension of individuality capable of degrees? Can there be individuals with more or less formal unity?

It is very clear that there may be more or less in *material* unity or cohesion: a crowd has less cohesion than a flock of birds, which itself has less cohesion than a galaxy. But when a compound is not one by aggregation but by its internal structure, can there be more or less in formal unity? In his *Metaphysics*, Aristotle says that "while in a sense we call anything one if it is a quantity and continuous, in a sense we do not unless it is a whole, i.e., unless it has unity of form; e.g., if we saw the parts of a shoe put together anyhow we should not call them one all the same (unless because of their continuity); we do this only if they are put together so as to be a shoe and to have already a certain single form" (Δ, 6, 1016b 12–17). This passage suggests that, for Aristotle, formal unity is a question of all or nothing. However, if the notion of formal unity can be analyzed into the two subnotions of formal indivisibility and of formal immiscibility, it is at least conceivable that there could be degrees of formal unity. It suffices that, alongside the individuals that are characterized by formal atomicity *and* formal immiscibility, we can conceive individuals that would be formally atomic *without being formally immiscible* and others that would be formally immiscible *without being formally indivisible*.

To illustrate this idea, let's introduce an ad hoc notation. Let F be, not any property, but a "form," for instance being human or being rabbit. Note a, b as individuals. Note +, the operation of physical agglomeration, and note $1/a$ and $1/b$, the result of the physical division of an individual.

(1) A prototypical individual is such that, if a and b are F, then $1/a$ and $1/b$ are not F and $(a + b)$ is not F.
(2) A divisible but immiscible individual is such that, if a and b are F, then $1/a$ and $1/b$ are F, but $(a + b)$ is not F.
(3) An atomic but miscible individual is such that, if a and b are F, then $1/a$ and $1/b$ are not F, but $(a + b)$ is F.

It seems clear that cases (1) to (3) represent a certain gradation in formal unity: compared to the maximum of formal unity of case (1), characterized by indivisibility *and* immiscibility, cases (2) and (3) are lower degrees of formal unity and, accordingly, lower degrees of individuality. But are there examples corresponding to cases (2) and (3)? It seems that some plants and some animals like the famous colonial ascidian tunicate *Botryllus schlossseri* correspond to case (3): although formally indivisible, they are formally miscible, so that the fusion of two individuals gives a larger individual of the same kind. And, corresponding to the case (2), individuals that would be formally immiscible while formally divisible, we could perhaps mention some varieties of planarians.

But even if one can remain hesitant about the interpretation of these examples, it is at least clear that, from the strict point of view of the logic of the concept of individuality, there is room for degrees of formal unity, which further strengthens the *scalar* dimension of the concept of individual.

2.2.3 Individuality and Singularity

Consider now what seems to be the third and final notional component of the ontological concept of an individual, a component whose connection with the concept of an individual is more complex. Imagine two notebooks coming out from the same production line. Let's look at the careers of these two notebooks and look at them some time later; cover one with the writings of Mr. X and the other with the writings of Mr. Y. These notebooks are individuals, according to the above criteria: they are *ens per se* and are formally immiscible and indivisible. These two features are present when we take these notebooks at the moment when their production is completed or after they have been used. However, there seems to be a sense according to which they have more individuality once covered with writings of their possessors than at the time they were manufactured.

What appears here is a third traditional dimension of the concept of individuality, the dimension of *singularity*.[17] Singularity, insofar as this notion is relevant to analyzing the concept of individuality, must not to be confused with the much more basic notion of *existential or ontological differentiation*. Any two items, whether or not individuals, must have a basic and primitive ontological differentiation allowing their accounts: they must be *two* existing items or they must have a *different existence*. But to say that *a* has more singularity than *b* is not the same as saying that *a* is ontologically different from *b*. The first judgment, despite appearances, is not *de individuo* but *de specie*. Saying that a human person has more individuality than a cat, and a cat more singularity than a fly, is saying, in effect, that there are more qualitative differences between two human beings than between two cats (of the same variety) and more qualitative differences between two cats (of the same variety) than between two flies (of the same variety). Similarly, saying that there is more singularity in a notebook covered with writings than in a freshly manufactured notebook is saying that there are more qualitative differences between two notebooks covered with writings than between two freshly manufactured notebooks. The singularity is, one might say, a *quantum of qualitative differences* within the same class, so that a judgment of the form "there is more singularity in *a* than in *b*" is actually of the form "there are more differences between two *A*'s than between two *B*'s" or "there are more differentiation among the *A*'s than among the *B*'s."

What is at stake with the notion of singularity is not *ontological discernibility* because, strictly speaking, discernibility applies to any pair of real items: where discernibility is lacking it is not only the *individuality* of the items that is rendered problematic, but it is more radically their simple *separate existence* that is problematic.[18] What is at stake in the notion of singularity is rather the unequal aptitude of individuals to differentiate *qualitatively* from each other intraspecifically. A carbon atom has, from this point of view, less singularity than a fly, because there are fewer qualitative differences (and even none) between two carbon atoms than between two flies.

Why, however, should singularity be a component of individuality? Why should an individual present singularity? The first difficulty is that singularity is not specific to individuals. It also applies to certain aggregates: there is, for example, more singularity in a cumulonimbus than in a stratus and more singularity in a basket of various fruits than in a basket of apples.

But not only is singularity not unique to individuals, it is also not necessary for individuality. If all individuals, regardless of the level of being at which they were taken and the species to which they belong, were unalterable clones of each other within each species, such that there would not be any singularity anywhere, but only individual discernibility and specific difference, there would not be fewer individuals in this world of cloned individuals than there is in the actual world.

Singularity is then rather what might be called a *booster of individuality*. From a strictly ontological point of view, there would be no contradiction in a world with specifically different individuals, but which would be qualitatively indistinguishable from each other within each species. But wherever individuals are not only ontologically distinguishable, but also qualitatively dissimilar, everything happens as if their singularity strengthened or emphasized their individuality, because singularity does not reflect degrees of unity, but rather degrees of complexity of individual unity. A carbon atom is simpler than a cell and a cell simpler than a fish. This scale does not reflect degrees of formal unity, since we have the same formal atomicity and formal immiscibility at each level. However, these individuals offer, metaphorically speaking, a growing *surface of individual differentiation* so that there are more differences between two fishes than between two cells (of the same lineage) and more differences between two cells than between two carbon atoms.

Singularity thus evidently plays a role in the natural multiplication of types of individuals, through the mechanism of natural selection. In addition, for the individuals that we are, we human persons, singularity is a value that completes the consciousness of our individuality. But if an individual, in the ontological sense of the term, is necessarily an *ens per se* formally atomic and immiscible, it is a singular being only because of its increasing complexity. Singularity strengthens individuality, but does not constitute it.

2.3 Individuals and Nonindividuals

Let's now try to complete this analysis by putting together the main ontological distinctions I have introduced. If the analysis is correct, it confirms that the concept of an individual is a scalar concept, a concept that defines a class whose members can be ordered following a triple scale: that of ontological autonomy, that of formal unity, and that of singularity. A source of the debates surrounding the notion of individuality stands, in my view, on the fact that the scalar dimension of the concept of individual is not always explicitly recognized, so that one tends to approach in terms of all or nothing a problem whose solution may lie in the identification of degrees of individuality.

Still, if the concept of an individual is capable of degrees, it does not follow that everything that exists is an individual. Instead there are entities that are not individuals or that lack of any individuality. Figure 2.1 summarizes the main categorical distinctions to which the previous analyses can lead.

I have represented in grayscale the main categories of entities to which the concept of individual can be applied.

The heart of the ontological concept of individual is constituted by the concrete beings that have an *ens per se* and that possess formal atomicity and formal immiscibility. However, at least three other classes of entities are eligible to the status of individuals:

2.3.1 Ideas

What we call "Ideas," in the Platonic sense, are entities separated from the sensible realm. It seems indeed fundamental to distinguish, by adopting the Platonic vocabulary, nonparticipated Ideas and the very same ideas as participated or multiply instantiated.[19] For instance, the Dyad, as a not participated Idea, is certainly a *separate nonsensible individual* because of its ontological autonomy and formal atomicity (although we may question its formal immiscibility). But the Dyad, as a universal instantiated by all pairs of concrete and sensible objects, is not an individual, according to the same criteria. If, like some philosophers of mathematics, we assume

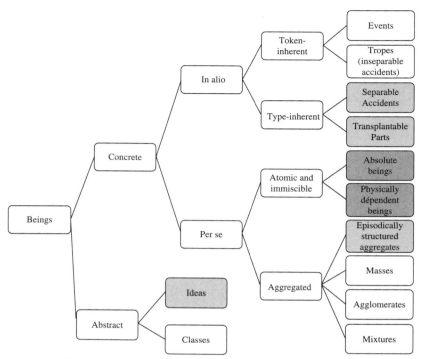

FIGURE 2.1 Schema of the main categories of entities to which the concept of individual can be applied

that there exists such intelligible entities, we have to hold them as intelligible individuals.

2.3.2 Transplantable Concrete Parts

The concrete beings we have called type-inherents, when they also meet the criteria of formal atomicity and immiscibility, can be recognized as individuals of a lower degree. All transplantable parts of living organisms fall into this category, the development of biomedical technologies being a sort of metaphysical tool, revealing the ontological joints of reality. I have also introduced, for the sake of completeness, the category of separable accidents. However, I am not sure that there are accidents that are *naturally* separable. And even if there were accidents *supernaturally* separable, as are the "species" of bread and wine in the Catholic doctrine of the Eucharist, it seems doubtful, despite the Ockham quote in note 12, that these accidents are really individuals, since they do not meet the criterion of formal atomicity.

2.3.3 Episodic Individuals

But the main consequence of the introduction of a narrow concept of individual is to bring out the varieties of the nonindividual.[20] The nonindividual includes categories as diverse as those of events, of tropes, of classes, and of aggregates. All these entities are *real* entities, real beings, but they are not individuals, even if the members of each of these categories can be treated as discrete particulars. From this point of view, the richest class is probably that of aggregates, and a metaphysics of aggregates would certainly be a natural and essential supplement to the metaphysics of individuality.

The distinctions I suggest between masses, agglomerates, and mixtures are probably not complete.[21] However, it is quite certain that among the aggregates, some may be granted the status of *episodic individuals*. Consider a crowd: it is a local agglomeration of people pursuing different ends. It is therefore not an individual. But a crowd can be structured by a common purpose or by a unity of command, which makes a group, for a few moments, an *episodic individual*. And what is true of an aggregate of human individuals is also certainly true of certain physical or biological aggregates, episodically structured by *emerging forms*, stable enough to impose a *formal episodic unity* to the whole aggregate.

2.4 Conclusion

The words have the meaning that we collectively agree to give them. But at the same time, these semantic decisions cannot be arbitrary: if we want to make use of a word, it is necessary that the meaning we give to it or the concept that we associate with it give it a *discriminating* power.

We can certainly be content to give the word "individual" one of its traditional meanings, the one that makes any real item that we can identify and isolate in the context of a demonstrative thought, an individual. But we must admit that such a concept of an individual has little discriminating power, since anything, or almost anything, can be counted as an individual in this sense. Individuality is indeed, then, secured on our *ability* to focus on pragmatically, phenomenologically, or linguistically salient portions of reality. The first Bambinette we come across is then an individual by all rights.

But there is another concept of individual, just as traditional as the other, but whose discriminating power is considerably stronger. According to this concept, an individual is a *being by itself, formally atomic and immiscible and possessing a degree of singularity*. This concept has some flexibility, since it admits that there may be degrees of individuality.

However, as flexible as it is, this ontological concept of the individual is significantly more discriminating than the logical-cognitive concept of an individual. If my analysis is correct, then an event, a state, a galaxy, and a train of waves are not individuals, although they are discrete particulars and possible objects of science. But are they, as Leibniz believed, "well-founded appearances" or "fictions of the mind"?[22] Nothing requires such a conclusion: if they do not have *ens per se* and are not formally atomic or immiscible, they *are* nevertheless, they have their laws, they have their effects: there are really events, states, galaxies, trains of waves.

One of the main consequences of the introduction of an ontological concept of individual is to show, not that individuality is rare, but that, ontologically, the material world is divided into individuals and aggregates. And an aggregate, if it perhaps has less being than an individual, less stability, less unity, is anyway a real being, so that a science of chemical, physical, or social aggregates is certainly not *de individuo*, but is not a science of appearances.

If individuality is important, it is probably because the ontological concept of an individual can reveal the joints of reality. But the joints of reality do not only pass at the boundaries of individuals. Individuals have an ontological stability that may explain the cognitive tropism in their favor. But they are not the whole of being. The world could not contain any individual or be itself the only major individual. There is probably nothing so dramatic if what a science studies cannot claim the ontological status of an individual.

Acknowledgments

I thank Thomas Pradeu and an anonymous reviewer for their comments and suggestions on an earlier draft of this chapter. I also thank Samuel Webb for his assistance in preparing the English version of this chapter.

Notes

1. There are two dimensions in the logical-linguistic division of the world: the individuative and the classificatory. I only deal in this chapter with the individuative division. For a systematic study of "The Division Problem," see Hirsch 1993.

2. In the text of *Categories* that we are discussing, Aristotle does not mention this second differentiation. However, his ontology makes room for massive beings or stuff, especially in the case of what he called "elements," such as water or fire. See especially Aristotle 1922, III, 3.

3. "By [Ideal] I understand the idea not merely *in concreto* but *in individuo*, i.e. as an individual thing which is determinable, or even determined, through the idea

alone" (Kant 1998, 551). The transcendental Ideal is "the concept of an *ens realissimum*" that is "the concept of an individual being" (556).

4. Cf. Aristotle, *De Caelo* 278a10–278b10 or Thomas Aquinas, *Summa theologiae*, I, 46, 1, ob. 1.

5. I have systematically studied this variety of individuals and the nature of the singular concepts we use to think to them in Chauvier 2010.

6. Simons (1987) argues that it is "counter-intuitive to suppose that there are degrees of being an individual" (290). Yet he says a little later (326) that there are degrees of ontological integrity or wholeness. But if the notion of an individual is defined by using concepts such as ontological integrity, then it is inevitable that the *definiendum* will inherit from the scalar dimension of the *definiens*.

7. A less fanciful example is that of the quaking aspen (*pando, populus tremuloides*).

8. We must distinguish abstract parts, such as a color or a shape, and concrete parts, such as a material piece of a certain color or of a certain shape. Abstract parts are universal and cannot be granted the status of individuals, except if we consider them, in a Platonic way, as separate intelligible entities that can exist without being "participated."

9. We use the notation of Simons 1987. "E" is a predicate of existence, distinct from the quantifier; "E!" "represents unique existence" and "<<" means "is proper part of."

10. The use of a modal operator is inevitable. If we did not use the modal operator, it would mean, for example, that a human being could not be part of a football team without losing his or her individuality.

11. We leave aside the case of a heart displayed in a butchery, which, in the words of Aristotle, is a heart by accident. See Aristotle, *Politics*, I, 2, 1253, 20–24 (about a sectioned hand).

12. Another classic example of parts that are type-inherent or transplantable are the accidents of bread and wine according to the Christian doctrine of the Eucharist. With this case in mind, Ockham wrote that, according to the theologians, "*accidens ... is individuum*," "the accident is an individual" (Ockham 1974, 66).

13. Taking the word "formal" according to the opposition of form and matter in Aristotle.

14. According to the criterion we envisage, a state is therefore not an individual since it is neither atomic nor immiscible: the division of a state may make two states (Sudan) and the fusion of two states can make one (West and East Germany). In addition, a state may be geographically dispersed in several unrelated pieces. States are then (more or less) stable aggregates of human individuals and territories.

15. For an analysis of the difference between internal and external boundaries, see Smith and Varzy 2000.

16. The problem of vagueness has therefore two aspects: the vague can have its source in the cognitive individuation, because the concept that we mobilize to individuate is vague. But the vague can also be located in the thing itself: a nonindividual as an aggregate is ontologically vague, because *it has no* ontological individuality. On vagueness, see Williamson 1994.

17. The concepts of singularity and of individuality are close. However, etymologically, the singular is not the atomic, the in-dividual, but the unique, the solitary, the isolated, or the fully differentiated.

18. It seems to me that the debate in philosophy of physics on the discernibility of elementary particles is not really about the individuality of particles, but, more basically, about their existence as entities really distinct from each other. The alternative is not either the particles are *individuals*, or they are *non*individuals (aggregate or events, for example); the alternative, rather, is either the particles are *separate beings*, or they are parts or aspects of a *single holistic being*.

19. In the *Parmenides* (132a), Plato speaks of the singular form (*eidos ekaston*) considered in its unity, in contrast with the multiplicity of participating things. This distinction is reflected in medieval philosophy by the difference between *universalia ante res* (universals before things) and *universalia in rebus* (universals in things). For a contemporary discussion of not instantiated universals, see Armstrong 1989, 75–82.

20. My approach to the nonindividual is quite different from that of E. J. Lowe in this volume, which mainly engages the concept that I call logical-cognitive.

21. On the concept of aggregate see Burge 1977, and Sharvy 1983 on the Aristotelian theory of mixtures.

22. Letter to Arnauld of 30 April 1687 (Leibniz 1960, 100–102).

References

Aristotle. 1922. *On the Heavens*. Ed. John L. Stocks. Oxford: Clarendon Press.
Aristotle. 1928. *The Works of Aristotle*. Vol. 8, *Metaphysica*. Trans. D. Ross. Oxford, Clarendon Press.
Aristotle. 1932. *Politics*. Ed. and trans. Horace Rackham. Cambridge, MA: Harvard University Press.
Aristotle. 1949. *Categoriae et liber de interpretatione*. Ed. L. Minio-Paluello. Oxford: Clarendon Press.
Armstrong, David M. 1989. *Universals: An Opinionated Introduction*. Boulder: Westview Press.
Burge, Tyler. 1977. A Theory of Aggregates. *Nous* 11: 97–117.
Chauvier, Stéphane. 2010. L'unique en son genre. *Philosophie* 106: 3–22.
Chauvinus, Stephanus. 1692. *Lexicon rationale sive thesaurus philosophicus*. Rotterdam: Petrus van der Slaart.
Hirsch, Eli. 1993. *Dividing Reality*. New York: Oxford University Press.
Kant, Immanuel. 1998. *Critique of Pure Reason*. Trans. Paul Guyer and Allen W. Wood. Cambridge: Cambridge University Press.
Leibniz, Gottfried W. [1879] 1960. *Die philosophischen Schriften von Gottfried Wilhelm Leibniz*. Ed. Carl Immanuel Gerhardt. Vol. 2. Reprint Hildesheim: Georg Olms.
Ockham, William. 1974. *Summa logicae*. Ed. Philotheus Boehner, Gedeon Gál, and Stephen F. Brown. *Opera philosophica*, vol. 1. New York: Franciscan Institute, St Bonaventure.
Plato. 1997. *Parmenide*. Ed. R. E. Allen. New Haven: Yale University Press.
Quine, Willard V. O. 1960. *Word and Object*. Cambridge, MA: MIT Press.

Quine, Willard V. O. 1969. *Ontological Relativity and Others Essays.* New York: Columbia University Press.
Sharvy, Richard. 1983. Aristotle on Mixtures. *Journal of Philosophy* 80: 439–457.
Simons, Peter. 1987. *Parts: A Study in Ontology.* Oxford: Oxford University Press.
Smith, Barry, and Achille Varzy. 2000. Fiat and Bona Fide Boundaries. *Philosophy and Phenomenological Research* 60(2): 401–420.
Strawson, Peter. 1959. *Individuals.* London: Routledge.
Thomas Aquinas. 1952–56. *Summa theologiae.* Ed. P. Coramello. Turin: Marinetti.
Williamson, Timothy. 1994. *Vagueness.* London: Routledge.

A Foreword to E. J. Lowe's Chapter

ALEXANDER BIRD

E. J. LOWE'S "NON-INDIVIDUALS" WAS written for the present volume, having been presented at the conference previously organized by the editors in Paris at the Sorbonne in May 2012. As readers will recall with sadness, Jonathan died on 5 January 2014. This was before the processes of commentary on and revision and editing of this chapter could be completed; it is included in this volume with the kind permission of his family. As readers will be able to tell, the chapter needed neither revision nor editing—it is a perfect example of the clear, precisely, and carefully but forcefully argued metaphysics that made Jonathan's work so greatly admired across the globe.

According to Lowe's view in this chapter, the notion of *individual* is a hybrid, being the conjunction of two conditions, *unity* and *identity*. Unity demands that for the entity in question to be an individual, it must be *one* thing; pluralities are not individuals. *Identity* requires that there be a determinate fact of the matter about which entity something is, if it is to be an individual. Lowe holds that unity and identity are distinct conditions that may come apart, so that an entity may fail to be an individual in more than one way: it may lack unity (although it may have identity) or it may lack identity (although it may have unity). This leads to a typology of entities:

Type I: Individuals. These have both unity and identity, for example, classical material bodies and [most] mathematical entities.
Type II: Pseudo-individuals. These have identity but not unity, for example, pluralities or quantities of stuff.
Type III: Quasi-individuals. These have unity but not identity, for example, particles that are members of assemblies of fundamental particles in certain quantum states.

Type IV: Sub-individuals. These have neither identity nor unity. Lowe says that it is difficult to think of real or imaginary examples. Nonetheless, one might imagine that there could be multiple pairs of electrons in an assembly such that each pair fails to have fully determinate identity relative to the other pairs. So each pair would have neither unity (it is a plurality) nor identity.

Now this typology would collapse into the distinction between unities and pluralities if there are no genuine failures of identity. A failure of identity would be provided by a case of two entities where there is no determinate fact of the matter which each entity is. Let us suppose that there is some property, φ, determinately possessed by one entity but not by the other. Then there is a determinate fact of the matter regarding which is which. One entity is that which possesses φ, and the other is that which lacks φ. So for there to be a Type III case, we want there to be entities that are determinately two while possessing all the same properties: there is no difference that could determine the identity of one (and therefore of the other). So finding such an exemplar of Type III requires that the identity of indiscernibles is violated.

The most famous counterexample to the identity of indiscernibles is Max Black's universe consisting only of two spheres, intrinsically alike in every respect. In this symmetrical universe nothing distinguishes the spheres, yet they are two—on the face of it, they are Type III entities. This counterexample suffers from the disadvantage that it is purely conceptual: metaphysical possibility is not guaranteed. And the case allows for an alternative analysis: perhaps it is only one sphere in curved space-time (this allows for the sphere to be a certain nonzero distance from itself). We do not need an imaginary example, however, for physics provides the cases we need where the relevant facts of the matter are perfectly symmetrical. Many examples exist where no facts distinguish the members of a pair of particles, for example, the two electrons in a helium atom in its ground state. In such a case we know for good physical reasons that the number of particles is two. Quantum chemistry tells us that nonionized hydrogen atoms have one electron, helium atoms have two electrons, lithium atoms three electrons, and so on. Furthermore, quantum theory tells us that the two electrons have different spins (which they couldn't if they were only one). Nonetheless this latter fact does not allow us to distinguish the two electrons, for nothing in the physics of the system allows us to pick out one electron as that which possesses spin up and the other as that which possesses spin down. All we know and can know about the electrons is perfectly symmetrical.

(Another example concerns "the" square root of minus one. We know that there are two such roots: if there is one such root there is another, distinct root with opposite sign. Yet all the facts concerning complex numbers are symmetrical with respect to the two roots.)

It is this symmetry that Lowe holds to be the source of trouble for the identity of indiscernibles. Yet, as he notes, some have argued that symmetry need not imply indiscernibility, so long as there is a relation that while symmetrical, is irreflexive. If such a relation applies to the two entities in question, then they are said to be *weakly discernible*. In the case of Black's spheres such a relation would be "_ is more than a mile away from _"; for the electrons the relation would be "_ has a different spin from _"; for the roots of minus one, the relation would be "_ has the opposite sign to _". If weak discernibility amounts to discernibility in the sense of the identity of indiscernibles, then it might appear that the entities in question do have determinate identity, or at least that they are not clear examples of entities that while determinately two in number lack determinate identity. Consequently, Lowe argues, with characteristic clarity and effectiveness, that what is taken for weak discernibility provides no ground for ascribing determinate identity to each entity in such cases. (While Lowe does not put things this way, the existence of weak discernibility shows only that the entities are indeed two in number; weak discernibility tells us nothing about identity.)

Jonathan Lowe's chapter is characteristically imaginative yet rigorous, trenchant yet respectful, and persuasive yet profound. His many readers around the world will regret the passing of a scholar who, with his masterpieces *The Possibility of Metaphysics: Substance, Identity, and Time* (1998) and *The Four-Category Ontology: A Metaphysical Foundation for Natural Science* (2006), led the world both as advocate and as practitioner of metaphysics. His most recent book, *Forms of Thought*, shows that he still had much to offer philosophy at the time of his early death. Those of us who had the good fortune to know and work with Jonathan will be reminded of his kindness and generosity, of his seriousness of purpose, and of his gentle good humor.

CHAPTER 3 | Non-individuals

E. J. LOWE

1. It may seem odd to present a chapter with the title "Non-individuals" in a volume entitled *Individuals Across the Sciences*. But it is often a good way to elucidate the scope of a concept by seeing to what, if anything, it *fails* to apply. Of course, some concepts have universal application, such as, plausibly, the concept of *self-identity*: plausibly, absolutely everything is self-identical. This doesn't mean that the concept of self-identity is trivial or vacuous. Some philosophers consider that the concept of an individual is likewise universal in its application: that absolutely everything is an individual. This likewise would not imply that the concept of an individual is trivial or vacuous. However, I do not agree with these philosophers: I think that the concept of an individual is not absolutely universal in its application. I believe that there are non-individuals, and certainly that there *could be* nonindividuals. But this requires me to explain what I understand by an "individual." This is a philosophical term of art, and to that extent its proper definition is a matter for philosophical debate and decision. It is not a matter to settle by lexicography.

2. First, we need to introduce some terminological distinctions. As I shall be interpreting the term "individual," it does not mean the same as any of the following terms, sometimes used interchangeably with it: "entity," "object," "particular," and "thing." "Thing" I am inclined to reject for philosophical use—because it has too many colloquial connotations—except as part of compound quantifier expressions such as "everything" and "something." I use "entity" as the term of broadest possible scope in ontology: everything whatever that does or could exist qualifies as an "entity" in my sense. By "object" I mean an entity that is a property-bearer but is not itself a property borne: in other words, I use it to mean something in the vicinity of the traditional metaphysical notion of an *individual substance*, although I shall avoid the latter expression for present purposes.

By a "particular" I mean an entity that is not a "universal," and I cash out the universal-particular distinction in the following way: universals always have (or at least *could* have) *instances*, but a particular could not have an instance. Any particular is, on my view, necessarily an instance of some universal (maybe of more than one), but it makes no sense to say that any particular is itself "instantiated" by something (at least, by anything other than itself). Thus, on my view, there may certainly be particulars that are not objects. So-called particularized properties, otherwise known as tropes or modes, are examples of such particulars, for they are not (on my view, anyway) bearers of properties and certainly are themselves properties. (Recall that, as I define it, an object is a bearer of properties that is not itself a property.) In what follows, I shall be concerned exclusively with individual (and nonindividual) *objects*, not individuals belonging to other ontological categories (for instance, individual *tropes*, or individual *events*, or individuals in the category of *universals*, such as individual *attributes* or *kinds*).

3. Here is how I propose to define "individual," perfectly generally:

[Def Ind] Something, x, is an *individual* if and only if (1) x determinately counts as *one* entity and (2) x has a determinate *identity*.

As I remarked above, I don't consider that *everything* is or must be an individual. That is to say, I believe that something may fail to count as *one* entity (failure of clause (1) above) and also that something may lack a determinate *identity* (failure of clause (2) above). As I see it, clause (1) may fail in two different ways, as follows. First, I believe that something may be *many* rather than one: that is to say, I believe in the existence of *pluralities*. Good examples would be *the planets of our solar system* and *the Tudor kings of England*. Second, I believe that something may *lack number altogether*. An example could be *the water in my bathtub*, or *all the gold in the universe*. Clause (2), in my view, may also fail. For instance: sometimes, I believe, *two* entities may be such that *there is no determinate fact of the matter as to which is which*. An example might be the two orbital electrons of a helium atom. There are certainly *two* electrons in this case, it seems, and one of them is spin-up while the other is spin-down: but there is, apparently, no fact of the matter as to *which* is spin-up and *which* is spin-down. I shall come back to this sort of example shortly, since it raises particularly interesting questions.

4. Before that, however, I need to expand on the notion of "oneness" or, as it is perhaps better denominated, "unity" (since "oneness" currently has some unfortunate mystical overtones). Many philosophers think that, trivially, *anything whatever* is just *one* entity (namely, *itself*): that there couldn't conceivably be *something* that failed to be *one* thing. (Or, we should perhaps say, "onething," observing my earlier recommendation to

avoid the freestanding use of the word "thing" for philosophical purposes.) But if we allow the use *plural* quantifiers and *plural* referring expressions, as I believe we should, this assumption must surely be rejected. The planets of the solar system are not *one* thing ("onething"), for they are *nine* in number (or eight, if one excludes Pluto). It would be wrong here to confuse the planets of the solar system with the *set* whose members are those planets. The set is undoubtedly *one* (it is one *set*), but they are *nine* (they are nine *planets*). Note here that ascriptions of number only make sense when they are, either explicitly or implicitly, associated with some suitable *sortal* concept: as, for example, *set* or *planet*. There is an incoherent view, espoused by Berkeley and flirted with by Frege, that *one F* might be (that is, be *identical* with) *many Gs*, where F and G are different sortal concepts: that, for instance, many *bricks* might *be* one *wall*. This is nonsense: many cannot be one. In the case of the bricks and the wall we should say not that many bricks *are* one wall, where "are" expresses identity, but that many bricks may *compose* one wall. Note, however, there is nothing wrong as such in using the plural form of the verb "to be"— "are"—to express identity, provided that it is flanked on *both* sides by plural expressions. We can happily say, for example, that the Jones brothers *are* Tom, Dick, and Harry. This is equivalent to saying that *each* of the Jones brothers is *one* of Tom, Dick, and Harry.

5. The notion of oneness or unity, and hence of cardinality quite generally, also fails of application where *mass* terms are involved, as in the sentence "The water that was in my bathtub is now in the River Thames." One can intelligibly ask *how much* water was in my bathtub, but not *how many*. Of course, it might be contended that the water in my bathtub is or was just *a number (a very large number) of water molecules*. This would imply that the water in my bathtub should really be classified as a *plurality*—a *many*. That would still mean, of course, that "it" is not a *one*. But I take this to be a proposed revision of the semantics of natural language, advanced for scientific or metaphysical reasons. Anyway, what seems clear is that if the atomic theory of matter had been *false*—that is, if matter had turned out to be "homogeneous" or "gunky" in nature—then it would not have been possible to regard amounts of material stuffs as pluralities in this way. In a "gunky" universe, we would have to be prepared to *quantify over* amounts of stuff without allowing that the domain of quantification for such purposes could meaningfully be ascribed a *cardinality*. Note that this is not to imply that the notion of *identity* cannot apply determinately to amounts of stuff, thus conceived. The aforementioned sentence—"The water that was in my bathtub is now in the River Thames," which makes perfectly good sense—serves as a counter to this thought. For there is nothing incoherent in supposing that the *very same* water that was in my bath-tub is now in the River Thames, and this is a

judgement of *identity*. The lesson is that the concepts of countability and identifiability are not necessarily always coapplicable. We can have determinate identity without determinate countability (as in the bathwater case) and we can have determinate countability without determinate identity (as in the electrons case). Note, incidentally, that although *amounts* of (gunky) material stuff are not individuals in my sense, because they lack oneness or unity, it by no means follows that *material bodies, composed* of such stuff, are not individuals in my sense. For it would be a mistake to *identify* a material body with the material stuff composing it. The body must be *unified*, not scattered or spread about arbitrarily, and must have both an *outer surface* and an *interior*. In topological terms, it must be *connected*, even if it has holes running through it. It must also be *maximal*, in the sense of not being a mere part of a larger such body. The search for a precise definition need not detain us here, since the idea is clear enough from familiar examples.

6. Having talked about oneness or unity, I now need to say more about *identity*. And the first thing to say here is that the term "identity" has two quite different, albeit related, uses in metaphysics. First, there is its use to express the logical *relation* of identity, formally signified by the equality sign, "=". The relation of identity may be exhaustively characterized (but I won't say *defined*) by two logical properties of that relation: its *reflexivity* and its governance by *Leibniz's law*. These logical facts are captured by the following two axioms of the logic of identity, where "φ" expresses any predicable condition whatever:

[Id 1] For anything, x, $x = x$.

[Id 2] For anything, x, and anything, y, if $x = y$ then φx only if φy.

I don't regard these axioms as providing even an implicit *definition* of identity, the notion of which I take to be primitive and indefinable. No one could *learn the meaning* of the term "identity" from grasping these axioms, because an understanding of identity is already required in order to grasp them (for instance, it must be grasped that "φ" in its two different—*nonidentical*—occurrences in **[Id 2]** should always be given the *same* interpretation). The second important use of the term "identity" is to express the notion of what is traditionally called *individual essence*. We use the term in this way when we speak of "the identity" of an entity, which is evidently a *non*relational use of the expression. But what *is* the "identity" of an entity, in this sense, it may be asked? I can do no better here than to quote John Locke's famous explication of the term "essence," in what he called its "proper original signification," namely, "the very being of any thing, whereby it is, what it is" (*Essay*, III, iii, 15). It is surely this sense of "identity" that is at issue in the electrons example. For there is no question that, in the *relational* sense of "identity," there are two *nonidentical* electrons orbiting the atom's nucleus. It's just that there is no fact of the matter as to *which* is (say) spin-up and *which* is spin-down, even though only *one* is spin-up and only *one* is spin-down. The clear implication appears to be that the electrons "lack a determinate identity" in the *second* sense of "identity," because it is apparently not true of either electron that there is something whereby it is "the very electron that it is." Indeed, the expression "the very electron that it is" seems to have *no application* to either electron. Such indeterminacy can certainly *lead to* indeterminacy in the identity relation as well, but is not to be confused with the latter. (It can, for example, lead to such indeterminacy in *diachronic* scenarios, such as the following: a positively charged helium ion, He^+, absorbs a free electron from the environment and later emits a free electron into the environment but, it seems clear, there is no determinate fact of the matter as to whether the emitted electron was the previously absorbed electron or the single electron that was originally possessed by the helium ion.)

7. In this context, some philosophers talk about different "strengths" of discernibility (or indiscernibility). For instance, with regard to Max Black's famous example of a world supposedly containing just two equally sized spheres of exactly similar composition one mile apart from each other, it has been said that the two spheres are merely "weakly" discernible for the following reason: each of them bears *to the other sphere* but *not to itself* the relation of being a mile away. Thus, in a recent paper, Adam Caulton and Jeremy Butterfield comment on Black as follows:

> Max Black's famous example of two spheres a mile apart ... is an example of two ... [merely weakly discernible] objects. For the two spheres bear the relation "is a mile away from," one to another; but not each to itself. The irony is that Black, apparently unaware of weak discernibility, proposes his duplicate spheres as a putative example of two objects that are qualitatively indiscernible (and therefore as a counterexample to the principle of the identity of indiscernibles). (2012, 50)

The supposed irony backfires, I think, on Caulton and Butterfield. For we need to distinguish between the following two properties that one of Black's spheres—call it "*a*"—might be supposed to *lack*: (1) the property of being such that it is a mile away *from itself* and (2) the property of being such that it is a mile away *from a*. *Both* spheres evidently lack property (1) and are therefore not discernible in this respect. But if we restrict ourselves, as Black does, to purely "qualitative" properties, *neither* sphere can apparently be said to lack the relevant version of property (2), because neither version appears to qualify as such property. If we arbitrarily label "one" of the spheres "*a*" and "the other" sphere "*b*"—acknowledging that this is done purely for convenience in describing the case—then the point may be made as follows, using the notation of lambda abstraction. Unproblematically, "~λx[x is a mile away from x]a" is true, as is "~λx[x is a mile away from x]b." These truths are unproblematic because only the variable "x" appears within the scope of the lambda-abstraction operator. However, what is thus said not to be true of "one" of the spheres (notionally, sphere a) is *exactly the same* as what is said not to be true of "the other" sphere (notionally, sphere b)—namely, that it has the (purely qualitative) property λx[x is a mile away from x]—so that the two spheres are indiscernible in respect. What is needed, it seems, for their so-called "weak discernibility" is that the following two sentences should *also* be true: "~λx[x is a mile away from a]a" and "~λx[x is a mile away from b]b"—for these two sentences putatively say of a and b respectively that they lack two *different* properties. However, *these* two sentences ostensibly invoke *non*qualitative properties, as can be seen from the appearance of the labels "a" and "b" *within* the scope of the lambda-abstraction operator. Those labels, though, were introduced only for convenience of description, and it cannot be assumed, without begging the question against Black, that there is any basis for regarding either of them as determinately designating one of the spheres rather than the other. Indeed, with the labels understood properly as mere descriptive conveniences, there is no non-question-begging reason to suppose that the property-expressions in the sentences "~λx[x is a mile away from a]a" and "~λx[x is a mile away from b]b" denote genuine properties at all.

8. Similar considerations apply in the electrons example. It cannot, for instance, be said that the two electrons are "weakly discernible" on the grounds that each has the same direction of spin as *itself* but not the same direction of spin as *the other*. This is *true*, but does not give us a genuine *discernible difference* between the two electrons, not even a "weak" one. And, indeed, the broader lesson is that the notion of "weak discernibility" seems to be a spurious one, born of a confusion concerning the logic and semantics of predication and property ascription. If we *don't* use lambda abstraction, the key point fails to emerge clearly. For the sentences, in the Black case, "~a is a mile away from a" and "~b is a mile away from b" can be subjected to lambda abstraction in two different ways, *both* to give the innocuous pair of sentences, "~λx[x is a mile from x]a" and "~λx[x is a mile from x]b," *and* to give the objectionable pair, "~λx[x is a mile from a]a" and "~λx[x is a mile from b]b." In short, and in plain English, "a is not a mile away from itself" appears to be *ambiguous*, as between "a has the property of not being a mile away *from itself*" and "a has the property of not being a mile away *from a*" (and likewise with regard to b). The latter is a contentious reading in the context of Black's example, however.

9. Perhaps it will be felt that I have been uncharitable toward the supporters of the notion of "weak discernibility." Perhaps it may be said that

their position amounts simply to this. *Whichever* sphere we label "*a*" and *whichever* "*b*" (so long as we assign the labels to *different* spheres) the following truths will hold, where "*R*" denotes the relation of *being a mile away from*: ~*Raa, Rab*, ~*Rbb*, and *Rba*. Of course, since *R* is a symmetrical relation, *Rab* and *Rba* are logically equivalent. Still, the fact remains that, for instance, ~*Raa* and *Rab* are both truths and so, by Leibniz's law, $a \neq b$. Recall that Leibniz's law was stated earlier in the following form:

1. For anything, *x*, and anything, *y*, if $x = y$ then φ*x* only if φ*y*.

Instantiating for "*x*" and "*y*" we get:

2. If $a = b$ then φ*a* only if φ*b*.

Contraposing, we get:

3. If φ*a* and ~φ*b* then $a \neq b$.

Now let φ be "~*Ra*ξ" to give (eliminating the double negation):

4. If ~*Raa* and *Rab* then $a \neq b$.

Finally, we can affirm the antecedent of this, "~*Raa* and *Rab*" and so, by modus ponens, detach the consequent, "$a \neq b$." Thus, it may be said, *Leibniz's law itself* recognizes a sense in which *a* and *b* are discernible and hence distinct (nonidentical). And this result holds *however* we choose to assign the labels "*a*" and "*b*"—a choice that can be regarded as completely arbitrary. What more can be demanded to vindicate the notion of "weak discernibility" as characterizing a genuine type of discernibility? But again I reply that this line of thought seems to involve sleight of hand. For *what* conjunction of truths can "~*Raa* and *Rab*" intelligibly and non-question-beggingly be taken to express in the present context? *One* such conjunction is this: "~λ*x*[*Rxx*]*a* and λ*x,y*[*Rxy*]*ab*." But *this* doesn't entitle us to infer, by way of Leibniz's law, "$a \neq b$." *This* conjunction just says of *a* that it does not have the property of being a mile away *from itself*—a property that *b* is equally said not have—and says of *a* and *b* that they stand in the relation of being a mile away *from each other*, which is perfectly symmetrical with respect to *a* and *b*. The *required* conjunction, however, is this: "~λ*x*[*Rax*]*a* and λ*x*[*Rax*]*b*," from which we could indeed derive, via Leibniz's law, "$a \neq b$," since here *a* and *b* are ostensibly said to *differ* in respect of their possession of a certain property. But, once again, for reasons explained above, we are not entitled to regard the lambda abstract "λ*x*[*Rax*]" as having any genuine meaning in the present context, because it could *only* be understood to express a nonqualitative property and to do so *only* on the condition that term "*a*" functions as a genuine name, not as an arbitrary label. Hence, it seems, there is no non-question-begging description of the situation in which Leibniz's law can be called upon to license an inference to the *distinctness* of *a* and *b*—their *nonidentity*—in virtue of some kind of (purely qualitative) *discernibility* between them Thus, I stand by my suggestion that the notion of "weak discernibility" is a spurious one. Black's honor, it seems, is upheld.

10. Here it may be objected that I have so far considered only *two* possible ways of construing the sentence "~*Raa* and *Rab*" in terms of lambda abstraction, when in fact there is also a *third*. The two that I have considered so far are these:

(A) ~λ*x*[*Rxx*]*a* and λ*x,y*[*Rxy*]*ab*.
(B) ~λ*x*[*Rax*]*a* and λ*x*[*Rax*]*b*.

But there is also this:

(C) ~λ*x,y*[*Rxy*]*aa* and λ*x,y*[*Rxy*]*ab*.

In (C), as in (A), neither of the labels "*a*" and "*b*" appears within the scope of the lambda-abstraction operator and hence (C) cannot be objected to in the way that I have objected to (B). The supporter of "weak discernibility" may now say that (C) unproblematically expresses a respect in which *a* and *b* are discernible, because its first conjunct denies that *a* and *a* stand in the relation denoted by the lambda abstract "λ*x,y*[*Rxy*]," whereas the second conjunct affirms that *a* and *b* stand in this relation. However, it seems that this really takes us no further forward than we were when contemplating the original sentence "~*Raa* and *Rab*," for (C) has *the same general logical form* as the latter. This can be seen clearly if we just replace both occurrences of "λ*x,y*[*Rxy*]" in (C) by "*R*," which immediately gives us "~*Raa* and *Rab*" once again. So now the question arises, just as it does with the original sentence "~*Raa* and *Rab*," whether we should construe the first conjunct of (C) as saying of *a* and *a* that "they" fail to stand in a certain relation, or instead as saying of *a* that it fails to stand in that relation *to itself*. An ambiguity remains and, as before, one way of disambiguating (C) provides us with nothing that signifies that *a* is discernible from *b*, while the other way is contentious. That *a* fails to stand in the relevant relation *to itself* is unproblematic and true, but it is equally true that *b* fails to stand in this relation *to itself*. But what about the claim that *a* fails to stand in this relation *to a*? Well, let us suppose, at least for the sake of argument, that this claim too is true. The question remains whether, and if so how, we can exploit it in order to deduce, via Leibniz's law, that "*a* ≠ *b*" is true. The problem is that the *only* way to do this, it seems, is by making the following substitution in line 3 of the proof set out in paragraph 9 above, namely, letting φ be "~λ*x,y*[*Rxy*]*a*ξ." Then we get, as our next line, "If ~λ*x,y*[*Rxy*]*aa* and λ*x,y*[*Rxy*]*ab* then *a* ≠ *b*," and we can proceed as before to apply modus ponens and detach "*a* ≠ *b*" as our conclusion. But this is *formally* no different from what was done in the original proof to get from line 3 to line 4—in that case by letting φ be "~$Ra\xi$"—and hence the objection that was raised concerning the original proof at that point just arises once again.

11. At this point it may be protested that we are stacking the cards against the supporter of "weak discernibility" because Leibniz's law, *as we have formulated it*, only enables us to infer a nonidentity, such as "$a \neq b$," from the fact that something (represented by the schematic letter "φ") is predicable of a that is not predicable of b, or vice versa. (Note, in this connection, that of (A), (B) and (C) in paragraph 10, *only* (B) has the form "~φa & φb," where "φ" is replaceable by *a lambda abstract*, namely, by "$\lambda x[Rax]$"; by contrast, (A) has the form "~φa & φ*ab," while (C) has the form "~φaa & φab.") This formulation of the law, it may be complained, doesn't make room for the notion of something's being predicable of a *pair* of objects, a and b—nor, hence, of its being predicable of *that* pair and not of "another," a and a. And so much the worse for the classical formulation of Leibniz's law, the defender of "weak discernibility" may say (adding, perhaps, that it builds in a prejudice against relations of which Leibniz himself was guilty). However, matters are not as simple as this. For we are entitled to ask what sense it makes to regard *a and a* as constituting a *pair*. How can there literally be a *pair* consisting of an object *and that very same object*? (Surely, it must really be a *singleton*, not a pair.) In the Black example, I am happy to agree that there is a *pair* of spheres, namely, the pair $\{a, b\}$. But that is the *only* pair of spheres that I am prepared to acknowledge in this situation. There is no "pair" $\{a, a\}$ and no "pair" $\{b, b\}$. That being so, even if we were to modify the formulation of Leibniz's law so as to make room for the notion of something's being predicable of a *pair* of objects, this would still be of no avail, it seems, to the supporter of "weak discernibility." For we can concede that a certain relation is predicable of the pair $\{a, b\}$—namely, the relation of *being a mile away from*—without conceding that this same relation is *not* predicable of some *other* pair of objects in the situation, since there *are* no other pairs in that situation, and in particular no "pair" $\{a, a\}$. What this leads to, once more, is the conclusion that, in the Black example, the *only* legitimate way to construe the formula "~Raa" is as saying of a that it is *not a mile away from itself*—not as saying of the "pair" $\{a, a\}$ that its "first" member is not a mile away from its "second." And, thus construed, the formula in question provides no basis for maintaining that there is some "qualitative" respect in which a is discernible from b. For, once again, it is equally true of b that it is *not a mile away from itself*.

12. Now, however—if not long before—it may be objected that my introduction of the names (or, better, *pseudo*-names) "a" and "b" was just a red herring all along, because the case for "weak discernibility" can be

made using a language altogether lacking singular terms and employing only quantifiers and variables. Thus, it may be said that the Black world is (partially) describable by the existentially quantified sentence "$\exists x \exists y(\sim Rxx$ & $Rxy)$," from which may be deduced, via Leibniz's law, "$\exists x \exists y(x \neq y)$." That is to say, the distinctness (nonidentity) of the spheres in that world may be deduced from a "purely qualitative" description of that world. But I *agree* that the terms "*a*" and "*b*" are superfluous, being used only for descriptive convenience, and would allow that they can just be regarded as being, in effect, free variables capable of being bound by quantifiers. *Still* I shall want to maintain that the sentence "$\exists x \exists y(\sim Rxx$ & $Rxy)$" is *three-way ambiguous*, and that the only acceptable disambiguation of it is the following: "$\exists x \exists y(\sim \lambda w[Rww]z$ & $\lambda w,z[Rwz]xy)$"—from which "$\exists x \exists y(x \neq y)$" cannot be derived via Leibniz's law. For the former sentence says only that there is something, x, and something, y, such that x lacks the property of *being a mile away from itself*, while x and y stand in the relation of being a mile away from each other. And again it is equally true that y lacks the property of being a mile away from itself, so that nothing distinguishes x qualitatively from y. The basic problem for the supporter of "weak discernibility," as I see it, is that the following (purely qualitative) description of the Black world seems to be *complete and fully adequate*: "There are two exactly similar spheres; each stands to the other in the relation of *being a mile away from*; and each lacks the property of *being a mile away from itself*." This description *leaves nothing out*, it seems. But the description is *perfectly symmetrical* with respect to the two spheres and hence, it seems, can encompass no qualitative respect in which either sphere is "discernible" from the other: they are, it seems, absolutely qualitatively indiscernible. The "weak discernibility" of the spheres that some philosophers purport to find in the Black world is, I strongly suspect, just an artifact of their preferred logical notation, which should not be regarded as sacrosanct nor as having a special revelatory power where the ontological content of the Black world is concerned.

13. Note that although I have denied that, in the Black example or in the electrons example, Leibniz's law can be appealed to in order to *infer* the nonidentity of the two spheres or the two electrons on the grounds of their supposed "weak" discernibility, I by no means want to *deny* the nonidentity (distinctness) of the two spheres or the two electrons (indeed, I robustly affirmed the nonidentity of the two electrons earlier). If there are indeed *two* spheres and *two* electrons, then the spheres must be *distinct* spheres and the electrons *distinct* electrons. For "x and y are *two Fs*" clearly entails "x is *not the same F* as y." At the same time, however, I want to say, at least in the electrons case, that here we have two Fs both of which *lack a determinate identity*, in my second sense of "identity." For, by my account, as I said before, there is no fact of the matter as to

which electron is spin-up and *which* spin-down, even though *just one* of them is spin-up and *just one* of them is spin-down. In this sort of situation, as I see it, we have a *plurality* of entities none of which has an "individual essence."

14. I am now in a position to use the foregoing considerations to formulate a general *typology* of entities, according to whether or not they satisfy clauses (1) and (2) of my general definition of individuality. This definition, recall, was as follows:

> **[Def Ind]** Something, x, is an *individual* if and only if (1) x determinately counts as *one* entity and (2) x has a determinate *identity*.

Let us call clause (1) the *unity* requirement on individuality and clause (2) the *identity* requirement on individuality. Then there are four possible types of entities, as follows. *Type I*: entities that satisfy both the unity and the identity requirements. These I call *individuals*. Examples would be classical material bodies and such mathematical objects as the natural numbers and sets of individuals. *Type II*: entities that fail to satisfy the unity requirement but do satisfy the identity requirement. These I propose to call *pseudo-individuals*. The most obvious examples are, on the one hand, pluralities of individuals and, on the other, amounts of material stuff, at least on a "gunky" conception of stuffs. In the biological domain, slime molds, for instance, might well be classified as pseudo-individuals, on the grounds that they are really pluralities of simple organisms. *Type III*: entities that satisfy the unity requirement but fail to satisfy the identity requirement. These I propose to call *quasi-individuals*. A good example would be that of the two orbital electrons in our helium atom case. There are also plausible examples in mathematics, such as in the domain of complex numbers, where similar cases arise. *Type IV*: entities that satisfy neither the unity requirement nor the identity requirement. These I propose to call *sub-individuals*. It is difficult, however, to think of plausible examples, either actual or imaginary. In my terminology, pseudo-individuals, quasi-individuals, and sub-individuals are all types of *non*individual, but differ importantly among themselves. It will be observed, incidentally, from my appeal here to mathematical examples, that I have no sympathy for attempts to characterize individuality, in its broadest metaphysical sense, in terms of spatiotemporal or causal features of entities. However, the typological status of an entity might certainly have a bearing on what spatiotemporal or causal characteristics it could have. We saw an example of this earlier, indeed, with the case of material bodies and how they differ, topologically, from amounts of material stuff.

15. It is my hope that the foregoing definition of individuality and the consequent fourfold typology of individuals, pseudo-individuals, quasi-individuals, and (even) sub-individuals may provide, at least for

some relatively modest purposes, a useful conceptual framework for the theoretical sciences, not just in the domain of natural science but also in the domains of the mathematical, human, and social sciences. Here I offer my services only in the capacity of Locke's humble "under-labourer" to the sciences, rather than with any hubristic desire to impose a priori metaphysical constraints on scientific theorizing.

Reference

Caulton, Adam, and Jeremy Butterfield. 2012. On Kinds of Indiscernibility in Logic and Metaphysics. *British Journal for the Philosophy of Science* 63: 27–84.

CHAPTER 4 | Individuality, Quantum Physics, and a Metaphysics of Nonindividuals

The Role of the Formal

DÉCIO KRAUSE AND JONAS R. BECKER ARENHART

The term thing (res) historically precedes the term object (which was introduced by the scholastics) and it is accompanied by a different conceptual meaning. The concept of object, unlike that of thing, implies an intentionality and activity of the conscious subject who, according to the circumstances, isolates and distinguishes, or links and fuses together, the elements of the real world. ... In my opinion, "objectuation" is a primitive activity; that is, it logically (and chronologically) precedes all other activities of thought. (Toraldo di Francia 1981, 220)

INTUITIVELY, AN INDIVIDUAL IS a unity, having identity and being such that it is possible at least in principle to discern it from any other individual.* But when we leave the intuitive realm and attempt a logical analysis, we are faced with a cluster of problems that are difficult to overcome within standard logico-mathematical apparatuses. In this chapter, we shall be concerned with some aspects of this intuitive concept of an individual and some related facts about individuation taken from recent discussions arising in quantum theory, pushing the discussion to a "logical" view. In particular, we propose a tentative characterization of individuals in a structure by means of invariance by automorphisms. We also propose a metaphysics where the notion of identity is substituted, for some objects, by a weaker notion of indiscernibility, and we try to justify such a move. In most of the uses of the expression "quantum theory," we shall not make explicit the distinction between the nonrelativistic and the relativistic approaches—although they of course are quite different—for we think that the problems we shall present appear in both versions. But, as the text goes, the context will distinguish them.

---

* Dedicated to Jonathan Lowe

The "dynamics" of this chapter will be as follows: we begin by discussing some forms of individuality typical of metaphysical discussions. We then describe how classical formalism of mathematics and logic encompasses some forms of those principles, in such a way that we may hold that classical mathematics obeys a theory of identity that endows its objects with some form of individuality. We then argue that quantum objects, given some sensible assumptions, fail to exemplify those features. Then we suggest that a distinct formalism should be employed to encompass the features of quantum objects, just as classical mathematics does encompass a traditional theory of identity. This formalism must take into account the fact that quantum objects are indiscernible and may come in pluralities, a feature not legitimately allowed by classical mathematics.

4.1 Individuals and Individuation

Intuitively speaking, by an individual we understand something that is considered as *one*, distinct from any other individual, and which at least in principle can be reidentified in a different situation (within the same context) as being *that* same item. Some authors leave reidentification (involving diachronic identity) out of the account, claiming that since it is an *epistemological notion*, it should not play a role in a legitimate definition of the *ontological notion* of individual (involving only synchronic identity). Here we shall only point to the fact that even without taking reidentification into account, some interesting metaphysical consequences may be drawn when we consider quantum entities (see Wiggins 2012 for a defense of reidentification as an important feature of individuality). Expressing the concept as we have done does not amount to providing a *definition* of an individual, due to the redundancies and vagueness of the characterization (indeed, most of the employed notions remain undefined, such as "distinct," "same," "reidentification," and so on). But we shall continue with this informal description, saying that an individual has *identity* and is *different* from any *other* individual (we shall take for granted that an individual belongs to a kind). In our view, individuals are entities of any sort, provided that they fall under the above very informal characterization.

Notice that some of the conditions for individuality may fail in some cases, making it clear that the item in question is not an individual. Thus, consider the example of an isolated cloud. Taken in isolation, it seems to be an individual: it seems to have a definite identity, and seems that it is one. However, the illusion is dissipated when we consider that our cloud sometime merges with other clouds to form a bigger cloud. In that case, it loses its identity, and we cannot claim that there is a determinate matter of fact as to whether there are, say, n clouds that form the bigger one. So there is only a *mock* individuality for clouds, as suggested by Toraldo di Francia (1985, 209). The same can be said of a portion of water, which at first can be thought of as an individual until merging with other portions in a lake, when its "individuality" is lost forever. So our notion of individual applies to some things and does not apply to others. Things may simply fail to be individuals, and that happens also when we address some questions concerning the quantum realm, as we shall see.

Our main concern in the cases of clouds and portions of water is not with transtemporal identity (even though that could be an issue too), but rather with the difference between one entity and a plurality of entities. A cloud does not seem to retain its unity in the presence of other clouds. The same goes for portions of water. The possibility of losing unity and identity does not hold for individuals. For instance, consider an ant. By itself, it is an individual. Once it enters the anthill, we cannot follow her any more (by hypothesis) and if some ants now leave the anthill, we may be in doubt whether some particular ant is *our* ant. Even so, in some sense the ant *has* individuality, and the problem is ours of being incapable of distinguishing it among other ants. We could mark the ant before it enters the anthill, say with a little mark and, so, when some ants leave the anthill, we could verify by inspection whether some particular ant is *our* ant. Contrariwise, we cannot mark either a cloud or the content of a portion of water, or a quantum object (we speak of *quantum objects* in a metaphysically neutral sense, not in the traditional sense of "object," which usually presupposes some form of identity condition); a long time ago, Schrödinger remarked that "we cannot mark an electron; we cannot paint it red" (Schrödinger 1953). According to the usual interpretation of quantum theory, he is right. Again, this fact is not grounded in limitations of our abilities or of our technology: we cannot do that because an electron is an entity of a kind such that in some situations any attempt of identification is rendered impossible.

As we shall discuss further in what follows, this impossibility of "marking" quantum objects, together with a plausible methodological assumption on the nature of the ontology of a scientific theory, makes it sensible to consider quantum objects as not being individuals. That is, when we adopt a methodological principle in ontology that quantum theory is the source of our information on quantum objects, along with some kind of prescription that we should not overstep what the theory informs us on what concerns ontological issues, we may conclude that the epistemological limitation is, at least by the moment, better accompanied by an ontological view of such entities as not being individuals. Attributing to them individuality, in the absence of any constraint by quantum theory to do so, would be a gratuitous attribution. Of course, this view only holds if we do not adhere to interpretations of the theory such as Bohm's, which does allow (or even seems to require) such an attribution of individuality (more on this issue soon). However, for a minimal interpretation of the theory, the general guidelines we are here invoking seem to be appropriate to grant that individuality does not apply in the quantum realm.

The search for a principle of individuation is well documented in the philosophical literature (see Quinton 1973 and Lowe 2003 for an overview), and refers to the question: "what is it that confers to an individual its individuality?" One may take a great variety of approaches to this problem, which may include, in one of the extremes, understanding individuality as a brute undefined fact, or as a consequence of some primitive undefined identity every individual is supposedly endowed with. In the same vein, in the other extreme we may hold that there is no such thing as individuality, so that no principle grounding it needs to be sought. In this line of reasoning, items may be taken as nonindividuals, understood as a brute undefined fact. We shall not pursue those kinds of approaches here (but see Arenhart and Krause 2014 for critical discussion of primitive individuality in quantum mechanics). Here we stick with the idea that individuality may be a defined concept, because this line of approach allows us to link identity, individuality, and indiscernibility in a very profitable way from a metaphysical point of view as well as from a logical point of view.

Seen as a defined concept, there are two basic standard answers to the problem of a principle of individuality; the first group can be unified by the term *transcendental individuality*, while the second group falls under the common denomination of *bundle theories*. The first group assumes that the individuality of a thing is provided by something lying *behind* (or *transcending*) its qualities, something that recalls the Lockean concept of substratum. Thus, despite the fact that *two* individuals may share all their qualities, the underlying substratum works for the purpose of granting numerical identity and individuality. Also, there is an easy answer to property change and transtemporal identity: an individual can change all its qualities, remaining the same because it is the substratum that makes the individual *that* individual. Obviously, there are other approaches to the problem that fall under the label "transcendental individuality," not only the substratum theory; for instance, *haecceities* and *primitive essences*. In general, a *haecceity* may be taken to be a *nonqualitative* property a particular item possesses, a nonshareable property lying beyond the item's qualities that grants it individuality (there are distinct versions of haecceitism, and the doctrine is in general related to the philosophical problem of identity in distinct possible worlds; even though this is an interesting problem, we shall not deal with it here). The metaphysical problems with those approaches are well known in the literature: the clarification of the nature of the underlying substratum (or quid, or haecceity) involves insurmountable difficulties. Besides, there are good arguments in the literature against forms of transcendental individuality both in general and in the particular case of quantum entities (for instance, see Teller 1998). For that reason, we shall be not concerned with this topic here.

Bundle theories, on the other hand, define a particular object as a bundle of coinstantiated qualities, so that only properties are needed to account for the nature of a particular object ("qualities" may involve also relations to other objects). Distinct versions of the theory arise accordingly as we specify how we should understand the "bundle" metaphor and also as we specify what it is that ties the properties together. The individuation is provided by some of the qualities or by a group of qualities. But this view also faces problems, such as: can *two* individuals be characterized by exactly the *same* bundle of qualities? If they can, there will be difficulties in explaining how numerical distinction is to be accounted for, and, as some have claimed, some resource to spatial location needs to be made (for a defense of such a view, see Demirli 2010). If they cannot, why is it so? Here too we encounter difficulties, for we need to assume some metaphysical hypotheses. Leibniz did it by means of his famous Principle of the Identity of Indiscernibles (henceforth, PII): indiscernible entities, that is, those sharing all of their qualities, are in fact not distinct entities, but just one. According to this view, there are no entities differing *solo numero*. Apparently, this idea accounts nicely for the individuality of the objects of our surroundings, for we never find two objects exactly alike. In classical mechanics it is also possible to follow this strategy; classical particles may share most of their properties, but they never share spatio-temporal location, for they are assumed to be *impenetrable*; thus space and time act as individualizing qualities in classical physics. This view squares nicely with the bundle theory of individuality because it grants, at least in principle, that no two bundles of properties will ever be the same: spatial location cannot be shared, so that even if most of the properties are the same, the bundles characterizing individuals are not. Of course, in order to maintain this position, we need to explain the nature of space and time without invoking the objects inhabiting it, a task engendering further difficulties. Furthermore, one must provide good grounds for the claim that spatial location is a legitimate individuating property. This is part of a general debate over what counts as a legitimate individuating property for particular objects, a debate the bundle theorist will surely have to face.

In classical logic (and mathematics), this view was captured, for instance, by Whitehead and Russell's definition of such Leibnizian identity in their *Principia Mathematica* (1999, 168), and standard logic and mathematics are Leibnizian in some way. In an updated language, their definition, which links identity (x = y, meaning "the same," "not two," etc.) with indistinguishability (agreement with respect to the attributes) can be put in a simple form as follows: $x = y := \forall F(Fx \leftrightarrow Fy)$, where

F is a variable ranging over the collection of the properties of individuals and x and y are individual terms (this definition can be extended for other levels on the type hierarchy). Part of the trick is to allow the property "is identical to y" in the scope of the properties quantified over in the definition. In that case, the notions of numerical identity and qualitative identity collapse. However, as we mentioned in the previous paragraph for the case of spatiotemporal properties, it is metaphysically doubtful whether properties such as "to be identical with y" may be legitimately employed in the individuation role, since they already presuppose that the item denoted by y is already individuated and available to somehow individuate x.

However, it is possible to conceive the existence of entities differing *solo numero*, that is, nonidentical objects that are qualitatively indistinguishable.[1] By the way, this was precisely the complaint by Ramsey against Whitehead and Russell's definition, qualifying it as a "serious defect in *Principia Mathematica* in the treatment of identity" (Ramsey 1968, 28; Wittgenstein made similar remarks in the *Tractatus*, proposition 5.5302). Classical logic and standard mathematics are typically Leibnizian theories; within their scope, there is no place for indiscernible (indistinguishable) distinct entities. In order to allow indiscernible things within a "standard" mathematical context, we need to do some mathematical tricks we shall see below (but let us remark that these moves just mimic the concept of indiscernibility). Let us now discuss classical identity theory.

4.2 The Standard Theory of Identity (STI) in a Nutshell

By the standard theory of identity we understand the theory of identity of standard classical logic and mathematics, and by this we understand that portion of present-day mathematics that can be built within first-order Zermelo-Fraenkel set theory. Of course we could admit alternatives, say by using other set theories, some higher-order logic, or even category theory, but limitations of space prevent us from discussing some peculiarities of those alternative systems, in particular category theory. In all these alternatives, in general, the particular STI could be a little bit different, but at the bottom all of them say the same: these frameworks are *Leibnizian*. Let us see in what sense this happens in the most usual frameworks.

To consider the case of a first-order setting first, we usually regard the binary predicate of identity "=" as a primitive concept subjected to the following postulates: (reflexivity) every object is identical to itself, or

$\forall x(x = x)$ in a standard first-order language, and (substitutivity): identical objects may be substituted *salva veritate*, or $\forall x \forall y (x = y \rightarrow (\alpha(x) \rightarrow \alpha(y)))$, where $\alpha(x)$ is a formula where x appears free and $\alpha(y)$ results from $\alpha(x)$ by substituting y in some free occurrences of x, y being a variable distinct from x. It results from those two postulates that the relation "=" is also symmetric and transitive. In an extensional set theory, such as ZF, these axioms are supplemented by the Axiom of Extensionality, which is a kind of converse of the substitutivity law, namely, $\forall x \forall y (\forall z (z \in x \leftrightarrow z \in y) \rightarrow x = y)$, which says that sets with the same elements are the very same set. If the set theory admits ur-elements, that is, objects that are not sets but can be members of sets, then the axiom of extensionality reads $\forall_s x \forall_s y (\forall z (z \in x \leftrightarrow z \in y) \rightarrow x = y)$, where S is a predicate saying that something is a set (and not an ur-element), and $\forall_s w\, \alpha(w)$ stands for $\forall w (Sw \rightarrow \alpha(w))$.

In higher-order logic, identity can be defined. The definition is Leibniz's law stated earlier, namely, $x = y := \forall F(Fx \leftrightarrow Fy)$.

Now, it is important to make some remarks about some of the consequences of these postulates and definitions.

(I) When a first-order language involves only a finite number of predicate constant symbols (with no individual constants or functional symbols), we can define identity by the exhaustion of the permutation of the terms in the predicates as follows: let P and Q be the only predicates of a first-order language, P a unary and Q a binary one. Then we can give the following definition:

$$x = y := (Px \leftrightarrow Py) \wedge \forall z \big((Q(x,z) \leftrightarrow Q(y,z)) \wedge (Q(z,x) \leftrightarrow Q(z,y)) \big)$$

This is essentially Quine's strategy (Quine 1986; see also Ketland 2006), but it is originally attributed to Hilbert and Bernays. It is important to note that even though this definition grants reflexivity and the substitution law, the result is only a definition of *indiscernibility with respect to the primitive predicates of the language*, and not identity properly speaking. In fact, it is easy to present a structure in which *two* (distinct) objects are "identical" according to this definition. Just think of a domain composed of straight lines in the Cartesian plane and let P stand for "to have null slope," while Q stands for "to be parallel to." Thus any two horizontal lines are "identical" according to this definition, for both have null slope and are parallel to one another, even though it is easy to see that they may be *distinct* straight lines. So, although the language cannot discern them, this fact does not make them *identical* (to be "the very same" object). This definition is perfectly consistent with the theory of identity given above, in the sense of satisfying the reflexivity law and the substitution law, but we prefer to call the defined relation *"indiscernibility with respect to the primitive predicates of the language."*

(II) Even Leibniz's law can be violated in certain interpretations. In fact, consider a second-order language with Henkin semantics. As is well known, in this kind of semantics, monadic predicate variables do not have as their possible range of values all the subsets of the elements of the domain; rather, we usually restrict the range to only some such subsets. Thus, we may envisage situations in which we have *two* objects that agree in all the chosen predicates, for they may belong to all the subsets of the domain taken in consideration in our semantics (for details, see French and Krause 2006, 257). From the semantical point of view, the only way to vindicate Leibniz's law is to take into account all the subsets of the domain, for in this case we need to consider all singletons, and they make the difference (see below). In that case, however, we lose some important results, such as completeness.

(III) Some philosophers say that even within a logical framework containing the above theory of identity it is possible to speak of *weakly discriminable* objects, that is, objects that obey an irreflexive and symmetric relation (see Muller and Saunders 2008). Furthermore, those philosophers claim that some objects, such as quantum entities, are at best weakly discriminable, so that we may not speak about absolutely discriminable objects, in the sense that there are no monadic predicates to distinguish them. A typical example from quantum physics would be the two electrons of a Helium atom in its fundamental state. The relation could be "to have spin opposite to." It is a physical fact that no electron has spin opposite to itself, and that the two considered electrons have opposite spins. So they are weakly discriminable. But can we be sure that there are no monadic predicates involved? The answer depends on what we call a "property" (see also **IV**). Within the framework of ZF, and by hypothesis we may assume it as our underlying logic, we can reason as follows. If we have a finite number of objects, say the two electrons, we can always enlarge the language of ZF with new constants a and b to name them. Thus, we can define the "properties" (formulas with just one free variable) $I_a(x) := x \in \{a\}$, $I_b(x) := x \in \{b\}$, and so on, depending on the objects we have (in a finite number). So it is a consequence of the axioms of ZF that a and b being two objects, $I_a(a)$ but $\neg I_a(b)$. Thus, they are discerned by a monadic property, and this happens for any object in ZF (see also point **V** below). In other words, the objects of ZF are absolutely discriminable.

It could be argued that the defined property is not a "property" at all, for membership is a relation. But once we accept our characterization above of a property as a formula (of the language of ZF) with one free variable, we have no reasons to refuse the given definition. We shall turn to this point below (see also Krause 2010 and Arenhart 2013 for further discussion on weak discriminability). Furthermore, one could argue that those are not legitimate properties; they are not truly "physical" properties. As the argument goes, one should concentrate on the physical properties and relations, not on the logical ones. However, once we present a language (for our physical theory) and its underlying logic, we must take them into account too; that is, the logical theorems are also theorems of the theory. Furthermore, in general one cannot separate sharply the syntaxes of mathematics (and logic) from that of physics, for the underlying set theory is an integral part of the physical theory. So it would be difficult to say which predicates would count as "physical." Second, we consider logic as the most general part of science and, as such, it cannot be completely a priori. We have no space to develop this view here, but it justifies our appeal to the underlying logic in cases such as the one we are discussing (see also da Costa 2002).

(IV) Another concern is with the notion of "property." Informally speaking, a property is an attribute of a thing. As we mentioned in **III**, some philosophers distinguish "physical" from "logical" properties. As we have argued, it is difficult to make such a distinction, and we propose here a distinction between a physical *approach* to properties and a logical one.

In quantum physics, the notion of property is very elusive, with the issue acquiring different meanings in distinct interpretations (and we shall not try to explore the problem in its whole extent here). The main problem seems to be that we cannot consider a naive realist view that quantum entities are property bearers in the same sense as everyday objects are. Depending on the interpretation we adopt, and in particular on how we see the eigenvalue-eigenvector link, the issue takes a distinct form. Some believe that a system has a property represented by the eigenvalue of a Hermitian operator *if and only if* the system is in its corresponding eigenstate. Others believe that the link is weaker, that is, that *if* a system is in the eigenstate of a Hermitian operator, *then* it has the property represented by the corresponding eigenvalue of the operator. This second option leaves open the matter as to whether a system may bear properties even when it is not in an eigenstate, as some modal interpretations of the theory seem to hold. Of course, even modal interpretations and others with a more realistic flavor than the Copenhagen interpretation will have to deal with issues like noncommuting observables, contextuality, and the like, which impose serious restrictions on a naive understanding of properties in quantum mechanics. Anyway, the theory allows us to calculate the probability that the value (an eigenvalue) of some observable (a Hermitian operator) for the system in a certain state lies inside a certain Borelian. And this is given by Born's rule, as is well known.

But in logic, for instance in stating the above Leibniz law, we regard a property simply as a formula with just one free variable. Assuming classical logic, as we do here (and as Muller and Saunders [2008] do), given such a property denoted by P, each object a either has the property or it has not. It is difficult to mesh this view with the above (very brief) discussion on properties in the quantum world. Worst, the issue becomes complicated if we aim at considering spatiotemporal "properties." Which kind of space and time are we speaking about? Would it be the standard absolute view, typical of nonrelativistic quantum theory, or perhaps the space-time of (special) relativistic quantum mechanics? Whatever move we choose, there are many difficulties facing the project of calling physical properties to testimony in favor of PII when it is stated in the logical fashion. As we mentioned, the representation of properties in logic does not do full justice to physics.

(V) Another problem with a framework such as ZF is that there may be individuals that cannot be defined by any formula of the language. But even so they are individuals in the above sense due to the underlying logic: they are one and have a well-defined identity as objects of ZF. Let us give an example. Due to the axiom of choice, we can prove in ZF that there is a well-ordering over any set A. A well-ordering is a partial order such that any nonempty subset of A has a first element relative to the well-ordering, that is, an element less than any other element of the set (in the given well-ordering). So, in particular there is a well-ordering W on the set $\Re$ of the real numbers (in fact, there are infinitely many of them). Thus, the interval $(0,1)$, described in the usual order, has a first element according to W, which is different from any other real number. The problem is that we cannot point to the difference, for neither can the well-ordering W on $\Re$ be described by a formula of the language of ZF, nor can the first element of any nonempty set be identified. Really, we could "name" it, say, t. But, what is t? Without W we cannot answer that, for we would need to stress that $\forall x(x \in X \rightarrow tWx)$, X being the nonempty subset, and W being the well-ordering; but, without a definition of W, this expression cannot be a theorem of ZF. Even so, as we have said, the first elements of the nonempty subsets X of $\Re$ are individuals, having an identity, as all real numbers have.

Thus, in considering Max Black's famous example of the two spheres one mile apart from each other (Black 1952; see Lowe 2012), we may say that if we model them within a framework such as ZF, there is no way out: they become individuals, and cannot be simply weakly discernible, as some want to say, independently of whether we can (qualitatively or not) discern them or not (Lowe 2012). Logic forces this conclusion. Thus, Lowe's argumentation, according to which "if there are indeed *two* spheres and *two* electrons [he is mentioning the two electrons of a helium atom], then the spheres must be *distinct* spheres and the electrons *distinct* electrons." This is true due to STI and the standard definitions of cardinal numbers. All of this shows that in considering foundational problems of physical theories, in particular of quantum theories, we should look to logic and consider logic as playing an important role in these studies.

Thus, to sum up, in order to correctly take into account nonindividuals, we conclude, the better strategy is to change the logical framework (we will present soon further arguments to back this up). But some still may say the language of ZF is so powerful that we can do the job within ZF in some way. This is correct, but then there will appear some mathematical tricks on the way to considering *legitimate* nonindividuals. Let us see why.

4.3 "Ersatz" Nonindividuals in ZF

Let us consider some conditions for something to be a nonindividual.[2] As we are insisting, the issue depends heavily on the logico-mathematical framework we are working in, and for certain purposes—as of course the ones concerning ontological disputes on identity and individuality—this framework must be rigorously specified.

To make clear what the proposal of rigorous specification of the underlying logic amounts to and to show how it bears on the issue of identity and individuality, let us proceed as follows. First, we shall see how we can "simulate" nonindividuals within ZF, thus defining what we may call *ersatz nonindividuals*. Then we turn to a characterization of *legitimate nonindividuals*.

Suppose we are working in first-order ZF. A structure in ZF is an ordered pair $E = \langle D, R_i \rangle$ ($i \in I$) comprising a nonempty domain D, and a collection of n-ary relations on D. Let us remark that this definition is quite general, and does not necessarily encompass just relations whose relata are individuals of D. The relations may be of a higher order, having as relata also subsets of D or yet more complicated entities (by the way, the structures relevant for mathematics and science—topological spaces, well-ordered sets—are not "first order" structures typical of standard first-order Model Theory). But we do not lose generality if we consider just these "first order" structures in what follows. An automorphism of E is a bijection h from D over D such that for any $x, y \in D$, $x R_i y$ iff $h(x) R_i h(y)$—this can be generalized to n-ary relations. If the only automorphism of E is the identity function (which of course is an automorphism of E), then, E is a *rigid structure*. Two individuals a and b of D are E-*indiscernible* if there is an automorphism h of E such that h(a)=b. Otherwise, they are E-discernible. For instance, consider the additive group of the integers, A=⟨Z,+⟩, where Z is the set of integers and + is a ternary relation on Z (the binary operation of addition). Thus A has two automorphisms, namely, the identity function and h:Z→Z defined by h(x)=−x. Then, for any n∈Z, we have that n and −n are A-indiscernible. In other words, *inside* the structure or, as we may say, from the *internal* point of view, nothing distinguishes between n and −n, for instance, 2 and −2 are A-indiscernible. But of course they are distinct! The problem is that their distinction cannot be seen *from the inside* of the structure A. For any structure E, let us call the E-discernible elements E-*individuals* and E-indiscernible elements E-*nonindividuals*. Then we can represent nonindividuality in ZF, regarding that we need to remain confined to a nonrigid structure, vindicating the above claim about the expressive power of its language.[3] So, by means of some mathematical tricks, we can simulate nonindividuality within standard logical frameworks. But in our opinion this is a philosophical sin if we strongly believe that there may exist *legitimate* nonindividuals (see below). Indeed, if we conveniently extend the structure, their distinctness may be taken into account. In the example above, we can add a new relation < to A, obtaining an extended structure A'=⟨Z,+,<⟩, and in this structure, the integers n ≠ 0 and −n are no more A'-indiscernible, for −2 < 2 but not the other way around. The structure A' is rigid. Furthermore, it is a theorem of ZF that *any structure in ZF can be extended to a rigid structure*. In other words, we can always "leave" a ZF structure E in order to discern the E-indiscernible elements, and convert nonindividuals into individuals. Furthermore, the whole ZF, seen as a structure, is rigid, a fact that sustains the above claim that *every object described in ZF is in fact an individual*.

The above technique of simulating indiscernibility (and nonindividuality) is exactly the strategy used in the *standard* quantum formalism. And of course it would be difficult, if not impossible, to do anything different, for our standard languages are made to speak of individuals (recall the epigraph of the chapter). Let us exemplify this fact with quantum physics. In order to deal with the two electrons in a Helium atom, we start by naming them *a* and *b*, and then we do a trick to veil this distinction; namely, we use antisymmetric functions (and symmetric functions in the case of bosons), such as (being the spin of the property being considered)

$$|\psi_{ab}\rangle = 1/\sqrt{2}\left(\left|\psi_a^{up}\right\rangle\left|\psi_b^{down}\right\rangle - \left|\psi_b^{up}\right\rangle\left|\psi_a^{down}\right\rangle\right).$$

This function changes sign if we permute the labels, but its square, $||\psi_{ab}\rangle|^2$, which gives the relevant probabilities, remains the same. Done! The indiscernibility of the electrons is brought to light. But, recall that, within ZF, being two, they are necessarily discernible and distinct. Thus, in order to speak otherwise, we need either to introduce some new kind of hidden variables, perhaps of a "logical nature," or to leave the standard formalism.[4] Due to difficulties with the former, we prefer to take the second option seriously into account.

4.4 Nonindividuals

Our intuitive view of nonindividuals can be illustrated as follows. They would be such as the Smiths in the film *Matrix*, when Mr. Smith, a computer program, multiplies himself in hundreds to capture the good guy, Neo. During the earlier teenages of one of us (DK), there was a TV cartoon series called *The Impossibles* where one of the good guys was the Multi-Man, who could multiply himself at will.

But, you could say, two replicas of the Multi-Man, or of Mr. Smith, occupy distinct locations in space and time, so they can be distinguished. Independently of the notions of space and time, we can say that this is right. But the problem is that two duplicates of the Multi-Man are to be indiscernible, so if they switch positions, nothing changes at all (at least this is the idea). The same happens with electrons or with quantum objects in general, though they sometimes (depending on the interpretation of the theory) cannot be considered as little balls, as the Multi-Man can). The very nature of quantum objects is not in discussion now; the only fact we want to fix concerning them is their permutation invariance, in addition to the fact that they may be (sometimes) considered as *ones* and *distinguishable* from others of the same species. It is important to remark that there is a theory of *multisets* where the collections (the multisets) may have several copies of *the same* object. For instance, the multiset {1,1,2,3,3,3} has cardinal 6, and not 3 as it would be in ZF. But in this theory STI also holds, so the two "1s" are really *the same* mathematical object, although counted twice in this case. But we do not regard a collection of indiscernible quanta as composed of replicas of *the same* entity (the noncloning theorem prevents that; see Wootters and Zurek 1982), though we can also find an interpretation of them in terms of multisets (for a discussion involving quasi-sets and multisets, see Krause 1991). A final remark concerning this last topic seems to be in order. The no-cloning theorem says that a quantum state cannot be copied while keeping the cloned object intact: the "old" state must be destroyed. So we cannot *produce* clones, hence, absolutely indiscernible objects. But our metaphysics does not avoid the existence of these entities, as Einstein's relativity theory does not avoid the existence of superluminal neutrinos, although we cannot *accelerate* quantum objects to produce them. Our nonindividuals were born nonindividuals.

Thus, let us informally characterize (metaphysically) nonindividuals as particulars having the following features (partially based on Lowe 2012): a nonindividual is an x such that (1) it counts as *one* entity, (2) it does not have a definite identity in the sense that x does not obey STI, (3) but x may be said to be *discernible* from nonindividual y in certain circumstances and *indiscernible* from y in others, and (4) any permutation of nonindividuals of the same kind does not conduce to a different state of affairs.

Indistinguishable quantum objects, *in the usual interpretation* of quantum theory, fit these conditions, whether they are described by nonrelativistic quantum mechanics or by the quantum field theories. Obedience to quantum statistics means exactly this, if we regard "state of affairs" as the relevant probabilities. But note that in (3) we have said that nonindividuals may be *discerned* from other nonindividuals, and not that they may be *different* or *distinct* from another nonindividual. This is consonant with item (2), for otherwise we would be committed to STI. So we prefer not to speak of *identity, difference, distinctness*, or whatever expression that connotes commitment to STI. The concept of "another" nonindividual here can be read as meaning "discernible from"; that is, they may form collections with cardinal greater than one.

Before we discuss how the idea of cardinality enters the stage, let us emphasize the role of the "usual interpretation" in our discussion of nonindividuals. Obviously, the quantum formalism has plenty of interpretations. Some of them, like Bohm's, do carry many more commitments that must be taken into account when it comes to discuss ontology, like the existence of a (hidden-variable accounting for) trajectory, leading to an ontology of individuals in a natural way. Other interpretations carry different kinds of commitments, resulting in distinct implications for ontology.

There is, however, a distinct kind of approach to quantum theory that does not carry such commitments. What we call the "usual interpretation" here is a minimal conception of quantum mechanics as providing for expectation values for measurements. The formalism of quantum mechanics is seen primarily as a tool for calculating probabilities for measurement outcomes. Adding this mathematical apparatus to a minimal realist view according to which quantum mechanics does provide for a description of some kind of reality, our intent here is to follow the lead of the theory and check whether the concept of individuals or nonindividuals (as discussed above) is more plausible when understood along the evidences provided by the theory. As we are arguing, nonindividuals seem to do better in this minimal setting.

However, one could argue that this ontological consequence does not follow from the theory. One may always shift to another interpretation or introduce a principle of individuation as being somehow "compatible" with the theory. After all, it could be said, ontology and epistemology are distinct matters. To answer this worry, we adopt explicitly a methodological principle, according to which we *are advised to put our metaphysics at the same level as our epistemology*. If the best we may know about quantum objects is provided by quantum mechanics, then, it seems fruitful to keep only with the information given by it, not introducing principles of individuation that go beyond the resources of the theory. Of course, other interpretations, such as Bohm's, would provide for another kind of information; the more we put in the theory, the more we get. Here we are concerned with the theoretical minimum, and we believe that nonindividuals do the job when quantum theory is understood in that way.

Having made that issue clear, let us move on to cardinalities. Concerning this topic, a natural question arises: can nonindividuals be counted? As usually understood, to count a collection with, say, five objects (we shall be restricted to the finite case) is to define a bijection from the collection in question on von Neumann's ordinal $5 = \{0,1,2,3,4\}$. But this entails that the considered collection must be a set of, say, ZF, that is, as put long time ago by Cantor, "a collection into a whole of definite, *distinct* objects of our intuition or of our thought" (italics added) (Cantor 1958, 85). In fact, if we cannot distinguish among the objects being counted, we cannot define the bijection (as Russell 1940, 102–103 had already pointed out). But, at least for the purposes of quantum theory, we do not need *to count* electrons in that way. All we need is to know *how many* electrons there are in a certain collection in some circumstances, say in the shell 2p of a sodium atom $1s^2 2s^2 2p^6 3s^1$. The six electrons are indistinguishable, although they obey Pauli's principle. Despite the differences in their quantum numbers, nothing tells them apart while in that shell. The important thing is the *cardinal* of the collection. But sometimes we have a sense in saying that we can count them. For instance, consider a neutral helium atom. By ionization, we can extract one of the electrons, and we can name it a. Thus, the electron that remains is electron b. So, some would say, $a \neq b$, but this again commits us with STI. What is the problem? The problem is that by ionization, we can make the atom neutral again, and the two electrons, b and the new one (is it a?) will be entangled and, according to the usual interpretation of quantum theory, nothing can tell us which is which. But, if they obey STI, they would retain their individuality and the distinction would exist in principle. Since we are not disposed to assume hidden variables (by hypothesis), the best thing is to say that there are no differences to be restored: the important thing is that the neutral atom has again *two* electrons, and we don't need to be worried about their individuality.

Let us give a further example taken from chemistry. In the combustion of methane, whose chemical reaction is $CH_4 + 2O_2 \rightarrow CO_2 + 2H_2O$, one methane molecule reacts with two oxygen molecules to produce carbon dioxide and water vapor. As we see, from the four oxygen atoms present in the two oxygen molecules, two go to the dioxide, and two to the water vapor. But which ones? It does not matter. Chemistry does not depend on this assumption. The individuality of these objects is not relevant to the theory; what really matters is the number!

4.5 A Positive Proposal

Identity and its metaphysical consequences could be suppressed from the philosophical discussion about quantum objects, and perhaps about objects in general. Identity is of course of fundamental importance in mathematics; if *my* number 2 is not the same as *yours*, we shall have troubles in discussing mathematics (suppose that we define the natural numbers as von Neumann did, by equating them to the set of the previous natural numbers in the usual order, and that you define natural number in Frege-Russell style, as "the class of all those classes that are similar to it [there is a bijection between them]" (Russell 1971, 18). Then we may arrive at different results, for we would tempted to form your union set ∪1 producing the universal set in your case, which does not exist in ZF (supposed consistent). But concerning people, chairs, clouds, and quantum objects, indiscernibility seems to be enough. People are discernible from themselves of some time ago, and when time is short, such as a few nanoseconds (the time of a blink), they look indiscernible from themselves, and all happens as if they were identical to the person of a moment ago. Quantum objects fall under this same category: suffice it to say that, in the above case of the Helium atom, the electron released from the atom by ionization and the electron that remains in the atom are discernible, but not necessarily *distinct*, which would commit us with STI (or with some other theory of identity).[5] Concerning the electron that was released and that which was captured by the ion, since the state of affairs does not change at all between the "first" neutral atom and the "second," we may say that they are indiscernible (and not that they are "identical," for they are—by hypothesis—not "the same" object).

Of course we do not need to change our terminology and ways of speaking; we can continue to say that the friend we meet today is the same person we met some years ago for the last time. But we should acknowledge that this is an abuse of language. If we apply STI, we will be in trouble. Concerning nonindividuals, they can be discernible or indiscernible, even if they are of a same kind, say electrons. Bosons in a BEC (a Bose-Einstein Condensate, see Ketterle 2007) could perhaps exemplify better the indiscernibility of nonindividuals, but entangled electrons do the job as well. Summing up, we think that the philosophical discourse would be much simplified if we simply dropped the notion of identity (as linked to STI) from our logic and assumed a weaker notion of indiscernibility instead. In this sense, all objects turn to be nonindividuals according to the given definitions, but they are still "objects of discourse." In a sense, there may be entities without identity.

How can we deal with nonindividuals, either discernible or indiscernible? How can we link our proposal to common language and reference, which demands uniqueness in cases where we speak about "the" electron released by ionization? Is it not the case that definite descriptions, as the standard notions of naming and reference to particular objects, need identity, at least self-identity? This is not the place for the development of a full account of how reference supposedly works in this case, but we believe that uniqueness conditions may be freed from identity by the very notion of cardinal. Indeed, in quantum mechanics we may be able to say that there are n items without needing to count them, so not needing (if someone asks us) to explain how to establish their identities (for the counting relation and cardinal attribution, see Arenhart and Krause 2014). The exchange of the underlying logic from one containing such a strong concept of identity to another one encompassing a weaker notion of indiscernibility would do the job quite well and perhaps may be closer to what physicists intend to say in the quantum domain at least.

Still concerning individuals and cardinals, it is important to say that, once nonindividuals in certain situations should count as one, as when they are isolated from others, they may also form collections with cardinal 1, but in this case we would not be able to distinguish between two unitary collections of indistinguishable nonindividuals of the same species. Such collections (which by their turn also count as two) can have a cardinal, but not an associated ordinal. These collections, which we term *quasi-sets*, may be also indiscernible, namely, when they have the same cardinal and the quantity (given by cardinals) of elements of one kind are the same. Furthermore, these collections must still remain indiscernible from an original one when one of its elements is exchanged (in some way) with an indiscernible one, as it happens with the atoms in the ionization processes mentioned above. Collections with such properties (*quasi-sets*) can be considered within *quasi-set theory*, but we shall not develop it here (see French and Krause 2006, 2010).

4.6 Conclusions

Identity is one of the invariants we use to construct a view of the world (Schrödinger 1964, Toraldo di Francia 1981, 220). Things *seem* to be individuals. Thus, when we face objects of some kind that appear to violate this condition, we become suspicious that something wrong is happening and try to accommodate what we face within our previous frameworks. Physicists are making these moves when they use standard mathematics and classical logic (encompassing STI) on what concerns the indiscernibility of quantum objects (recall the use of the antisymmetric function as exemplified above), even if they are not aware of that. We propose something different. In believing that our logic and our mathematics should be compatible with our metaphysics, and in accepting a metaphysics where nonindividuals are possible, it seems that we have strong grounds to favor the idea of looking for a different formalism to cope not only with our intuitions, but also with observed quantum facts, a formalism that could cope with discernible nonindividuals (playing the role of individuals in certain circumstances), individuals properly speaking (such as mathematical objects like numbers), and of course indiscernible nonindividuals. We can do it by eliminating identity for some objects of our discourse, although it can be maintained for certain objects. Thus, as a primitive concept, we may use a weaker notion of indiscernibility, or indistinguishability, so exchanging "=" by "≡" (an equivalence relation), but contrary to identity, the substitutivity rule doesn't hold in general. That is, indiscernible objects cannot be substituted for each other *salva veritate* in whatever context (just in some of them).

But, you may say, in refusing substitutivity, we are just eliminating the very intuition regarding quantum objects, namely, that the expectation value (roughly, the probabilities) does not change when a quantum object, say an electron, is substituted by "another" one! Of course, this is the right conclusion *by the moment*. In fact, to suppose the failure of substitutivity is the way we found to make indiscernibility distinct from identity. Indiscernibility is only an equivalence relation. Concerning permutations, we can reason as follows: quasi-set theory encompasses certain quasi-sets of objects that can be indiscernible (no identity conditions exist for them) but that have a certain cardinal. It is a theorem of the theory that if we "exchange" (by the set-theoretical operations) an element of the collection with another one that is not in the collection (it is not a *different* one, but a *distinguishable* one), the new collection remains indistinguishable from the original one (yes, the indistinguishability relation applies also for collections—see French and Krause 2010, Th.3.1). This way of expressing the permutation invariance, anyway, seems to be in closer connection with quantum mechanics. If we regard the collection of the electrons of the outer shell of a certain neutral atom as being represented by a quasi-set, the exchanging of electrons by ionization may be represented by the (quasi-)set-theoretical operations in accordance with quasi-set theory.

Furthermore, in eliminating identity, we can maintain Lowe's first condition of his characterization of nonindividuals, namely, that the considered object (a nonindividual) is a unity (see Lowe 2012). Really, we wish to make reference to, and speak of, *the electron* being released from an atom by ionization. The problem is that we cannot keep its identity meaningful after it becomes entangled with other electrons, say in the environment. If we regard it as an individual, we need to conform it to the STI, and then all the above undesirable consequences enter again by the back door. It is amazing that today, when most philosophers see no evil in questioning ancient "principles" like those of noncontradiction and the excluded middle, identity remains untouched. Perhaps it's time for this last dogma to go too.

Notes

1. It does not matter, for the intended application of these ideas to the quantum realm, if our "objects" are particles or fields. In any case, physical theories deal with *things* of some kind, and it is for those *quantum things* we address our considerations (Auyang 1995, 152).
2. The term "nonindividual" is used here for historical reasons (see French and Krause 2006).
3. Of course if the structure is rigid, the only E-indiscernible from an individual x is x itself.
4. By "hidden variables of a logical nature" we mean some form of distinction provided by the underlying logic. As far as we know, the philosophical discussion on hidden variables has never considered such a possibility, which looks quite "natural": the underlying logic imposes a distinction from the start.
5. That is, "discernible" would not entail difference. Since it looks "natural" to suppose that identity entails indiscernibility, the counterpositive is possible only by modifying the underlying logic, as we are proposing: our logic does not contain identity as a primitive relation and it is defined only for certain objects—see French and Krause 2006 for a full discussion.

References

Arenhart, Jonas R. B. 2013. Weak Discernibility in Quantum Mechanics: Does It Save PII? *Axiomathes* 23: 461–484.

Arenhart, Jonas R. B., and Décio Krause. 2014. Why Non-individuality? A Discussion on Individuality, Identity, and Cardinality in the Quantum Context. *Erkenntnis* 79(1): 1–18.

Auyang, Suny Y. 1995. *How Is Quantum Field Theory Possible?* New York: Oxford University Press.

Black, Max. 1952. The Identity of Indiscernibles. *Mind* 61: 153–164.

Cantor, Georg. 1958. *Contributions to the Founding of the Theory of Transfinite Numbers*. New York: Dover.

da Costa, Newton C. A. 2002. Logic and Ontology. *Principia* 6(2): 279–298.

Demirli, Sun. 2010. Indiscernibility and Bundles in a Structure. *Philosophical Studies* 151: 1–18.

French, Steven, and Décio Krause. 2006. *Identity in Physics: A Historical, Philosophical and Formal Analysis*. Oxford: Oxford University Press.

French, Steven, and Décio Krause. 2010. Remarks on the Theory of Quasi-sets. *Studia Logica* 95(1–2): 101–124.

Ketland, Jeffrey. 2006. Structuralism and the Identity of Indiscernibles. *Analysis* 66(4): 303–315.

Ketterle, W. 2007. Bose-Einstein Condensation: Identity Crisis for Indistinguishable Particles. In *Quantum Mechanics at the Crossroads: New Perspectives from History, Philosophy and Physics*, eds. James Evans and Alan S. Thorndike, 159–182. Berlin Heidelberg: Springer.

Krause, Décio. 1991. Multisets, Quasi-sets and Weyl's Aggregates. *Journal of Non-classical Logic* 8(2): 9–39.

Krause, Décio. 2010. Logical Aspects of Quantum (Non-)Individuality. *Foundations of Science* 15(1): 79–94.

Lowe, E. J. 2003. Individuation. In *The Oxford Handbook of Metaphysics*, ed. Michael J. Loux and Dean W. Zimmerman, 75–95. Oxford: Oxford University Press.

Lowe, E. J. 2012. Non-individuals. Prepublication.

Muller, Fred A., and Simon Saunders. 2008. Discerning Fermions. *British Journal for the Philosophy of Science* 59: 499–548.

Quine, Willard V. O. 1986. *Philosophy of Logic*. 2nd ed. Cambridge, MA: Harvard University Press.

Ramsey, Frank P. 1968. *Los fundamentos de la matemática y otros ensayos sobre lógica*. Collected by R. B. Braithwaite. Trans. Emilio del Solar Petit y Wilfred Reyes Scantlebury. Santiago: Universidad de Chile.

Russell, Bertrand. 1940. *An Inquiry into Meaning and Truth*. London: George Allen and Unwin.

Russell, Bertrand. 1971. *Introduction to Mathematical Philosophy*. New York: Simon and Schuster.

Schrödinger, Erwin. 1964. *My View of the World*. Cambridge: Cambridge University Press.

Teller, Paul. 1998. Quantum Mechanics and Haecceities. In *Interpreting Bodies: Classical and Quantum Objects in Modern Physics*, ed. Elena Castellani, 114–141. Princeton, NJ: Princeton University Press.

Toraldo di Francia, G. 1981. *The Investigation of the Physical World*. Cambridge: Cambridge University Press.

Toraldo di Francia, G. 1985. Connotation and Denotation in Microphysics. In *Recent Developments in Quantum Logic*, ed. Peter Mittelstaedt and E. W. Stachow, 203–214. Mannheim: Bibliografishes Institut.

Whitehead, Alfred N., and Bertrand Russell. 1999. *Principia Mathematica.* Cambridge: Cambridge University Press.

Wiggins, David. 2012. Identity, Individuation and Substance. *European Journal of Philosophy* 20(1): 1–25.

Wootters, William K., and Wojciech H. Zurek. 1982. A Single Quantum Cannot Be Cloned. *Nature* 299: 802–803.

PART II | Puzzles about Individuals in Biology and Physics

CHAPTER 5 | Individuality and Life Cycles

PETER GODFREY-SMITH

5.1 Introduction

The living world appears to us as a collection of individual organisms. This individual-based view of life probably runs deep in our "folk-biological" habits of thinking. Indeed, if we imagine the scene confronting our ancestors in prehistory, organisms must have been some of the most clearly bounded and conspicuously countable things around, especially before people started making artifacts. When Aristotle wanted examples of *primary substances*, the most basic kinds of things that exist, in his *Categories*, the examples he gave were individual horses and individual men.[1]

From at least the late 18th century onward, there has been a keen interest in puzzle cases—those where individuality is unclear, or as Darwin described one marine example, "not yet completed." Recent work in biology and the philosophy of biology now contain an extensive literature on what individuals are, how individuals of different kinds arise, and how to think about intermediate and enigmatic cases.[2]

In retrospect, the majority of recent discussion has been concerned with what collects the parts—the spatial parts—of a system into a living individual. People discuss superorganisms, genetic chimeras, modular organisms (such as trees and corals), and symbiotic consortia. But individuals are also unified in time; you are made up of both spatial parts (arms and legs) and stages or temporal parts (you yesterday, you today). Individuals like Aristotle's man and horse persist through time, as well as extending in space.

One term in metaphysics for an object that persists over time is "continuant." This term has narrower and broader uses. In its narrower sense, a continuant is something that is wholly present at various different times. In the broader sense, a continuant is something that persists, but where this "persistence" might be a matter of different stages or temporal parts existing at different times, and a living thing would be seen as a connected succession of these temporal parts. In this chapter I use the broader sense of the term, not taking a stand on metaphysical debates about the nature of persistence itself. Familiar living organisms, then, seem to be *reproducing continuants*. They come into being, persist for a while, may reproduce, and eventually die. While they are alive they grow and develop, and when they reproduce they make more things of the same kind.

This chapter is about some questions about individuality in biology that involve time, especially some that put pressure on the idea of a "reproducing continuant." These are cases of complex life cycles, featuring an "alternation of generations," between physically different forms. I'll introduce the problem using a particular example, briefly consider a range of solutions that involve relatively minor adjustments to standard frameworks, and then outline a new way of approaching these issues, in which the familiar notion of "reproduction" is treated as a special case of a broader category.

5.2 Individuals and Evolution

I began above by noting the role of individuality in an intuitive view of living organisms. To begin the move to scientific contexts, I'll introduce a standard summary of evolution by natural selection, due to the geneticist Richard Lewontin (1985, 76).

> A sufficient mechanism for evolution by natural selection is contained in three propositions:
>
> 1. There is variation in morphological, physiological, and behavioral traits among members of a species (the principle of variation).
> 2. The variation is in part heritable, so that individuals resemble their relations more than they resemble unrelated individuals and, in particular, offspring resemble their parents (the principle of heredity).
> 3. Different variants leave different numbers of offspring either in immediate or remote generations (the principle of differential fitness).
>
> [A]ll three conditions are necessary as well as sufficient conditions for evolution by natural selection. ... Any trait for which the three principles apply may be expected to evolve.

Lewontin's summary can be applied to individuals that are smaller or larger than familiar organisms, but it's certainly supposed to be applicable to organisms as well. In other work I've argued that there is a bifurcation in the "individual" concept as it plays a role in modern biology (2013).

A distinction can be made between *organisms* and *Darwinian individuals*. Organisms, in this sense, are metabolic units, which may or may not reproduce. Darwinian individuals are reproducing entities, which may or may not have the metabolic features of organisms. Both are important kinds of "individuals" from a biological point of view. Within mainstream views of reproduction and metabolism, entities such as people and pigeons are examples of both. Viruses, in contrast, are Darwinian individuals without the metabolic features of organisms, and some symbiotic collectives might be organisms without being Darwinian individuals.[3]

Most of this chapter is about Darwinian individuals, because I'll be discussing evolution rather than metabolic and physiological units. But let's focus here on cases that seem, at least initially, to be in both categories. The Lewontin summary has it that species evolve as a result of patterns in the survival and reproduction of many individual organisms. Individual organisms come to exist, often reproduce, and die. Some reproduce more than others. An important feature of some traits of organisms is *heritability*, a tendency for offspring to resemble their parents more than they resemble unrelated individuals. When a trait is helpful in reproduction and is heritable, it tends to become more common in the species (though there are many exceptions and special cases). Through the accumulation of many of these events, new forms of life can arise. The main ideas here are due to Darwin, a thinker who focused on observable phenomena as much possible. He knew a lot about breeding—of pigeons and farm animals, for example—and drew on these cases when developing his theory. So far, the requirements in Lewontin's summary have a comfortable relationship to the folk-biological view of organisms (though the same probably cannot be said about the folk-biological view of *species*). There are no immediate tensions.

A scientifically oriented person might next note that below these conspicuous countable objects are the entities of basic physics. Modern physics has an uneasy relation to the idea of an individual object (see other chapters in this volume, such as the chapter by Saunders). If we imagine zooming in on a living organism, we find it is made of cells, of molecules, of atoms ... and below the atomic level, the notion of a persisting and countable object becomes problematic. In this chapter I'll remain at a more macroscopic level, though there are interesting questions about how the themes I'll discuss are related to those arising in physics.

As well as organisms, biology recognizes individuals at different scales in time and space. Some relationships between these scales are represented in a famous diagram due to Willi Hennig (1966, figure 5.1 here). Circles represent individuals at different scales, and time goes up the page. Species are the largest circles. Within them are smaller circles for individual organisms, connected by sexual reproduction. Below that level, an individual

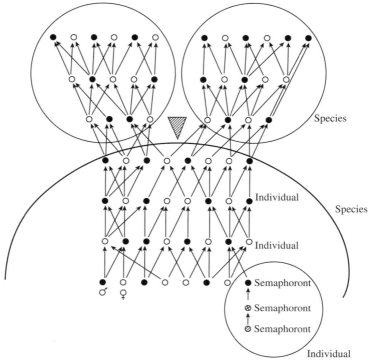

FIGURE 5.1 Change at three scales, in a diagram simplified from Willi Hennig (1966)

organism is broken down by Hennig into stages, which he called *semaphoronts*. Hennig, an entomologist, was very attentive to changes in form within the lifetime of an individual. A juvenile can look entirely different from an adult. In some cases the "stages" are only artificially bounded, as change is more or less continuous. In other cases there is a sharper and nonarbitrary divide between some stages—a metamorphosis step.

I take it that "folk biology" is not greatly troubled by change within a lifetime. A caterpillar changes into a butterfly. Aristotle's horse and human also change a great deal, though more continuously than the caterpillar. Someone might insist that the butterfly is a new object, a different individual, from the caterpillar: when the divide between stages is sharp and the forms are very different, the result is a new organism. But if someone else disagrees, and says it's the same individual though it has changed some of its properties, there does not seem a lot to argue about. The view that the caterpillar and butterfly are stages of a single individual is certainly an available position here.

It's also possible to note something like an inconsistency in Hennig's diagram. Organisms are drawn as, roughly speaking, spatial and temporal parts of a species. At each time, a species-stage has many organisms that are its spatial parts, and at different times in the species' life, different organisms are alive. That is how Hennig handles species and their parts. But at the next level down, an organism usually contains parts of something like the same kind—its cells. It's possible to redraw the Hennig diagram, with the structure within his "individuals" drawn in a way analogous to the within-species structure. In my philosophy of biology textbook, the Hennig diagram was redone to make part-whole relationships more consistent (2014, 29). The smallest scale in the drawing then looks quite a lot like the largest scale; in cell division, each cell gives rise to two, and species do the same thing. Between them, at the level of individual organisms, there is a different pattern: two parents give rise to each new organism, and each organism can enter into many of these interactions and contribute to many new organisms. So far, the theoretical machinery of evolutionary biology and the picture of individual organisms seen in folk biology, filled out with some metaphysics, fit together fairly well.

5.3 Alternation of Generations

This section begins taking a closer look at reproduction, change, and time. The Lewontin summary I quoted does not explicitly mention reproduction, or say what it must involve, but two of its clauses make implicit reference to it: the clause about fitness differences and the clause about heritability. Further, this summary presupposes reproduction in a sense that is stronger than mere *recurrence*; there must be parent-offspring lineages. Each new individual is mapped to one or a few *parent* individuals, who have special causal responsibility for it. Only if reproduction in that sense is occurring can we ask Lewontin's Darwinian questions: do some individuals reproduce more than others, and to what extent are the traits of successful individuals inherited by their offspring?

I'll now introduce some phenomena that put pressure on standard ways of thinking. These phenomena involve the *alternation of generations*. They interfere with the ordinary idea of reproduction in the following manner: rather than *A*s making more *A*s, *A*s make *B*s, which in turn make *A*s.[4]

Alternation of generations is common; it is seen in plants, fungi, protists, and some animals. In some cases there is compression of one generation, so the alternation is barely noticed. In other cases there is more symmetry. Alternation of generations is often part of a *haploid-diploid cycle*: genomes double and halve in size.

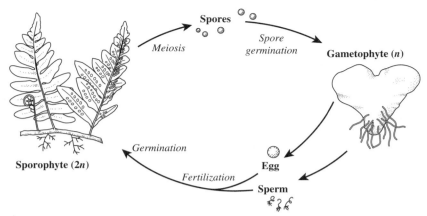

FIGURE 5.2 Fern life cycle
Drawn by Eliza Jewett-Hall.

Ordinary ferns are a dramatic case. The *sporophyte* (familiar fern-shaped plant) is *diploid* (with two sets of chromosomes). It produces *spores* by *meiosis*. Spores have one set of chromosomes. Each spore can grow up into a haploid plant, a *gametophyte*. These are visible plants, with photosynthetic capacity. These produce haploid *gametes* (sex cells). Male gametes swim to female gametes on gametophytes, and fuse with them. A new *sporophyte* grows up there, attached to the gametophyte. So sporophytes produce gametophytes asexually. Gametophytes produce sporophytes sexually. (See figure 5.2 for a summary of the pattern.)

Is this a case of metamorphosis, which we encountered above with butterflies and caterpillars? Perhaps the sporophyte to gametophyte step is metamorphosis? No, because each *A* can make *many B*s, just as each *B* can make many *A*s. If only one *B* can come from each *A*, then this step could be seen as metamorphosis. Here, though, at both stages there is multiplication, and at both stages there is also a *bottleneck*, a narrowing to a single-celled stage.

To think about the ferns, imagine that humans were all the same sex, and produced sperm-like stuff, each cell of which could grow up by cell division into something that looked like a giraffe, and the giraffe-like things mated with each other and produced humans.

What are the "individuals" in this case? It is straightforward to say that both the gametophytes and the sporophytes are organisms, metabolic units in their own right. More generally, I don't suggest that there is a problem saying that both gametophytes and sporophytes are continuants, persisting individual objects. The problem arises around the idea of a *reproducing* continuant—something that persists and makes more things of the same kind. It is the Darwinian individual that is not so clear.

Let's look more closely at cases like the fern. *A*s produce *B*s, and *B*s produce *A*s; who, if anyone, *re*produces? You might say that the new *A* is made indirectly by the old *A*. But the new *A* has also been made directly by a *B*. There are fitness-like properties on both sides: a sporophyte can be good at making gametophytes. A gametophyte can be good at making sporophytes.

If we want to think about a case like this in a way that applies a familiar notion of reproduction, there seem to be five ways to do it:

Priority: Make the *A*-to-*A* process (or *B*-to-*B*) the reproduction step. The other step is part of the reproductive machinery. This is due to some objective distinction between them. Perhaps one stage is larger than the other, or one stage is phylogenetically prior in some relevant way.

Conventionalism: We freely choose which thing is seen as the reproducer; we foreground one and background the other, and the choice is based on convenience. We might switch to and fro if that is helpful.

Reduction or **subsumption**: Treat neither *A* nor *B* as reproducers. Find something lower-level, or higher-level, that does not have alternation: perhaps genes, or cells. If there is reduction to a lower-level reproducer, then *A* and *B* are treated as nonreproducing products of these other objects. If there is subsumption to a higher-level reproducer, *A* and *B* are both treated as mere parts of the reproducing object.

Strange heredity: *A* making *B* is reproduction, even though they are so different.

Entanglement: Both *A* and *B* are reproducers. There are two Darwinian processes tied together. *A* makes *A* with *B* as a way station; *B* makes *B* with *A* as a way station.

In the case of ferns, some might initially choose *priority*, and say the sporophyte reproduces while a gametophyte is a mere way station. The sporophyte is bigger and longer-lived (not so in mosses). But gametophytes sometimes resist the idea that they are not genuine reproducers. They can *break out* of the cycle of alternation, either entirely or partially (Farrar 1967, 1990). About 10% of fern species have gametophytes that reproduce vegetatively, by means of "gemmae"—little propagules with one "apical" cell, plus other material, that give rise to more gametophyes. In some cases gametophytes shift between allocating resources to asexual reproduction of gametophytes, and allocating them to sexual reproduction of sporophytes. In other cases they have given up alternation completely; in the Appalachian mountains of the United States there are at least four species of ferns that have become gametophyte-only. Sporophytes have not been seen for a long time. There are cases of sporophyte breakout as well.

Many biologists might opt for *reduction*. Genetic reproduction (replication) has no alternation, and both sporophytes and gametophytes might be treated as among the varied products of gene action. Parent-offspring relationships between these larger objects are not important—those objects are like clouds in the sky, or dust storms in the desert. Alternation of generations might be used to motivate the replicator-interactor framework of Dawkins (who made the analogy with dust and clouds, 1976) and Hull (1980).

This reductionist approach is a genuine possibility, but in this chapter I'll set it aside without arguing against it. My aim is to use the alternation of generation cases to introduce and motivate a different approach, one involving abstraction rather than reduction, but in this chapter I won't argue that abstraction is superior.

A more unorthodox view of reproduction, which might be applied here, holds that *the cycle itself* is the reproducing object. One token or instance of the cycle gives rise to others (Griffiths and Gray 1994). I think there is a question about the causal claims required by such a view even in simple cases—rather than a cycle-token producing another cycle-token, each stage produces the next stage, in a way that eventually gives rise to a return to earlier forms. A cyclical pattern is better seen as the upshot of stage-to-stage causation, rather than a cause itself. In cases like the fern there is a further problem. When a sporophyte (for example) gives rise to several gametophytes, each of those gametophytes initiates a chain that can give rise to several new sporophytes. If a cycle-token is taken as a concrete material entity here, something that reproduces more things of the same kind, then many of these cycle-tokens will overlap, sharing parts. If sporophyte A_1 produces gametophytes B_1 and B_2, which (sexually) give rise to more sporophytes, those gametophytes are part of many cycle-tokens, depending on how many sporophytes arise downstream. Perhaps this is not a bad problem for the view, but in the next section I'll develop a framework that avoids consequences of this kind.

The other options listed above, *strange heredity* and *entanglement*, seem ad hoc as responses to the problem when considered individually. In some ways, the view I'll sketch below combines elements on both of these. More importantly, although I used to think we had to choose between the options above, now I think cases of the alternation of generations motivate a different kind of move, a generalization in which we treat the existence of objects that allow standard Darwinian forms of counting as a special case of something broader.

5.4 Forms of Production

My analysis begins with a very broad—almost indefinitely broad—concept of *production*.[5] This is seen when one object, or an arrangement of objects or a larger system, gives rise to another, which might be anything at all. This product may in turn give rise to something else.

Within this broad category, some patterns of production exhibit *cycles*, with recurrence of forms seen earlier: *A* makes *B*, *B* makes *C* ..., and eventually this sequence gives rise to a new *A*. Such a cycle can be short—*A* makes *A* directly—or long and indirect. The amount of cycling can change; in the fern "breakout" cases discussed above, a larger cycle (*A* to *B* to *A*) has been reduced to a smaller one (*A* to *A*).

Any form in a sequence might be seen as the "beginning" of a cycle, and whether a cycle is seen at all will depend on the scale at which the system is viewed. There might be recurrence at a microscopic level, but where collections of these low-level objects always give rise to macroscopic novelty, never returning to the same state. In this discussion I'll assume we are dealing with multicellular organisms, and looking at them on a scale where that fact is apparent.[6] If one looks closely, it will also be evident that a "cycle" is never perfect; a new object of the *A* type will be a bit different from earlier ones.[7]

A further distinction can then be made, dividing cases of cyclical production into those that involve *multiplication* and those that do not. From one *A* or a small number, many might arise, directly or indirectly. If there is no cycle—if *A* makes *B* which makes *C*, and onward—this distinction concerning multiplication still *seems* available, but it will (sometimes? always in principle?) be unclear whether there is multiplication or not, because of questions about how to count the products (is that one *B* or several?) If a cycle is present, the distinction between multiplicative and nonmultiplicative cases is clearer: either more objects of the *A* type appear than were present before, or not. The earlier *A* gives us a "reference object" for assessment of multiplication.

Adding another distinction, some chains of production include a *bottleneck* step and some do not. A bottleneck is a dramatic narrowing in the amount of structure present, as in the human reproductive bottleneck to a single-celled stage. As in the case of multiplication, the question of whether bottlenecks exist in a sequence *seems* applicable in cases without cycles, as well as cases with them. But perhaps questions about "more versus less" arise that are analogous to those that arose with multiplication? In any case, from here onward I'll only consider cases with cycles, and will also only discuss narrowings that reduce multicellular stages of organisms to single-celled stages, though the distinction could be handled in a broader way.

Lastly, there may or may not be *sex* at any stage. I understand sex very broadly here, as any fusion of contributions from two productive lineages. Whether sex is present depends, again, on the scale at which the system is viewed.

"Biological reproduction" in the broadest possible sense is cyclical production in a living system, that includes *some* sort of event that differentiates the situation from one of simple persistence with change. Sex, bottlenecks, and (especially) multiplication are all events that might be seen as playing this role, but the cases of reproduction that we are familiar with have a particular arrangement of these features.

Figure 5.3 represents the human life cycle in a way that makes this clear. In this life cycle there is sex, multiplication, and a bottleneck, and they take place once before recurrence at roughly the same place. Humans make many gametes, which can each fuse with a gamete of the other sex to make a zygote, which can grow up into a new human. In figure 5.3, I mark the "multiplication" step as the production of gametes, as this is the point at which many future humans may have their histories converge.[8] Small circles indicating bottlenecks are drawn at both the gamete and zygote stages, but I count this as a single narrowing. The diagram is not drawn as a *cycle*; as noted before, it's good to make explicit the fact that "recurrence" is to a state that is similar, not qualitatively identical, to what was present before.

A case like this is naturally described in terms of reproduction by a temporally extended object, reproduction by a continuant. A new human appears at conception and continues to death, with reproduction along the way. Humans are reproducing continuants. Once we have isolated the features that make this fact clear, we see that the arrangement of them seen in familiar cases is just one of many ways these features can be arranged, even when all of them are present. Figure 5.4 gives a diagram of the same kind for ferns. In each turn of the cycle there are now two multiplication steps, both with a bottleneck, one with sex. (Note that the spores and gametes, here represented as little circles that look similar, are different sorts of things, so there is no tighter cycle from one-celled stage to one-celled stage within the main cycle.) Sex is present, the narrowing of a bottleneck is present, and multiplication is present, but not in the same places they are in the human cycle.

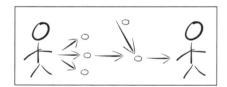

FIGURE 5.3 The human life cycle, drawn in a way emphasizing the location of multiplication (divergent arrows), single-celled "bottleneck" stages (small circles), and sex (convergent arrows)

FIGURE 5.4 The fern life cycle, drawn in a way emphasizing the location of multiplication (divergent arrows), single-celled "bottleneck" stages (small circles), and sex (convergent arrows)

The feature that causes the most disruption for standard ways of thinking about reproduction, as I see it, is the second multiplication step. Without that step, you could see a fern as a reproducing continuant, a single Darwinian individual, despite a lot of change through its life. If each sporophyte could make at most one gametophyte, this could be seen as a metamorphosis step. A sporophyte-gametophyte continuant would then make, by means of sex, more things of the same kind. The bottleneck within the life of the sporophyte-gametophyte would look like a partial "fresh start," but this could be treated as an ontogenetic change of the caterpillar-butterfly kind, though with a more dramatic narrowing.[9] In any case, in ferns there are two steps, not one, with bottlenecks and multiplication before recurrence, and that makes it more difficult to identify a reproducing continuant in the system. There are plenty of continuants, and plenty of *producing* ones, but no reproducing ones, at least in the familiar sense.

I'll introduce a case with yet another arrangement of these ingredients. Figure 5.5 represents the life cycle of a Scyphozoan jellyfish (the most familiar kind of jellyfish) in a redrawn version of a 19th-century diagram

FIGURE 5.5 Life cycle of a scyphozoan jellyfish. Stages 1–8, planula attachment and metamorphosis to scyphistoma stage; 9–10, scyphistoma strobilation; 11, ephyra release; 12–14, transformation of the ephyra into an adult medusa.

Redrawn by Christine Huffard from a figure by M. J. Schleiden (*Die Entwicklung der Meduse*, 1869).

by Mattias Schleiden. Here there is alternation between *polyp* and *medusa* forms. The medusa swims, and the polyp, formed by development of a sexually produced larva, lives anchored to a surface. In this case there is a bottleneck only at the sexual production of gametes by the medusa, but there is multiplication at two stages: polyps produce medusae multiplicatively but without a bottleneck (stage 11 in figure 5.5).

Figure 5.6 gives the linear diagram for the jellyfish. As the figure shows, before recurrence there are two multiplication steps, one bottleneck, with sex at the multiplication step with the bottleneck.[10] In *Cubozoan* jellyfish, in contrast, each polyp metamorphoses into a single medusa. But polyps themselves can multiply by budding, so again there are two multiplication steps with one bottleneck. And in the Cubozoan case, there is a subcycle from polyp to polyp within the larger cycle from medusa to medusa.

Many other ways of arranging the features are possible. Some, as far as I know, are hypothetical. Suppose one bottleneck step and one multiplication step are present before recurrence, but these are not the same step. Then two features of familiar "reproduction" would appear once in the cycle, but at different points. For me, the multiplication step would be in a sense the more important one, with respect to identifying the beginning of the life of a continuant, but I don't see this as something to argue about. The important point, rather, is that this dissociation of the multiplication and bottleneck steps would be a departure from the structure that makes thinking in terms of reproducing continuants natural and convenient.

One interesting empirical case has recently been revised in response to ongoing work. For some time it appeared that in *rhizocephalan* barnacles, which parasitize crabs, the barnacle is able to cross into the interior of the crab's body by injecting a single-celled stage through crab's shell (Glenner and Høeg 1995). The single-celled stage, it was thought, then grows up into a multicellular structure, which eventually reproduces sexually, having made its way again to the exterior of the crab. Assuming that the normal outcome would be a single multicellular stage inside each crab, this would be a case where there are two bottleneck stages in the cycle, one multiplicative and one nonmultiplicative, where the multiplicative one includes

FIGURE 5.6 The scyphozoan life cycle, drawn in a way emphasizing the location of multiplication (divergent arrows), single-celled "bottleneck" stages (small circles), and sex (convergent arrows)

sex. However, more recent work indicates that there is no reduction to a single-celled stage after all, and the invader remains multicellular in its passage into the interior of the crab (Glenner et al. 2000).[11] So this is not, as it seemed to be, a case with a nonmultiplicative single-cell bottleneck. I don't know whether there are other cases.

It would be interesting to multiply examples in a longer discussion (see Herron et al. 2013, and Haber, this volume, for some good and difficult cases), but instead I'll take stock.[12] The kind of cyclical production seen in familiar cases of reproduction, the kind that enables us to recognize reproducing continuants—Darwinian individuals—is a subset of a larger set of phenomena, actual and possible. The simplest cases (simple for us to think about, not simple in their biology) have a cycle with one multiplication step, one bottleneck, and sex, all at roughly the same location in the cycle. If those features are in place, a good deal of metamorphosis can be present without interfering with the folk-biological impression of a series of reproducing continuants. Sometimes there is no sex at the step with a bottleneck and multiplication, and that has prompted some to recognize distributed "evolutionary individuals" who *grow* through the bottleneck rather than reproducing (Janzen 1977). Others, including myself (2009), have seen these asexual cases as reproduction. As we move further and further from a situation with a single multiplicative bottleneck, we reach phenomena that are harder and harder to think of as ordinary reproduction by a continuant. In a case like the jellyfish, with two multiplication steps, there is considerable disruption of the familiar pattern. When both multiplication steps have their own bottlenecks, as in the fern, there is more disruption still.

To the extent that standard concepts of heritability and fitness depend on simple and familiar modes of reproduction, they need to be generalized in some way to cover all cases. Some would see this as a reason to shift Darwinian description down to a genetic level—to reduce rather than abstract. An alternative is to see the form of multiplication and recurrence present at the genetic level as one of the variants. What we call "persistence" of sequence in genes across replication events is a very tight cycle.[13]

It would be interesting also to link this treatment of production and reproduction to recent literatures on symbiosis, and the role of symbionts in the evolution of multicellular collectives. It can be argued, for example, that because of their metabolic dependence on cellulose-digesting bacteria, horses like Aristotle's are *recurring* objects that arise through the repeated fusion of animal and bacterial parts, without being things that stand in parent-offspring relations of the familiar kind.[14] The part of a horse comprised of eukaryotic animal cells does not make a new horse (even sexually). Instead, this object and its sexual partner produce the eukaryotic animal part of a new horse, and that provides a context in which the prokaryotic part of the new horse becomes established and grows up. I don't yet accept, or reject, this analysis, but if we work within it provisionally for a moment, it combines with the ideas above in an interesting way. There are two sets of phenomena at odds with the simplest "like-makes-like" picture. One is comprised of cases where there is *production* without each object giving rise to similar things; organisms make, but don't make "like." Another is comprised of cases where the recurrence of organism-like units occurs through the fusion of many productive lineages, so that "like" recurs, but not through its production by a small set of parent individuals of similar form. In two ways, then, the elements of an intuitive notion of reproduction—recurrence and production—can be separated from each other. These are two coins, rather than two sides of the same coin, because the phenomena that involve recurrence without simple parent-offspring relations can (for all that's been said so far) involve relatively simple like-makes-like phenomena at the level of the parts that come together, and the phenomena that involve cyclical production without like directly making like need not include the fusion of many diverse productive lineages. Each can, in principle, occur alone, but they also occur together.

5.5 Closing Thoughts

I'll conclude by indicating some themes that might be developed in further work. Individuality of the familiar horse-and-man kind is a derived trait, especially in collectives. The evolution of individuals of this kind involves organization of biological material in space and time. Here I've looked specifically at organization in time. What evolutionary factors tend to give rise to life cycles that support familiar notions of individuality?

Applying adaptive thinking here is complicated by the fact that we are talking about the evolution of arrangements that help make adaptive evolution possible. Bottlenecks, for example, make collective lineages more "evolvable," but that may not be the reason they exist. As organisms evolve, forms of reproduction change, and then so does the mode of evolution. With those cautions in mind, though, here are some thoughts.

The idea of "division of labor" is usually applied to spatial parts, but it can also be applied to stages, including metamorphosing stages and alternation of generations. Over space, division of labor is a form of heterogeneity that does not compromise collective individuality; it is heterogeneity with "common purpose." The same thinking can be applied to time. Familiar cases of metamorphosis often have a division of labor between feeding and reproduction (and other tasks such as dispersal).

A temporal division of labor is associated with special expenses of breakdown and rebuilding, however, and a temporally organized division of labor can sometimes be replaced by a spatially organized one. Complex life cycles are often associated with simpler (i.e., spatially simpler) organisms. Simplification of the life cycle is seen in some conspicuous lineages, such as flowering plants and vertebrates. One evolutionary path has been toward a large, physiologically *complex life-form* that has a *simpler life cycle* than various of its relatives (algae and ferns, cnidarians) and likely ancestors. This is part of what gave rise to the paradigm folk-biological organisms—Aristotle's individual man and individual horse—the sorts of objects, back before Aristotle, that perhaps most induced us to think and talk in subject-predicate structure in the first place, a structure that exerted such strong effects on the philosophical tradition.

Acknowledgments

Ancestors or earlier forms of this chapter were given at "Philosophy of Biology at Dolphin Beach 4" (Moruya, Australia, 2010), at the "Individuals Across the Sciences" conference (Paris, 2012), and at the 2013 meetings of the European Society for the Philosophy of Science in Helsinki. Thanks to Thomas Pradeu and Maria Cerezo both for their organizational work at the latter two meetings and for many helpful comments on this material. Thanks also to Eliza Jewett-Hall for figure 5.2 and Christine Huffard for figure 5.5. For comments on drafts I am also grateful to Matt Haber, Maureen O'Malley, Ron Planer, and an anonymous referee.

Notes

 1. Aristotle, *Categories*, 1a20, 2a13 (Ackrill 1963).
 2. See Buss (1987), Maynard Smith and Szathmary (1995), Dupré and O'Malley (2009), and the papers in Bouchard and Huneman (2013). For the Darwin quote, see *The Voyage of the Beagle* (1839, 128).
 3. See Dupré and O'Malley (2009), Pradeu (2012) for relevant arguments, and Godfrey-Smith (2013) for discussion.
 4. Some pressure on a simple view of the role of reproduction in evolution comes from sexual dimorphism, especially in organisms where the sexes are more different than they are in horses and men. In anglerfish, barnacles, and other animals, females can be hundreds of times the size of the males, which may be little parasites clinging to the female. Sometimes the two fuse altogether. Setting aside questions arising from fusion, in any mating in a dimorphic species one parent has offspring very different from itself, and this puts pressure on the idea of heritability.

5. The framework I'll outline here has been influenced in places by Griesemer's work (2005). Appendix A6 of my 2009 book also complements this section by giving a formal treatment of some features of production in a generalized sense.

6. See O'Malley (forthcoming) for a discussion of unicellular forms of reproduction. It would be interesting to consider using something like the framework developed here to these cases, but it would probably have to be modified.

7. Here I generally treat symbols like "A" as referring to types, which may be coarse or fine-grained. Some sentences in this discussion will not differentiate types and tokens, in a way that I hope will be clear in context.

Various special cases may arise that are not considered here. It would be possible, for example, for a "cycle" to exist with respect to one form in a sequence but no others—A recurs but the intervening steps are always different.

8. They may also converge at the zygote stage, as in monozygotic twinning, but I'll set those cases aside to keep things simple.

9. Ben Kerr (personal communication) suggested that this move can be applied to the ferns even with the second multiplication step: all the asexually produced material (including all the gametophytes coming from one sporophyte) could be seen as parts of a continuant initiated by sex. In the terms of the taxonomy introduced earlier, this is a "subsumption" move. One feature of this view that some might regard as problematic is the fact that the gametophytes are genetically different from each other, because of meiosis, as well as being different from their originating sporophyte. On these issues see also Godfrey-Smith (2009, chapter 5) for discussion of the evolutionary importance of multiplication.

10. Earlier I mentioned the possibility of using phylogenetic criteria as a basis for "priority" claims in these cases (as suggested by an anonymous referee). In the jellyfish case, it is thought fairly likely that the medusa stage is a later addition to the life cycle in which the adult form was a polyp (Marques and Collins 2004).

11. This case was discussed by Stephen Jay Gould (1996) and picked up as a problem for Dawkins's analysis of reproduction (1982) by Sterelny and Griffiths in *Sex and Death* (1998). I made use of the case in *Darwinian Populations and Natural Selection* (2009) but without checking more recent empirical work. So this is not, as I said in that book, a case where a metamorphosis stage without multiplication includes reduction to a single-celled bottleneck. I am grateful to Jens Høeg for correspondence about this case.

12. Herron et al. (2013) and Haber (this volume) both emphasize cases where life cycles have alternative paths (A can give rise to A via B ..., or via C ...). This is seen also in some of the fern "gametophyte breakout" cases mentioned earlier. This is an important complication that a fuller development of these ideas would have to consider in detail.

13. See Godfrey-Smith (unpublished) for development of this point.

14. Herron et al. 2013 discusses the connection between the two sets of issues, and see Dupré and O'Malley (2009) and Pradeu (2012) for arguments bearing on the horse.

References

Ackrill, John. 1963. *Aristotle: Categories and De Interpretatione*. Oxford: Clarendon Press.

Bouchard, Frédéric, and Philippe Huneman, eds. 2013. *From Groups to Individuals: Perspectives on Biological Associations and Emerging Individuality*. Cambridge, MA: MIT Press.

Buss, Leo. 1987. *The Evolution of Individuality*. Princeton, NJ: Princeton University Press.

Darwin, Charles. 1839. *Journal and Remarks, 1832–1836 (Voyage of the Beagle)*. London: Henry Colburn.

Dawkins, Richard. 1976. *The Selfish Gene*. Oxford: Oxford University Press.

Dawkins, Richard. 1982. *The Extended Phenotype: The Gene as the Unit of Selection*. Oxford: W. H. Freeman.

Dupré, John, and Maureen O'Malley. 2009. Varieties of Living Things: Life at the Intersection of Lineage and Metabolism. *Philosophy and Theory in Biology* 1: 1–24.

Farrar, Donald. 1967. Gametophytes of Four Tropical Fern Genera Reproducing Independently of Their sporophytes in the Southern Appalachians. *Science* 155: 1266–1267.

Farrar, Donald. 1990. Species and Evolution in Asexually Reproducing Independent Fern Gametophytes. *Systematic Botany* 15: 98–111.

Glenner, Henrik, Jens Høeg, Jack O'Brian, and Tim Sherman. 2000. Invasive Vermigon Stage in the Parasitic Barnacles Loxothylacus texanus and L. panopaei (Sacculinidae): Closing of the Rhizocephalan Life-cycle. *Marine Biology* 136: 249–257.

Godfrey-Smith, Peter. 2009. *Darwinian Populations and Natural Selection*. Oxford: Oxford University Press.

Godfrey-Smith, Peter. 2013. Darwinian Individuals. In *From Groups to Individuals: Perspectives on Biological Associations and Emerging Individuality*, ed. Frederic Bouchard and Philippe Huneman, 17–36. Cambridge, MA: MIT Press.

Godfrey-Smith, Peter. 2014. *Philosophy of Biology*. Princeton, NJ: Princeton University Press.

Godfrey-Smith, Peter. (Unpublished). Complex Life Cycles and the Evolutionary Process. To be presented at PSA 2014.

Gould, Stephen J. 1996. Triumph of the Root-Heads. *Natural History* 105: 10–17.

Griesemer, James. 2005. The Informational Gene and the Substantial Body: On the Generalization of Evolutionary Theory by Abstraction. In *Idealization XII: Correcting the Model, Idealization and Abstraction in the Sciences*, ed. Martin Jones and Nancy Cartwright, 59–115. Amsterdam: Rodopi.

Griffiths, Paul, and Russell Gray. 1994. Developmental Systems and Evolutionary Explanation. *Journal of Philosophy* 91: 277–304.

Hennig, Willi. 1966. *Phylogenetic Systematics*. Trans. D. Dwight Davis and Rainer Zangerl. Urbana: University of Illinois Press.

Herron, Matthew, Armin Rashidi, Deborah Shelton, and William Driscoll. 2013. Cellular Differentiation and Individuality in the "Minor" Multicellular Taxa. *Biological Reviews* 88: 844–861.

Hull, David. 1980. Individuality and Selection. *Annual Review of Ecology and Systematics* 11: 311–332.

Janzen, Daniel H. 1977. What Are Dandelions and Aphids? *American Naturalist* 111: 586–589.

Lewontin, Richard C. 1985. Adaptation. In *The Dialectical Biologist*, ed. Richard Levins and Richard Lewontin, 65–84. Cambridge, MA: Harvard University Press.

Marques, Antonio, and Allen Collins. 2004. Cladistic Analysis of Medusozoa and Cnidarian Evolution. *Invertebrate Biology* 123: 23–42.

Maynard Smith, John, and Eors Szathmáry. 1995. *The Major Transitions in Evolution*. Oxford: Oxford University Press.

O'Malley, Maureen. 2014. *Philosophy of Microbiology*. Cambridge: Cambridge University Press.

Pradeu, Thomas. 2012. *The Limits of the Self: Immunology and Biological Identity*. New York: Oxford University Press.

Sterelny, Kim, and Paul Griffiths. 1998. *Sex and Death: An Introduction to Philosophy of Biology*. Chicago: University of Chicago Press.

CHAPTER 6 | What Biofilms Can Teach Us about Individuality

MARC ERESHEFSKY AND MAKMILLER PEDROSO

6.1 Introduction

This chapter uses the example of biofilms to examine Hull's (1980) and Godfrey-Smith's (2009) accounts of biological individuality and to explore the nature of individuality more generally. Biofilms are single or multispecies communities of microorganisms. Biofilms are useful for examining accounts of biological individuality because they fail to satisfy commonly suggested properties of biological individuals, such as having reproductive bottlenecks, forming parent-offspring lineages, and being composed of members of the same species. Nevertheless, biofilms have many of the other qualities associated with biological individuals. Biofilms are embedded in a self-produced extracellular substance that prevents predation, captures nutrients, and allows the cells of a biofilm to communicate and share genes. Biofilms have repeatable life cycles, and they have biofilm-level adaptations that vary among biofilms. On some accounts of reproduction, they are reproducers. Whether there is inheritance between earlier and later biofilms is an open question. Nevertheless, there is trait transmission between earlier and later biofilms, even though biofilms form via aggregation.

Biofilms provide a good case for studying biological individuality. The nature of biofilms, for example, suggests that Godfrey-Smith's account of biological individuality is too restrictive. Biofilms fare poorly on Godfrey-Smith's account of individuals because they fail to have reproductive bottlenecks, they do not stand in the appropriate parent-offspring relations, and their reproductive division of labor is not high. Despite violating Godfrey-Smith's criteria for paradigmatic or middling individuals, biofilms fulfill many (if not all) of his underlying desiderata for individuals in natural selection, which gives us reason to think they are in fact individuals.

In this chapter we support Hull's interactor account of individuality. We do so for two reasons. One is that according to Hull's account of individuality biofilms are good candidates for interactors in natural selection and hence individuals in natural selection. The other reason is that Hull's interactor account of biological individuality can be placed in a more general interactor account of individuality, one that extends beyond biology. In what follows we develop an interactor account of individuality in the tradition of Hull's, but we renovate Hull's theory in several ways. First, we place it in a general sortal framework, along the lines of Wiggins's (2001) sortal account of identity. There are different sorts of individuals in the world, and when asking if an entity is an individual we need to specify the sort of individual under consideration. Second, we dive into the metaphysics of individuality, articulating the different types of processes (internal interaction versus external forces) that cause entities to be parts of an individual. Third, we depart from Hull's commitment to replicator theory and allow for a more liberal account of reproducers. In the end, we offer an interactor account of biological individuality embedded in a more general theory of individuality.

6.2 Biofilms: A Primer

Let us start by introducing our case study. Biofilms are found throughout the environment. They grow on the rocks of rivers, on the surfaces of stagnant water, and on our teeth. The bacteria of a biofilm collectively produce, and are embedded in, an extracellular polymeric substance (EPS). EPS matrices hold the cells of a biofilm together. More interestingly, they are digestive systems that trap nutrients in the environment and break those nutrients down with extracellular enzymes (Flemming and Wingender 2010). EPS matrices also protect biofilms with molecules that bind to antimicrobial agents and prevent their access to biofilm cells. In addition, EPS matrices are media for cell communication among the bacteria of a biofilm, and they foster the exchange of genetic material through lateral gene transfer (see below).

The life of a biofilm proceeds through a series of stages (Hall-Stoodley, Costerton, and Stoodley 2004). For example, a multispecies oral biofilm begins its life cycle with first colonizers, *Streptococcus gordonii*, attaching to tooth surfaces. Then secondary colonizers from the species *Porphyromonas gingivalis* coaggregate with the cells already attached (Kolenbrander et al. 2010). Coaggregation is "a process by which genetically distinct bacteria become attached to one another via specific molecules" (Rickard et al. 2003, 94). It is common for biofilm formation to be a sequential process involving different species at different stages

(Kolenbrander et al. 2010). Once a biofilm is fully colonized it matures. Then, in its last stage, dispersal cells are produced and released to the environment. A biofilm life cycle, thus, consists of four stages: planktonic lifestyle (unattached single cells); attachment; colonization; and dispersal.

The cells of a biofilm interact in numerous ways. Quorum sensing, for example, is a cell-to-cell signaling system that enables bacteria within a biofilm to regulate cellular density. Quorum sensing occurs through the secretion and detection of molecules called "autoinducers." When the concentration of autoinducers reaches a certain threshold, cell differentiation in a biofilm is affected (Davies et al. 1998). Another signaling system, called "molecular signalling," affects a biofilm's lifecycle. For example, low concentrations of nitric oxide produced by *P. aeruginosa* trigger biofilm dispersion (Stewart and Franklin 2008).

Another type of biofilm interaction is lateral gene transfer (LGT). LGT is gene transfer among bacterial cells that is not due to reproduction. It occurs among conspecific strains and strains in different species. Biofilms provide favorable conditions for LGT. Consider two LGT mechanisms: transformation and conjugation. Transformation consists of the uptake of free DNA from the environment by a bacterial cell. Transformation requires extracellular DNA. In biofilms this prerequisite is met because environmental DNA is a major constituent of biofilms. The other mechanism for LGT, conjugation, occurs via cell-to-cell junctions or bridges. Such bridges allow the transfer of mobile genetic elements, usually plasmids. The physical stability caused by EPS matrices reduces the chance of conjugal bridges breaking (Ehrlich et al. 2010). In short, lateral gene transfer occurs within biofilms for several reasons: the occurrence of extracellular DNA, high cell density, and the physical stability EPS matrices provide.

Stepping back from these details, we see that biofilms have repeatable life cycles. Those cycles are caused by various types of interactions within biofilms, such as quorum sensing, molecular signaling, aggregation, and lateral gene transfer. In addition, EPS matrices serve as digestive systems, defense mechanisms, and media for communication. Biofilms are not mere agglomerations of organisms but groups of organisms with multiple finely tuned interactions. Though some biofilms contain only conspecifics, many biofilms are composed of organisms from multiple species.

6.3 Godfrey-Smith's Account of Biological Individuality

Godfrey-Smith's (2009, 2013) account of biological individuality starts with Lewontin's (1985) characterization of natural selection. According to Lewontin, natural selection occurs when three necessary conditions are met: there is variation among individuals; that variation is heritable; and that variation results in differential fitness among individuals. Godfrey-Smith explores all three of Lewontin's conditions for selection, but he pays special attention to the role of reproduction in selection. As Godfrey-Smith (2009, 86) writes, "The link between 'individuality' and reproduction is ... inevitable. Reproduction involves the creation of a new entity, and this will be a countable individual." Individuals in natural selection must be reproducers: those entities that not only vary and have differential fitness, but also have countable descendants.

Godfrey-Smith's discussion of reproducers focuses on what he calls "collective reproducers": "reproducing entities with parts that themselves have the capacity to reproduce, where the parts do so largely through their own resources rather than through the coordinated activity of the whole" (2009, 87).[1] Multicellular organisms are collective reproducers. Godfrey-Smith measures such reproduction using three parameters. Those parameters come in degrees. The higher an entity scores on the parameters, the closer it is to being a paradigmatic individual. The first parameter is reproductive bottleneck. According to Godfrey-Smith, paradigmatic cases of reproduction require a bottleneck, such as when a zygote develops from a small propagule. Human reproduction involves such bottlenecks. No bottleneck occurs when a new structure is formed by the aggregation of cells, for example when free living *Dictyostelium* cells aggregate and form a slime mold (2009, 95). Godfrey-Smith's second parameter measures the degree of reproductive division of labor within a reproducer. Humans score high because we have distinct germ and soma lineages, where the first type of lineage is responsible for reproduction. Sponges, on the other hand, score low on this parameter when they reproduce asexually because any fragment of a sponge can start a new sponge (2009, 92). Godfrey-Smith's third parameter, integration, concerns the boundary between an individual and its environment, and the mutual dependence of its parts. Mammals have high integration, buffalo herds have low integration, and sponges somewhere in between.

Are biofilms reproducers and individuals on Godfrey-Smith's account? Let's start with the bottleneck parameter. A bottleneck occurs when a new individual develops from a small propagule. A bottleneck does not occur when a new individual is the result of the aggregation of numerous cells. As we saw earlier, biofilms form by aggregation. Consequently, they lack bottlenecks. Though biofilms fail Godfrey-Smith's bottleneck criterion for reproduction, they nevertheless satisfy his reason for positing bottlenecks as a condition for paradigmatic individuality. Bottlenecks foster biological individuality because when a mutation occurs in the germ line of an organism, a bottleneck spreads that genetic change to an individual's somatic cells. As Godfrey-Smith writes, "Because a bottleneck forces the process of growth and development to begin anew, an initially localized mutation can have a multitude of downstream effects" (2009, 91). Biofilms have an alternative process for doing this—lateral gene transfer. Biofilm evolution is due in no small part to the introduction of new genetic material in a biofilm, and then the transfer of that material to other parts of a biofilm. For example, Ehrlich and coauthors (2010) posit the Distributed Genome Hypothesis (DGH) to explain the abundance of biofilms that cause chronic disease by citing the role of LGT in biofilms. According to DGH, new disease strains are caused by novel combinations of genes that are spread throughout a biofilm by LGT. The example of DGH illustrates how LGT distributes genes in a collective that lacks a bottleneck. It illustrates how Godfrey-Smith's motivation for requiring bottlenecks for paradigmatic or marginal individuals can be satisfied when there is no bottleneck present.

While biofilms score poorly on Godfrey-Smith's bottleneck parameter, they have an intermediate score when it comes to division of reproductive labor. Recall that humans score high on this parameter: few of our parts are passed on, just our gametes. Sponges, when they reproduce asexually, score poorly on reproductive division of labor because any part can start a new sponge. Slime molds score in the middle: they have "some reproductive specialization," though more of their parts can reproduce than the parts of a mammal (Godfrey-Smith 2009, 95). Biofilms are in the same boat as slime molds when it comes to division of reproductive labor. More parts of a biofilm can start a new biofilm than the parts of a mammal can start a new mammal. Nevertheless, there are specialized reproductive cells in a biofilm, in contrast to the cells in a sponge. For example, a biofilm's planktonic cells are dispersal cells and the source of new biofilms, whereas sessile cells primarily perform the function of helping biofilms adhere to surfaces. Biofilms, thus, score somewhere near the middle on Godfrey-Smith's criterion of division of reproductive labor.

Biofilms do well on Godfrey-Smith's third parameter for reproduction, integration. Godfrey-Smith measures integration by how effectively an entity maintains its boundary between itself and the environment, and how much its parts depend on each other for their viability. Biofilms are distinct from their environments. The cells of a biofilm are molecularly bonded through aggregation and bounded within an EPS matrix. An EPS matrix catches and digests nutrients from the environment and protects a biofilm's cells from predators. Furthermore, the cells of a biofilm share genetic material via LGT, and there is intercellular communication within a biofilm that regulates a biofilm's development. These interactions set a boundary between a biofilm and its environment (more on this below). Biofilms also score high on Godfrey-Smith's other measure of integration, the degree to which the parts of an individual rely on each other for their viability. There are a number of biofilm-level processes that cause bacteria to have a significantly higher survivorship when they are part of a biofilm than when they live on their own (Costerton 2007). For example, a biofilm's EPS matrix contains chemicals that protect its component bacteria, and it contains mechanisms for catching and digesting nutrients.

Stepping back from these details, we see that biofilms lack bottlenecks, they score somewhere in the middle when it comes to division of reproductive labor, and they score high on integration. For Godfrey-Smith, paradigmatic reproducers, and consequently paradigmatic individuals, need to score high on all three parameters (2009, 94). Biofilms do not—they fail to have bottlenecks and are middling on division of reproductive labor. Nevertheless they score high on integration because they have a number of processes that promote their stability and demarcate them from the environment. Moreover, though biofilms fail to have bottlenecks, they satisfy Godfrey-Smith's motivation for requiring that paradigm individuals have bottlenecks: to spread mutations among the parts of an individual. Biofilms do this through LGT. Biofilms satisfy this desideratum of Godfrey-Smith's account, yet they fail to satisfy his specific criteria for being either paradigmatic or middling individuals. This gives us reason to think that biofilms are individuals in natural selection, and that Godfrey-Smith's account, as it stands, is too restrictive an account of individuality.

Biofilms reveal a further problem with Godfrey-Smith's account, namely his view of what sort of parent-offspring lineages can form individuals. Some biofilms are multispecies. Godfrey-Smith allows the existence of multispecies individuals so long as the different species lineages within an individual run in tandem (2013). He cites the case of aphids and their symbiotic bacteria to demonstrate this. These bacteria and their host aphids have the same reproductive cycle: an aphid mother transfers bacteria to its offspring through its ovary. Biofilms, however, do not consist of lineages that run in tandem. The bacteria that form a biofilm are scattered in the environment and they come from different sources. Furthermore, their coaggregation occurs at different stages of biofilm formation. In other words, the different bacterial lineages that comprise a biofilm do not run in tandem, and they fail to form a unified parent-offspring lineage in Godfrey-Smith's sense. More generally, biofilms serve as a counterexample to the requirement that an individual be composed of lineages that have the same beginnings and endings. We hasten to add that the occurrence of LGT in biofilms and not in aphid-symbiont combinations (Nikoh et al. 2010) makes biofilms better candidates for biological individuals than aphid-symbiont combinations.

Before leaving Godfrey-Smith's account of individuality, we should address several possible objections to biofilms being biological individuals. First, one might worry that biofilms are ecological communities and not individuals. We have tried to address that concern above. To emphasize that biofilms are individuals and not merely communities, contrast biofilms with common examples of symbiotic complexes, such as the symbiotic relation between ants and acacias (Godfrey-Smith 2011). Bacteria in a biofilm exchange genetic content; ant/acacia symbionts do not. Bacteria within a biofilm build and employ EPS matrices. Such matrices, as we have seen, defend a biofilm's bacteria from predators, capture and digest nutrients, and facilitate communication among component bacteria. The sorts of interactions and interpenetrations that occur among the bacteria of a biofilm outstrip symbiotic and other kinds of ecological relations.

One might grant that biofilms are more organized than ecological units, but nevertheless maintain that biofilms are not individuals in natural selection, that they are something in between. For example, Godfrey-Smith (2013) discusses metabolic organisms. A metabolic organism is a system of entities that collectively work together using environmental resources to maintain that system. For Godfrey-Smith, some organisms fail to be individuals (in natural selection) because they do not form reproductive lineages. Godfrey-Smith explains his individual/organism distinction using the example of squid-bacteria symbiotic complexes. Godfrey-Smith argues that such complexes are not individuals in natural selection because each complex is "a metabolic knotting of reproductive lineages that remain distinct" (2009, 30). Using this concept of organism, one might object that biofilms are not individuals (in natural selection) but merely organisms as Godfrey-Smith defines them. We respond by pointing out that biofilms are not simply organisms *sensu* Godfrey-Smith. Bacterial lineages within a biofilm do not remain distinct, as do the lineages of squids and their symbiotic bacteria. Lateral gene transfer genetically blends the different species lineages of a biofilm. Then there are the other interactive processes among the bacteria of a biofilm that we have discussed, such as the production of a common EPS matrix that protects biofilms from predators, captures and digests nutrients, and allows cells to communicate for biofilm growth. The interactions and interpenetrations among the bacteria of a biofilm outstrip the interactions among the members of symbiotic consortia. A biofilm is more than a mere metabolic knotting of bacteria.

Finally, one might worry that if biofilms are not reproducers given Godfrey-Smith's multifaceted account of reproduction, then biofilms are not reproducers at all. And if biofilms are not reproducers, then they are not individuals in natural selection. We address this concern in section 6.5.

6.4 An Interactor Account of Individuality

In this section we turn to an interactor account of individuality. Our aim in this section is threefold. First, we offer a general interactor framework for individuality—general in the sense that it applies to biological and nonbiological individuals. Second, we discuss Hull's (1980) interactor account of individuality and demonstrate that biofilms are individuals under that account. Third, we discuss how biofilm biology raises the question of what sort of interaction is required among the parts of an individual.

The general interactor framework we suggest has two components. First there is a sortal component: when asking if X is an individual we need to ask if X is an individual of sort S. Here we follow Wiggins's (2001) sortal account of identity. When asking if two entities are the same entity, we need to place that entity under a sortal and enquire about the identity conditions for the sort of entity in question. The guiding idea is that different sorts of entities have different identity conditions. We follow a similar route when it comes to individuals. When asking whether something is an individual we need to specify the sort of individual under consideration. Evidence for this sortal approach to individuality is found in the different sorts of individuals in biology and their varying identity conditions. There are individuals in natural selection (the focus of this chapter), individuals in systematics (species and other taxa, Hull 1978), metabolic individuals (Godfrey-Smith 2013), immunological individuals (Pradeu 2010), and perhaps other types of biological individuals.

The second component of this interactor framework concerns the interactions necessary for an entity to be a certain sort of individual. Individuals of different sorts have different outcomes (functions, states, products). Consider two examples: individuals in natural selection require processes that allow them to vary and pass on that variance; individuals in biological systematics require processes that cause them to be distinct lineages. Once we determine the sort of individual under investigation and the outcomes necessary to be that sort of individual, our focus turns to the types of interactions required of individuals of that sort. We need to ask if the parts of an entity appropriately interact among themselves or with the environment to form the sort of individual in question. This framework for individuality is quite general. It applies to various kinds of individuals, both in and outside of biology.

How this framework applies to individuals in biology will be demonstrated shortly. But first let us briefly mention how it applies to individuals outside of biology. Armstrong (1980) provides a causal theory of individuality that is an interactionist account. He uses the example of individuating a cup to illustrate his account. Suppose the function of cup is to be a solid material that holds liquid. In order for cup parts to form an individual that holds liquid, those parts must be appropriately attached. According to Coulomb's law, objects are bound into a solid that can hold water only if there are certain electrostatic forces among the molecules in those objects (Halliday, Resnick, and Walker 2013). Consequently, cup parts are parts of an individual cup only if there are certain electrostatic forces among molecules in those parts. This is a simple example and offered merely to indicate that an interaction account of individuality is applicable beyond biology.

Returning to biology, numerous philosophers adopt an interactor account of individuals in natural selection (for example, Hull 1980, Sober and Wilson 1998, Lloyd and Gould 1993, Dupré and O'Malley 2009). Here we will discuss Hull's (1980) interactor account of individuality, which is embedded in his interactor-replicator framework for natural selection. In that framework, interactors and replicators are both necessary for natural selection. According to Hull, replicators "pass on their structure largely intact from generation to generation" (1980, 315). Genes and asexual organisms are replicators for Hull. Though some asexual parents and offspring may not be genetically identical, they are similar enough to pass Hull's standard. Sexual organisms, colonies, and more inclusive units are not replicators. Recombination in sexual reproduction, for instance, reshuffles the genetic contributions of parents so that sexual offspring fail Hull's criterion for replicators. Turning to interactors, an interactor is "an entity that directly interacts as a cohesive whole with its environment in such a way that replication is differential" (1980, 318). Hull suggests that organisms and perhaps colonies are interactors, but he is suspicious of more inclusive entities being interactors (1980, 325). In Hull's theory of natural selection, both replicators and interactors must be present for selection to occur, but they need not be the same entities. In fact, very few entities are both replicators and interactors. Hull suggests that genes fulfill both roles (1980, 320). In the majority of cases, interactors and replicators occur at different hierarchical levels, for example, some organisms are interactors and their genes are replicators.

Are biofilms replicators or interactors on Hull's account? Let us start with replicators. According to Hull (1980, 315) replicators must pass on their structure "largely intact." Hull tells us that the entire genome of an asexual organism is a replicator because "in asexual reproduction, the entire genome is transmitted" (1980, 321). In contrast, the genomes in sexual organisms may be altered by recombination, so only portions of the genomes in sexual organisms are replicators. Just as not all of the genes of a sexual parent make it into an offspring, not all of the strains of a biofilm make it into a descendent biofilm (Kolenbrander et al. 2010, 478). Furthermore, according to Ehrlich and coauthors (2010, 270), lateral gene transfer "among the component strains (and species) [of a biofilm] leads to the continuous generation of a cloud of new strains with a novel combination of genes." In other words, LGT can cause a biofilm to genetically change over time. Biofilms vary too much to be replicators. Nevertheless, biofilms contain replicators, their genes and bacterial cells.

Are biofilms interactors on Hull's account? Hull places two restrictions on interactors. First, an interactor must be a cohesive whole. Second, an interactor's interaction with the environment must have a unitary effect on its constituent replicators. Starting with the notion of cohesive whole, Hull contrasts such wholes with mere groups. Mere groups are spatiotemporally localized, but that is merely due to their being at the same location (Hull 1980, 314). Interactors should be more because "[a mere] group can be selected only incidentally—e.g., because all its members happen to be in close proximity to each other" (314). Hull also describes interactors as "functionally organized systems" and "organized wholes" (325). He adds that populations are interactors only if they have "populational adaptations, properties characteristic of the population as a whole that allow it to interact with the environment as a whole" (325). Biofilms satisfy the cohesive whole requirement for being an interactor. Biofilms are not mere groups of organisms that happen to be at the same location at the same time. *P. gingivalis*, for example, does not become part of an oral biofilm simply because it happens to be near other colonizers. *P. gingivalis* becomes part of a biofilm through coaggregation (Hojo et al. 2009), and coaggregation mechanisms restrict which species can bind together (Rickard et al. 2003). Then there are mutualistic interactions among the species of a biofilm in which the byproduct of one species' metabolism is utilized by another species within that biofilm (Stewart and Franklin 2008). In general, the orchestra of cell-to-cell interactions within a biofilm shows that biofilms are not mere groups but cohesive wholes *sensu* Hull.

What about Hull's second condition for being an interactor, namely that an interactor's interaction with the environment has a "unitary effect" on its constituent replicators? Does a biofilm's interaction with the environment have a unitary effect on its constituent replicators? If "unitary effect" means that the failure or success of a biofilm affects the survivorship of its constituent cells (replicators) in a uniform way, then biofilms meet this condition. Bacterial cells typically have higher survivorship when they are parts of a biofilm than when they are in their planktonic state (Costerton 2007). There are multiple reasons. EPS matrices help protect bacteria from antimicrobial agents (Flemming and Wingender 2010). Biofilms allow bacteria to withstand shear stresses in flowing environments (Hall-Stoodley, Costerton, and Stoodley 2004, 99). And by forming biofilms, bacterial cells form "opportunistic self-mobilizing communities" capable of surviving environments, such as the deep ocean, that bacteria by themselves could not survive (Costerton 2007, 64ff.). The capacity to form biofilms is central for explaining the proclivity of bacterial cells. A biofilm's interaction with the environment *as a biofilm* has an important unitary effect on its bacterial cells: it increases their survivability.

One might worry that biofilms are no more interactors whose interactions have a unitary effect on their constituents than symbiotic consortia or ecological groups. We agree that symbiotic consortia and ecological groups have interactions that affect their constituents. In the squid-bacteria consortium discussed earlier, a selection force against the bacteria will negatively affect the squid that depend on those bacteria for camouflage. Similarly, selection that affects a segment of a buffalo herd and reduces the available gene pool for the herd may decrease the fitness of all. Generally, the number of interactions a group experiences and the extent to which those interactions affect all the members of a group come in degrees. We contend that the interactions among the members of a biofilm and the degree to which they affect its members surpass the interactions among the members of an ecological group. As discussed throughout this chapter, the bacteria of a biofilm have a robust number of interactions: from EPS formation to quorum sensing, from shared defensive and nutrient-gathering mechanisms to the sharing of genetic material. The interactions and interpenetrations are more numerous than symbiotic interactions. Moreover, the sharing of genes through LGT demonstrates that the interaction among a biofilm's members has a more significant genetic downstream effect than found in symbiotic consortia or ecological groups. The interactive effects among the constituent replicators of a biofilm are more uniform than similar effects among the replicators of an ecological group.

The upshot is that biofilms are good candidates for interactors and consequently individuals on Hull's natural selection account of individuality. On Hull's account, biofilms are individuals in natural selection, despite their lacking reproductive bottlenecks, despite their being multispecies individuals, and despite their not forming single parent-offspring lineages. Given this result and a result of the previous section (that biofilms satisfy many of Godfrey-Smith's underlying desiderata for individuality), we believe that biofilms are good candidates for individuals in natural selection. We also offer further reasons for thinking that biofilms are individuals in natural selection in the next section. Given how well Hull's account of individuality captures biofilms as individuals in natural selection, we prefer his account to Godfrey-Smith's.

Before leaving this section we would like to discuss the metaphysics of the cohesiveness or interaction required of the parts of an individual. The nature of biofilms suggests that such interaction is less intuitive than one might think. Recall Armstrong's example of the sort of interaction required among the parts of a cup for it to be parts of a cup: electrostatic interaction among the different parts. That sort of interaction is internal to the individual in question. We would like to suggest that the sort of interaction required among the parts of individual could be due to external forces acting on those parts. When Hull tells us that an individual in natural selection needs to be a cohesive whole whose interaction with the environment has a unitary effect on its replicators, does he require that the parts of the individual causally interact with each other, or can acting in a unity fashion due to environmental forces suffice for being an individual?

Hull does not address this question in his work on selection, but he does in his work on species. He writes that the sort of cohesion needed for a species to be an individual lies in its organisms having "fairly consistent, recognizable phenotypes" (1978, 343). He also tells us that such cohesion can be the result of processes that require the members of species to causally interact, or the result of processes that act independently on those organisms (1978, 343–344). Gene flow among the members of a species is an example of an interactive process among the members of a species. Genetic homeostasis and exposure to common selection regimes are examples of cohesifying mechanisms that act independently within or on members of a species. Genetic homeostasis is internal to each developing organism. Exposure to common selection regimes is the effect of the environment on each organism. So for Hull, external forces acting on organisms, not just interactions among a species' organisms, contribute to the cohesiveness of an individual.

What about interactor individuals in natural selection? Can they be individuals due to external processes acting on the parts of an individual rather than interactive processes among those parts? We will not provide a definitive answer to this question, but suggest that just like the parts of a species, the parts of an interactor in selection may be parts of that interactor due to external forces. As an example, consider cooperation among the parts of a biofilm. The literature on cooperation among the cells of a biofilm centers on the notion of public goods. Public goods are costly products manufactured by some cells that benefit other cells in a biofilm, such as signaling molecules (for quorum sensing) and EPS compounds. The existence of public goods in biofilms poses the problem of what prevents the spread of cheats within a biofilm—those cells that benefit from the products of other cells but do not themselves produce public goods. There are several suggested mechanisms that foster cooperation among the cells of a biofilm. We discuss one.

One set of experiments focuses on the bacterial species *Pseudomonas fluorescens* (Brockhurst, Buckling, and Gardner 2007). A strain of *P. fluorescens*, the wrinkly spreaders, produces a public good, cellulosic polymer, which improves access to oxygen by enabling the construction of biofilms at the surface of liquids. However, biofilms with the wrinkly spreader strain are susceptible to invasion by another strain of *P. fluorescens*, the smooth spreaders: they reap the benefits of being part of a biofilm without paying the cost of building the biofilm. Brockhurst, Buckling, and Gardner (2007) show that some forms of ecological disturbance promote biofilms with cooperating cells. In their work, they varied the degree of ecological disturbance affecting the biofilms under study. Brockhurst, Buckling, and Gardner (2007) found that under frequent disturbance, the density of cells is below the level at which biofilm formation is beneficial. Under intermediate ecological disturbance, the proportion of cooperators (wrinkly spreaders) peaks. When there is infrequent disturbance, the number of cheaters increases significantly and biofilms produce fewer public goods. Thus, in cases of intermediate ecological disturbance, selection favors biofilms with higher proportions of cooperators. Brockhurst, Buckling, and Gardner's (2007) work shows that ecological disturbance can be an external force that keeps cheaters in check.

The metaphysical implication of this finding is that a group of organisms can, at least in part, be caused to be an interactor due to external forces acting on those organisms. Biofilms of *P. fluorescens* vary in their proportions of cheats and noncheats. Those with higher proportions of noncheats pass on more cells that form biofilms than biofilms with lower proportions of noncheats. That variation in proportions of cheats and noncheats among biofilms is at least in part caused by environmental causes. According to Hull, an interactor's interaction with the environment should have a unitary effect on its replicators, and among interactors that replication should be differential. That is happening among *P. fluorescens* biofilms due to external forces acting on the cells of those biofilms. This is an interesting metaphysical result. This biofilm example and Hull's species example raise the question of whether the sort of interaction required for individuality must be internal to an individual or if it can be due to external forces acting on that individual. At the very least, these examples show that one type of interaction that contributes to the individuality of some individuals is the environment acting on the parts of those individuals.

6.5 Reproducers, Inheritance, and Natural Selection

A concern raised in section 6.3 is that if biofilms are individuals in natural selection, then they need to be reproducers.[2] Recall that biofilms score poorly on Godfrey-Smith's (2009) parameters for reproducers. Biofilms lack bottlenecks, even though they perform the desideratum for bottlenecks—the spreading of genetic novelty within an individual. Biofilms are at best middling on his criterion of division of reproductive labor. And biofilms do not form the sort of parent-offspring lineages Godfrey-Smith attributes to individuals. We could at this stage say that we do not care if biofilms are reproducers and point out that in Hull's (1980) interactor-replicator framework it does not matter whether biofilms are reproducers. Biofilms just need to be interactors with appropriately related replicators, and in the previous section we suggested that biofilms are such interactors.

Nevertheless, we think a case can be made for biofilms being reproducers using Griesemer's (2000a, 2000b) account of reproduction. According to Griesemer, reproduction is the "multiplication with material overlap of mechanisms conferring the capacity to develop" (2000a, 361). There are two parts to this account. First, parents and offspring must have a genealogical relationship caused by material overlap. Second, entities capable of reproducing must develop or have life cycles. Griesemer describes development as the acquisition of the capacity to reproduce (2000a, 360). For Griesemer, "The realization of a reproduction process entails the realization of a developmental process. The realization of development entails reproduction" (2000b, 74). This interdependence between reproduction and development forms a hierarchical structure. The reproduction of a multicellular organism requires an organism to develop from cells; cell reproduction requires cells to develop from organelles and chromosomes; and so on. This hierarchy bottoms out at the level of "null development," which is a case of reproduction of offspring that lack the capacity to develop (Griesemer 2000a, 362).

Biofilms satisfy Griesemer's account of reproduction. Once a biofilm matures, it releases cells to the environment, either as individual cells or as clumps of cells. For example, *P. aeruginosa* biofilms produce motile cells that swim out of a biofilm, and *S. aureus* biofilms shed clumps of hundreds of nonmotile cells (Hall-Stoodley, Costerton, and Stoodley 2004). The released cells, or their descendants,[3] aggregate with other cells and form new biofilms. New biofilms, thus, are built using material contributed by old biofilms. Furthermore, that material provides new biofilms with the capacity to develop. As we saw in section 6.2, biofilms have four developmental stages in their life cycles: planktonic lifestyle, attachment, colonization, and dispersal. Biofilm development also has the reproduction-developmental hierarchy that Griesemer proposes. The reproduction of a biofilm requires a biofilm to develop from cells; and cell reproduction requires cells to develop from organelles and chromosomes.

Biofilms reproduce according to Griesemer's account. Furthermore, we have seen that biofilms are interactors with appropriately related replicators. Still, one might be committed to Lewontin's (1985) framework for natural selection and wonder if biofilms satisfy his requirements for individuality. Recall that according to Lewontin, entities are individuals in natural selection if three conditions are met: there is variation among individuals; that variation is heritable; and that variation results in differential fitness among individuals. Similarly for Griesemer, being a reproducer is not sufficient for being an individual in natural selection. Griesemer writes that more is needed. "So long as the components of development by which the capacity to reproduce is acquired result in component transmission fidelities greater than zero there is scope for evolution to operate among them. Populations of reproducers have the capacity to evolve, insofar as the pieces of development that realize their reproductive capacities themselves have heritable properties that vary" (Griesemer 2000a, 363). For the remainder of this section we explore the question of whether there is heritable, adaptive variation among biofilms.

Biofilms reproduce by aggregation, and at first glance that seems to undermine their having heritable variation. The cells that aggregate to form a biofilm often come from different biofilms, and they often come from different species. In addition, the particular species compositions of earlier and later biofilms can vary (Kolenbrander et al. 2010). For example, oral biofilms usually contain the species *S. mutans*, but such biofilms can be formed without that species. It seems that the transmission of genetic information, or any information, among earlier and later biofilms is too diffuse, too disorganized for there to be heritable variation among biofilms. However, we think that this conclusion is too hasty given current empirical work on biofilms. Consider oral biofilms that usually contain the species *S. mutans*. That species can be absent in a new oral biofilm, yet that biofilm still forms and is "caries inducing" without *S. mutans* (Kolenbrander et al. 2010, 478). In other words, the biofilm phenotype "caries inducing" occurs even when there is variation in the species composition of an oral biofilm. There is a general point here. Oral biofilms and other biofilms reliably exhibit a number of adaptive traits across biofilm generations, such as quorum sensing, EPS production, mutualist interactions, and other life cycle traits (see section 6.2). That seems uncontroversial. The pressing question is, how do these traits get reliability transmitted through aggregation? Or to put it differently, if biofilms reproduce via aggregation, how do they have "transmission fidelities greater than zero" (Griesemer 2000a, 363)? Empirical work on this question is in its early stages, but there seem to be at least two mechanisms that cause such transmission.

First, recall that not all species can aggregate with each other to form a biofilm. Bacteria do not form a biofilm because they happen to be near each other. There are specific molecular mechanisms that determine which bacteria can aggregate with each other. For instance, *F. nucleatum* acts as bridge between early and late colonizers in the formation of oral biofilms (Hojo et al. 2009). Mechanisms that regulate how biofilms form by aggregation also regulate which cells and genes get transmitted between older and newer biofilms. Another transmission mechanism in biofilms is lateral gene transfer (LGT). We have discussed how LGT transfers mutations within a biofilm. LGT also transfers genetic material among biofilms and is a mechanism for biofilm inheritance. Consider work on the mechanisms that foster cooperation among the bacteria in a biofilm. One set of experiments focuses on mobile genetic elements (MGEs)—genes that can move among prokaryotic genomes via LGT. MGEs are akin to infectious agents, capable of benefiting or harming their bacterial hosts. Smith (2001) hypothesizes that if cooperation is coded in MGEs, then the lateral transfer of these mobile elements may infect noncooperative bacteria, causing them to become cooperative and produce a public good. Nogueira and coauthors (2009) provide empirical evidence for Smith's hypothesis by studying the genes that code for the protein secretome. Such proteins are costly to produce, yet they benefit neighboring bacteria. Nogueira and coauthors (2009) found that the genes coding for secretome are overrepresented in MGEs and are laterally transferred among and within biofilms. LGT in this case keeps the number of cheats in check. This example illustrates how transmission of a trait can occur among biofilms even though biofilms reproduce by aggregation. It is also an example of a transmitted trait that increases the fitness of a biofilm: biofilms with fewer cheaters do better than biofilms with more cheaters (see section 6.4 and Brockhurst, Buckling, and Gardner 2007).

Stepping back from these details, we believe that biofilms may indeed satisfy Lewontin's three criteria for individuals in natural selection. Work on biofilms indicates that they do have mechanisms for nonzero transmission fidelity. Furthermore, the suggestion that biofilms are connected by inheritance seems plausible given that biofilms have traits that affect the fitness of whole biofilms and their component bacteria, and those traits occur over and over again. When it comes to the general question of whether biofilms are reproducers with the sort of inheritance needed to be individuals in natural selection, we can say the following. Biofilms are individuals in Hull's framework because they are interactors with appropriately related replicators. There is promising evidence that biofilms may be individuals in natural selection according to Lewontin's and Griesemer's frameworks. However, biofilms are poor individuals according to Godfrey-Smith's account.

6.6 Conclusion

In this chapter we have offered an interactor account of biological individuality embedded in a more general theory of individuality. That general approach to individuality employs a sortal framework: the world consists of different sorts of individuals, and whether or not an entity is an individual depends on whether that entity's parts interact (among themselves or with its environment) in a sortal-specific way. We see the inclusiveness of this framework as a virtue—it allows for multiple theories of individuality corresponding to the multiple kinds of individuals in the world.

When it comes to biofilms, we have seen that their nature teaches us several things about individuality. First, it teaches us that some standard ideas about individuals in natural selection should be abandoned. Individuals in natural selection need not have bottlenecks or a high division of reproductive labor. Such individuals can be composed of lineages from different species, and those lineages need not run in tandem. Second, biofilms teach us that a proper theory of reproduction should be more inclusive than commonly conceived. Aggregation is not normally seen as reproduction, but through aggregation biofilms may pass on heritable variation. Third, biofilms teach us that common intuitions about the type of relations required among the parts of an individual may be wrong. Perhaps some individuals owe their individuality to external rather than internal processes. The lowly biofilm does teach us a thing or two about the metaphysics of individuality.

Acknowledgments

We thank Ellen Clarke, David Crawford, Peter Godfrey-Smith, and Thomas Pradeu for their very helpful comments on an earlier version of this chapter. Financial support for work on this chapter was provided by the Canadian Institutes of Health Research, and the Social Sciences and Humanities Research Council of Canada.

Notes

1. Godfrey-Smith (2009) also discusses two other types of reproduction: simple and scaffolded. A simple reproducer reproduces using its internal machinery, but the parts of a simple reproducer cannot reproduce using their internal machinery. Scaffolded reproducers are reproduced by mechanisms external to them. Godfrey-Smith's discussion of reproduction focuses on collective reproduction, which we simply call "reproduction."
2. Bouchard (2010), however, questions this assumption. He suggests that in some cases we should count the fitness of an individual in terms of differential growth rather than differential reproduction.
3. Some bacterial cells multiply through binary fission during their planktonic stage.

References

Armstrong, David. 1980. Identity through Time. In *Time and Cause*, ed. Peter Van Inwagen, 67–78. Dordrecht: D. Reidel.

Bouchard, Frederic. 2010. Symbiosis, Lateral Function Transfer and the (Many) Samplings of Life. *Biology and Philosophy* 25: 623–641.

Brockhurst, Michael, Angus Buckling, and Andy Gardner. 2007. Cooperation Peaks at Intermediate Disturbance. *Current Biology* 17: 761–765.

Costerton, John. 2007. *The Biofilm Primer*. Berlin: Springer.

Davies, David, Matthew Parsek, James Pearson, Barbara Iglewski, John Costerton, and E. Greenberg. 1998. The Involvement of Cell-to-Cell Signals in the Development of a Bacterial Biofilm. *Science* 280: 295–298.

Dupré, John, and Maureen O'Malley. 2009. Varieties of Living Things: Life at the Intersection of Lineage and Metabolism. *Philosophy and Theory in Biology*. Online. http://hdl.handle.net/2027/spo.6959004.0001.003.

Ehrlich, Garth, Azad Ahmed, Josh Earl, N. Luisa Hiller, John Costerton, Paul Stoodley, Christopher Post, Patrick DeMeo, and Fen Ze Hu. 2010. The Distributed Genome Hypothesis as a Rubric for Understanding Evolution In Situ during Chronic Bacterial Biofilm Infectious Processes. *FEMS Immunology and Medical Microbiology* 59: 269–279.

Flemming, Hans-Curt, and Jost Wingender. 2010. The Biofilm Matrix. *Nature Reviews Microbiology* 8: 623–633.

Godfrey-Smith, Peter. 2009. *Darwinian Populations and Natural Selection*. Oxford: Oxford University Press.

Godfrey-Smith, Peter. 2011. Agents and Acacias: Replies to Dennett, Sterelny, and Queller. *Biology and Philosophy* 26: 501–515.

Godfrey-Smith, Peter. 2013. Darwinian Individuals. In *From Groups to Individuals*, ed. Frederic Bouchard and Philippe Huneman, 17–36. Cambridge, MA: MIT Press.

Griesemer, James. 2000a. Development, Culture, and the Units of Inheritance. *Philosophy of Science* 67: 348–368.

Griesemer, James. 2000b. The Units of Evolutionary Transition. *Selection* 1: 67–80.

Hall-Stoodley, Luanne, John Costerton, and Paul Stoodley. 2004. Bacterial Biofilms: From the Natural Environment to Infectious Diseases. *Nature Reviews Microbiology* 2: 95–108.

Halliday, David, Robert Resnick, and Jearl Walker. 2013. *The Fundamentals of Physics*. Hoboken: Wiley and Sons.

Hojo, Kenichi, Seiji Nagaoka, Tomoko Ohshima, and Nobuko Maeda. 2009. Bacterial Interactions in Dental Biofilm Development. *Journal of Dental Research* 11: 982–990.

Hull, David. 1978. A Matter of Individuality. *Philosophy of Science* 45: 335–360.

Hull, David. 1980. Individuality and Selection. *Annual Review of Ecology and Systematics* 11: 311–332.

Kolenbrander, Paul, Robert Palmer Jr., Saravanan Periasamy, and Nicholas Jakubovics. 2010. Oral Multispecies Biofilm Development and the Key Role of Cell-Cell Distance. *Nature Reviews Microbiology* 8: 471–480.

Lewontin, Richard. 1985. Adaptation. In *The Dialectical Biologist*, ed. Richard Levins and Richard Lewontin, 65–84. Cambridge, MA: Harvard University Press.

Lloyd, Elisabeth, and Stephen Gould. 1993. Species Selection on Variability. *Proceedings of the National Academy of Sciences* 90: 595–599.

Nikoh, Naruo, John McCutcheon, Toshiaki Kudo, Shin-ya Miyagishima, Nancy Moran, and Atsushi Nakabachi. 2010. Bacterial Genes in the Aphid Genome: Absence of Functional Gene Transfer from Buchnera to Its Host. *PLoS Genetics.* doi:10.1371/journal.pgen.1000827.

Nogueira, Teresa, Daniel Rankin, Marie Touchon, Francois Taddel, Sam Brown, and Eduardo Rocha. 2009. Horizontal Gene Transfer of the Secretome Drives the Evolution of Bacterial Cooperation and Virulence. *Current Biology* 19: 1683–1691.

Pradeu, Thomas. 2010. What Is an Organism? An Immunological Answer. *History and Philosophy of the Life Sciences* 32: 247–268.

Rickard, Alexander, Peter Gilbert, Nicola High, Paul Kolenbrander, and Pauline Handley. 2003. Bacterial Coaggregation: An Integral Process in the Development of Multi-species Biofilms. *Trends in Microbiology* 11: 94–100.

Smith, Jeff. 2001. The Social Evolution of Bacterial Pathogenesis. *Proceedings of the Royal Society B* 268: 61–69.

Sober, Elliott, and David Wilson. 1998. *Unto Others: The Evolution and Psychology of Unselfish Behavior.* Cambridge, MA: Harvard University Press.

Stewart, Philip, and Michael Franklin. 2008. Physiological Heterogeneity in Biofilms. *Nature Reviews Microbiology* 6: 199–210.

Wiggins, David. 2001. *Sameness and Substance Renewed.* Cambridge: Cambridge University Press.

CHAPTER 7 | Cell and Body
Individuals in Stem Cell Biology

MELINDA BONNIE FAGAN

7.1 Introduction

Organisms like us are considered paradigmatic biological individuals. Such entities share a number of properties that furnish common-sense criteria for biological individuality: spatiotemporal continuity, physical boundaries, participation in biologically significant causal processes (e.g., reproduction, evolution, and development), physiological integration, and functional autonomy.[1] One question that has concerned many philosophers is whether biological individuality extends to other levels of biological organization: genes, cells, groups, species, and so on. Of these, the cell has arguably the strongest claim to biological individuality by "organism-centred" criteria (Wilson and Barker 2013). The idea that cells are "autonomous living units" or "elementary organisms," while the organism is a "society of cells" or "cell state," dates back to mid-19th-century formulations of cell theory (Conklin 1939, Hall 1969, Virchow [1859] 1958; for recent discussions see Nicholson 2010, Reynolds 2007). The opposing view is that cells are subordinate to the organism, harmoniously organized by a top-down division of labor. Tension between these two perspectives contributes to several long-running debates about biological individuality and agency. In engaging these debates, philosophers of biology have investigated the cell-organism relation primarily in terms of evolution by natural selection. This chapter takes a different perspective, focusing on cell development. The following sections examine the question: are stem cells biological individuals?

This question may seem merely a special case of a more general (and philosophically significant) question about cells. But that impression is misleading. Questions about stem cells go to the core of contemporary ideas about biological individuality. Consider the following simple argument:

(1) Cells of multicellular organisms are biological individuals.
(2) Stem cells are cells of multicellular organisms.
(3) Therefore, stem cells are biological individuals.

At first glance, the conclusion follows deductively, by inference from the general to the particular. Premise (1) asserts that cells that make up the body of a multicellular organism (neurons, skin cells, muscle cells, and so on) qualify as biological individuals. Premise (2) asserts that stem cells are among these cells. Given the two premises, (3) follows ineluctably. The conclusion appears so obvious as to require no further defense: if stem cells are a kind of cell, and cells are biological individuals, then stem cells must be biological individuals. But the argument's simplicity is deceptive. I will argue, in what follows, that stem cells are *not* biological individuals *in the same way* as more familiar cell types that make up the body of a multicellular organism. The simple argument for (3) therefore fails due to equivocation.

This approach targets a particular interpretation of premise (2). On another interpretation, this premise is correct: stem cells are cellular entities, and they are parts of (or derived from) multicellular organisms. But premise (1) asserts more than this. The way in which stem cells are cells of multicellular organisms does not coincide with the cells of multicellular organisms that are plausibly taken to be biological individuals according to familiar criteria. This equivocation between premises (1) and (2) blocks the conclusion that stem cells are biological individuals. Another approach would be to reject premise (1). However, there is considerable, though not decisive, support for the thesis that cells of multicellular organisms are biological individuals (section 7.2). So we should provisionally accept premise (1).

The charge of equivocation rests on two contrasts between stem cells and cells of multicellular organisms in the sense of premise (1). These two contrasts are demonstrated in sections 7.3 and 7.4. First, the stem cell concept is not a purely cellular notion, but involves the organismal level as well (section 7.3). So the question of biological individuality for stem cells is more complicated than for cells that are parts of multicellular organisms but can be conceived as purely cellular entities. The second contrast is that the stem cell concept (understood as an abstract model) cannot be experimentally shown to correspond to single cells in biological systems. That is, stem cells cannot be experimentally individuated at the single-cell level, as cells of multicellular organisms can (section 7.4). More precisely, while specialized mature cells of a multicellular organism can be defined and experimentally characterized as individual cells, stem cells cannot. Though it sounds paradoxical, stem cells are not cells—at least, not in the same way as cells that are plausibly taken to be biological individuals.

One might accept both contrasts and yet reject premise (1). On such a view, cells of multicellular organisms and stem cells both fail to qualify as biological individuals, although for different reasons. From a developmental perspective, however, it is more interesting to consider stem cells as transformative entities, mediating between two different levels of biological individuality: cell and organism. Section 7.5 sketches one idea along these lines, which offers independent support for conclusion (3). I argue that at least some stem cells can be considered biological individuals, by analogy to multicellular organisms. These stem cells are *model organisms* for investigating mammalian development. Finally, the evidential challenges of stem cell research reveal a surprising parallel with physics: uncertainty relations in the context of Bohr's account of complementarity (section 7.6).

7.2 Cells as Biological Individuals

This section argues that there is strong support for the thesis that cells of multicellular organisms are biological individuals, clarifying the baseline with which stem cells will be contrasted in later sections. The basic tenets of cell theory offer a starting point:[2]

(1) Every organism begins as a single cell, which, in multicellular organisms, gives rise to all the body's cells.
(2) Cells reproduce by binary division.[3]
(3) The life of a cell begins with a division event and ends with either a second division event yielding two offspring, or cell death (and no offspring).
(4) Generations of cells linked by reproductive division form a lineage.[4]

Together, these four tenets establish cells as plausible candidates for biological individuality. Tenet (1) relates cells to organisms, the paradigmatic biological individuals. Every organism begins as a single cell; most continue that way. Single-cell organisms such as bacteria and yeast "dominate life on this planet" numerically, historically, in biodiversity, and in biomass (O'Malley and Dupré 2007, and references therein). So in the majority of cases, questions of cell and organismal individuality coincide.[5]

Single-cell organisms satisfy important criteria for biological individuality. Individuals in general are thought to "have three-dimensional spatial boundaries, endure for some period of time, are composed of physical matter, bear properties, and participate in processes and events"

(Wilson and Barker 2013). Single-cell organisms evidently satisfy these general criteria. They are physically contiguous, with clear spatial and temporal boundaries. Indeed, the physical boundary of a semipermeable membrane separating a cell from its external environment is among the clearest in all of biology. Single-cell organisms also bear a number of properties, including size, shape, and morphology, and function as agents within an environment. Beyond the general case, further criteria for biological individuality are diverse and somewhat contentious. Among the most familiar and widely accepted are participation in biologically significant causal processes (notably reproduction and evolution), physiological integration, and functional autonomy. Single-cell organisms clearly satisfy these criteria. Tenets (2) and (3) state the basics of cell reproduction, a process that yields cell lineages consisting of generations of parents and offspring (tenet 4). As members of potentially ongoing reproductive lineages, single-cell organisms are "evolutionary individuals" (Hull 1992, Godfrey-Smith 2009).[6] Moreover, "the cell" is an exemplar of physical and physiological individuality for the biological world. Cellular metabolism, including respiration, energy storage, and excretion, comprises a self-regulating system that unifies heterogeneous components into a "one discrete and cohesive entity" (Pradeu 2012, 227). Overall, single-cell organisms qualify as biological individuals, according to most criteria.[7]

For cells of multicellular organisms, matters are less clear-cut. Because multicellular organisms are paradigmatic biological individuals, it is tempting to suppose that their parts are not. Yet a strong case can be made for the biological individuality of these cells, if the term "part" is understood in a minimal, common-sense way. To say that cells are parts of a multicellular organism, in this sense, is just to say that a cell-level inventory of an organism would result in a list of cells that make up that organism. Such an inventory is not a philosophical fantasy, but grounded on more than a century of biological practice yielding robust and detailed characterizations of cells extracted from multicellular organisms. Properties of these cells do not vary continuously, but cluster into discrete groups sharing a suite of structural and functional traits. These groupings are referred to as "cell types." Major cell types such as neurons, muscle, and blood cells exhibit considerable robustness across organismal taxa, and can be clearly distinguished from one another by microscopic examination, biochemical tests, molecular techniques, or some combination thereof. Characteristic features of cell types are typically stable across cell generations—liver cells divide to produce more liver cells, fibroblasts beget more fibroblasts, and so on.

Cells of multicellular organisms can be experimentally individuated and characterized independently of their organismal context. Because of this, many of the points made above about single-cell organisms apply to them as well. This is not to say that experimental practices are somehow constitutive of these cells' biological individuality. Rather, experimental practices of cell biology reveal facts about cells of multicellular organisms (e.g., clear physical boundaries, heritable properties) from which a strong case for their biological individuality can be built. Cells of multicellular organisms evidently satisfy minimal criteria for individuality: they are physical entities with clear spatiotemporal boundaries, which bear properties and are involved in processes. In addition, tenets (1)–(4) apply, just as for free-living single-cell organisms. Cells of multicellular organisms also satisfy physiological and evolutionary criteria, each having a metabolism, heritable structural and functional traits that vary in response to mutation, and clear criteria for reproductive success. Though most cells of a multicellular organism are "evolutionary dead ends" in the long run, this does not preclude their being units of selection within an organism's lifetime.[8]

The primary objection to premise (1) is that cells of a multicellular organism fail to satisfy a criterion of "minimal functional autonomy" (Wilson and Barker 2013). Unlike single-cell organisms, so the objection goes, cells that make up the body of a multicellular organism lack the ability to "act for themselves." Instead, they are controlled from the top down by the integrated system that defines the organism as a biological individual. This view is the oppositional counterpart of the bottom-up view that cells are individuals that make an organism. So we are back to the intuition that cells of multicellular organisms cannot be individuals because they are parts of individuals, but with the added assumption that the term "part" implies lack of autonomy. I think we should not accept that assumption. There is currently no consensus, among biologists or philosophers, about how control is transmitted across levels of biological organization, or in what direction. Furthermore, it is not obvious that biological individuality must be a zero-sum game; that is, that an organism's functional autonomy must be at the expense of that of the cells that compose it. Though these considerations do not decisively rebut the functional autonomy objection, it would be premature to rule out biological individuality for cells of multicellular organisms on its grounds. At the very least, we can conclude that cells of multicellular organisms that can be isolated and characterized (the specialized neurons, muscle cells, blood cells, and so on) satisfy some important criteria for biological individuality. Premise (1) above therefore has strong support.

7.3 Defining Stem Cells

I now turn to the stem cell case. The term "stem cell" itself seems to imply that stem cells must be a kind of cell, and thus fall under the scope of the arguments above. However, close examination of the concept reveals two important contrasts. In this section, I show that the stem cell concept cannot be understood solely in terms of single cells isolated from their context. In the next section, I argue that stem cells cannot be experimentally individuated at the single-cell level. The overall result is that stem cells can neither be defined in terms of nor justifiably identified with individual cells. In both respects, they contrast with cells of multicellular organisms, which can be experimentally identified and characterized as single cells, analogous to free-living single-cell organisms (section 7.2). Therefore, stem cells are not individual cells in the way as the specialized cells of multicellular organisms. The argument for this conclusion rests on careful analysis of the stem cell concept.

Stem cells are defined as undifferentiated cells that self-renew and give rise to differentiated cells (e.g., Ramelho-Santos and Willenbring 2007, 35; Melton and Cowan 2009, xxiv). This general definition suggests that a stem cell is a kind of cell. But in fact there are many kinds of stem cell: adult, embryonic, pluripotent, multipotent, induced, mesenchymal, neural, hematopoietic (blood-forming), embryonal carcinoma, and so on. Any general definition must cover this bewildering variety of cases. I have elsewhere defended a modeling approach to this problem (Fagan 2013a). The basic idea is to conceive the general definition of "stem cell" as an abstract conceptual model. Here I briefly sketch the main points and results of this analysis. The model represents the two defining capacities of stem cells: self-renewal and differentiation. Both are reproductive processes that involve comparison across cell generations. Self-renewal is cell reproduction in which parent and offspring resemble one another, while differentiation is the process by which parts of a developing organism acquire diverse, specialized traits over time. So the two processes are complementary. It is important to note that stem cells are defined as being *capable* of participating in both. Whether those defining capacities are realized depends on environmental context (see section 7.4). The stem cell concept can be represented as a model combining these two reproductive processes, with tenets (2)–(4) serving as background assumptions.

A minimal model of a stem cell consists of one cell undergoing a division event, such that one offspring cell is similar to the parent, while the other is more specialized (figure 7.1). This model consists of three objects ("cells") standing in two reproductive relations, which define a structure: cell lineage L. L is not simply a genealogical structure, but includes cross-generation comparisons. However, no two cells are similar or different in *every* respect. The comparative concepts of self-renewal and differentiation are only scientifically useful if they are understood as relative to a set of variable characters. So the model requires a set of characters C, specific values of which can be assigned to objects in the model. Characters of interest for stem cell biology are attributes of individual cells, such as

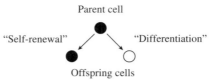

FIGURE 7.1 Minimal stem cell model: asymmetric cell division

size, shape, or concentration of a particular molecule. The minimal model of figure 7.1 is also restricted to a single cell division event and two cell generations. Generalizing to allow for any number of cell division cycles introduces a further variable (n). When estimates of cell division rate are available, this number can be converted to an interval of calendar time, which in practice ranges from hours to decades. These three variables (L, C, and n) suffice to define self-renewal: parent and offspring cells in lineage L resemble one another with respect to some character set C for time interval (number of cell cycles) n.

Differentiation is more complicated, because not all changes across cell generations count as differentiation, but only those in the direction of increased specialization.[9] So an additional parameter needs to be included, alongside cell lineage, set of characters for parent-offspring comparison, and time interval. This additional parameter refers to the end of development: mature cells that have completed the process. A cell *specializes* over some time interval just in case its character values are more similar to those of mature cells at the end of that interval than the beginning. The characters relevant for differentiation are those of a mature cell type of interest (character-set M). Mature cells have specialized morphology, functions, and molecular traits (an array of values for M), and therefore correspond very closely to cells of multicellular organisms as discussed above. But they have a developmental aspect, which is not presupposed by the cell types discussed in section 7.2. A mature cell has reached the endpoint of development; the process is complete. We do not have a cell-level concept of complete development. Characters of mature cells (M) are defined in terms of their location and functional role within a fully developed organism; that is, an organism that has reached reproductive maturity. Insofar as the character set and range of alternative values for M (and C) are selected with reference to a fully developed organism, the latter concept is implicated in the process of differentiation. In this sense, the stem cell concept presupposes that of an organism.

Putting all this together, this general stem cell model consists of three or more objects ("cells") with characters C or M, values of which vary over some time interval n, organized by two reproductive relations (self-renewal and differentiation) into cell lineage L (figure 7.2).

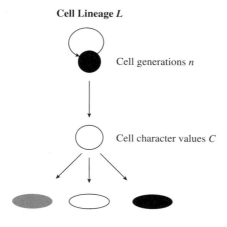

FIGURE 7.2 Abstract stem cell model: the general concept

A stem cell is defined by its position as the unique stem of L, with maximal self-renewal and differentiation potential relative to C, M, and n. These reproductive relations that define L are in turn defined in terms of comparisons of character value across cell generations. The comparisons implicate not only individual cells, but also multiple cell generations arranged in a lineage, as well as the whole organism and its specialized cellular constituents. The general stem cell concept, therefore, cannot be understood solely in terms of individual cells. Cell lineage relations and whole organisms are also implicated. So the stem cell concept, unlike that of cells of multicellular organisms, is not purely cellular.

7.4 Experimental Uncertainty

The above argument shows that stem cells are not defined as single cells; the concept is more complicated. However, if that concept were shown to apply to a single cell, then an argument like that of section 7.2 could be made for stem cells. That is, experimental individuation of stem cells could show that objects in the stem cell model correspond to single biological cells, and therefore have clear physical boundaries, physiological integration, and so on.[10] If the same or similar points that support the claim of biological individuality for cells of multicellular organisms could be made for stem cells, the simple argument in section 7.1 would not be equivocal. So the next task is to consider how the abstract stem cell model relates to the biological world. As the previous section shows, a cell can qualify as a stem cell only relative to a lineage, set of characters, and time

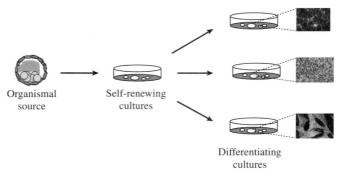

FIGURE 7.3 Stem cell experiments: general design

interval of interest. These parameters are specified by experiments that aim to isolate and characterize stem cells. The stem cell concept applies to different entities, depending on how the model's parameters are specified. So application of the concept is experiment-relative.

However, experiments that aim to isolate and characterize stem cells share a basic pattern: cells are first removed from an organismal source (which establishes the cell lineage of interest), then placed in a context where their character values can be measured, and finally moved to another context to measure self-renewal and differentiation (figure 7.3). Experiments with this design specify parameters the abstract model leaves open, either as materials and methods or measured results. For example, for human embryonic stem cells (hESC), the source organism is an early human embryo and the specific site of extraction is the inner cell mass. Character values of interest include "undifferentiated morphology" and a long list of molecular traits shared with embryonic or cancer cells.[11] The duration of interest is ≥50 cell divisions, with no upper limit.[12] Subtle changes in cell culture environment can (in principle) induce differentiation of hESC to any of the mature cell types that make up an adult human body: pluripotency.

Diverse experimental methods give rise to the current variety of stem cells: adult, embryonic, induced, pluripotent, multipotent, blood, neural, muscle, and many more. Pluralism about stem cells is compounded by the diversity of representational assumptions that are in principle available to link the model to biological objects: single cells, cell populations, or cell types. Cells in the model could conceivably correspond to any of these, and possibly other biological entities as well. But the representational assumption of concern here is the simplest and most intuitive: that objects in the model correspond to single cells in the biological world. Granting that the latter are biological individuals (as argued in section 7.2), if this representational assumption holds, then the stem cell model picks out biological individuals. But that consequence is blocked by a severe evidential constraint. The stem cell model interpreted via the representational assumption that its objects correspond to single cells cannot be supported by experimental evidence. Hypotheses of the form "Cell A is a stem cell" face intractable evidential problems.

As noted above, stem cell experiments involve two sets of measurements, both of which provide data about characters of single cells. *But no single cell persists through both sets of measurements.* Because cells reproduce by division, descendants and ancestors do not coexist. A single stem cell, therefore, can only be identified retrospectively. Stem cells are defined as being capable of both self-renewal and differentiation. To experimentally establish a candidate stem cell's differentiation potential, that cell must be placed in an environment conducive to differentiation, and character-values of progeny that indicate differentiation measured later. To establish self-renewal, a candidate stem cell must be maintained in an environment that blocks differentiation, and its progeny after n cell cycles measured for character-values that indicate similarity to the original. It is not possible to perform *both* experiments on a single cell.

Even if this could be done, self-renewal on its own presents a problem. This process is defined as production of offspring cells with the same traits and capacities as the parent. An offspring cell with the same capacities as a stem cell parent has the same potential for differentiation and for self-renewal. To establish that it has the same self-renewal capacity as the parent, the offspring cell must be maintained in the same environment as the dividing parent, and *its* offspring measured. Whether the offspring cell has the same self-renewal capacity as its parent depends on the reproductive capacities of the "second generation" cell. And so on. Establishing that a single cell is capable of self-renewal is infinitely deferred. Nor can a single cell's differentiation potential be definitively established. Experiments can show what happens when a cell is placed in a particular environment, but not what would have happened if that cell had been placed in a different environment. If we have pure populations of identical stem cells, we can learn what happens to a stem cell from that population in a range of environments. But the antecedent assumes what these experiments are designed to discover: cells that share features had by all and only the stem cells of a particular lineage. If we cannot demonstrate that a single cell is a stem cell, how can we assume a population of identical ones? Even if that assumption were justified, there is no basis for generalizing claims about a cell's differentiation potential beyond the range of environments used in experiments. Inferences about stem cell capacities are therefore relative to the environments examined—that is, to an experimental method, with a specific organismal source, characters measured, and manipulations of cell environment.

It follows that attribution of stem cell capacities to any individual cell is necessarily uncertain. No matter what technological advances are made in tracking and measuring single cells, experiments cannot decisively resolve stem cell capacities at the single-cell level. This "uncertainty principle" is an unavoidable evidential constraint for stem cell biology (see section 7.6). To measure both self-renewal and differentiation potential for a single cell, and to elicit the full range of a cell's potential, multiple copies of that cell are needed—a homogeneous (with respect to characters of interest) *population* of candidate stem cells.[13] So a strictly single-cell stem cell model cannot be experimentally confirmed. In other words, no direct mapping between objects in the stem cell model and individual biological cells can be supported by experimental evidence. To summarize the results of this and the previous section: stem cells, as currently defined and experimentally identified, do not correspond to single cells of multicellular organisms. Therefore, I conclude that stem cells are not biological individuals in the same way as cells of multicellular organisms.

7.5 Model Organisms, Model Individuals

A further conclusion would be that stem cells are not biological individuals at all. But this judgment would be hasty. The above arguments suggest that stem cells mediate between different modes of biological individuality: cell and organism. Their mediating role offers a fresh developmental perspective on the cell-organism relation. The conceptual challenge of understanding this relation is a fundamental problem in biology. Paul Weiss (1940) presents the challenge incisively:

> At the end of development we are confronted with a unitary organized system, called an "organism," which, at the same time, is a collective of cells. At the beginning of development we find just one primordial cell—the egg ... Now, there arises a dilemma. Either the egg already possesses supra-cellular organization of the same order as the later body—then it is not just another cell, but an uncellulated organism; or it is merely a cell like others ... [and] development would create organization of a higher order. (1940, 37–38)

Weiss's own solution is to acknowledge that the cell has "a dual character," as both an active biological individual and passive component regulated by top-down organismic control. A more sophisticated understanding of development as a collection of interrelated mechanisms, some operating from the "bottom up," others "top down," he argues, would dissolve the dilemma. Stem cells, I propose, are a crucial part of such an understanding of development. Moreover, in their mediating role between cell and

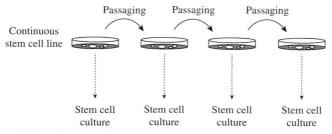

FIGURE 7.4 Cultured stem cells and stem cell line

organism, stem cells can exhibit biological individuality in a way analogous to a multicellular organism. I here argue that cultured pluripotent stem cells qualify as biological individuals in this way. We can see this, and account for contrasts between cultured stem cells and multicellular organisms, by recognizing the former as *model organisms* for investigating organismal development.

As described above (section 7.3), stem cell experiments produce new biological entities, each derived from a particular organismal source and individuated by a set of measured character values. Cultured pluripotent stem cells are one such entity. They are produced by removing cells from an organismal source and placing them in artificial culture, with chemical factors added to prevent differentiation. Cells that rapidly divide under these conditions form colonies, which are selected and put into new cultures (figure 7.4). This "passaging," repeated every few weeks, maintains a continuously growing lineage of undifferentiated cells. So self-renewal is imposed by the experimental design. Differentiation potential is demonstrated by moving small clumps of cells from a culture to a new environment that includes chemical factors that induce differentiation into a certain cell type: cardiac muscle, neurons, bone marrow, and so on. Continuously growing lines of undifferentiated cells that can differentiate under appropriate culture conditions qualify as stem cell lines. Because cells derived from very early (~5d) embryos tend to show greater self-renewal and differentiation potential than cells from older organisms, the paradigmatic pluripotent stem cell lines are embryonic stem cells.[14] Cultured pluripotent stem cells are entities that exist between "passages" from one culture to another.

Cultured pluripotent stem cells exhibit many of the individuating features of organisms. They have clear spatial boundaries in three dimensions, determined by the size of cell culture flasks: a transparent artificial body. Cultured pluripotent stem cells also endure over time; indeed, pluripotent stem cell lines have (apparently) unlimited lifespans. They have molecular, cellular, morphological, and physiological properties (see note 11) and participate in causal processes of self-renewal, experimental manipulation, and—under suitable conditions—cell development. So cultured pluripotent stem cells satisfy minimal criteria for individuality. They are also evolutionary individuals, subject to artificial as well as natural selection. Cycles of passaging, which select for cells "with a uniform undifferentiated morphology," impose a version of the germ/soma distinction (figure 7.4; Thomson et al. 1998, 1147 n. 6).[15]

However, cultured pluripotent stem cells differ from multicellular organisms in two key respects: (1) they exhibit much simpler organization and (2) they are artificially produced. I discuss each in turn, and argue that these contrasts do not warrant rejecting the thesis that cultured stem cells are biological individuals. First, unlike organisms, cultured pluripotent stem cells exhibit minimal physiological organization and functional integration. But in appropriate environmental conditions, they can and do transform into mature tissues and organs—indeed, a whole organism, at least in principle.[16] Even in artificial culture that encourages self-renewal, pluripotent stem cells are liable to differentiate spontaneously—particularly when they contact or "overgrow" one another. Such spontaneous differentiation is far less elaborate and organized than unmanipulated embryonic development in mammals. So pluripotent stem cell cultures could perhaps be considered minimal or borderline organisms.[17] They embody early mammalian development in drastically simplified form: a layer of undifferentiated, dividing cells, selected so as to minimize variation. Under certain environmental conditions, they exhibit cell differentiation processes in a very simple and manipulable way. I return to this point below.

A second contrast is that cultured stem cells, unlike naturally occurring organisms, do not form lineages of transparent bodies on their own, but only as a result of experimental intervention. So they do not reproduce "autonomously," in the way of biological organisms. This is obviously an important difference, but it does not, I think, rule out cultured stem cells as biological individuals tout court. Obviously, cultured stem cells are artificial products. Though derived from organisms, pluripotent stem cells have no direct organismal counterparts. But artificiality per se does not entail lack of individuality. Many biological fields concentrate research efforts on model organisms: inbred fruit flies, mice, yeast, and zebrafish. Though not identical to their naturally occurring counterparts, these entities are clearly organisms, as their moniker implies. Many cannot survive outside a laboratory environment. For example, certain mouse strains, widely used in biomedicine, lack an immune system and must be maintained in a sterile room. But such mice are clearly organisms, and therefore biological individuals. Therefore engineered obligate inhabitants of a laboratory can qualify as biological individuals. Cultured pluripotent stem cells, I suggest, are a new kind of model organism, derived from parts of an organism rather than a whole.

Cultured pluripotent stem cells are evidently not biological entities *found in nature*. It follows that they are not biological entities that have evolved by natural selection, strictly speaking. So one might object that they do not qualify as evolutionary individuals, being insignificant from a broad evolutionary standpoint. However, such a standpoint is not the only one relevant for questions of biological individuality.[18] In stem cell research, the developmental perspective takes priority. By taking this perspective and its epistemic goals seriously, we can better understand the biological individuality of stem cells. The modeling role clearly applies here. Cultured pluripotent stem cells are laboratory artifacts created to study mechanisms of organismal development. By studying them, scientists hope to understand and manipulate developmental pathways in more paradigmatic biological individuals (like ourselves). The notion of a "model organism" accounts for the contrasts between cultured pluripotent stem cells and multicellular organisms. Indeed, this distinctive modeling relation exactly captures the relation between these two biological entities.

Classic model organisms (e.g., *E. coli, C. elegans*, and *D. melanogaster*) are constructed to be tractable for laboratory use and to exhibit phenomena of interest in a way accessible to controlled manipulation. Model organisms for studying development, in particular, share several characteristic features: small body size, large offspring number, rapid growth rate, robust processes of reproduction and development, tractability, simplicity, and accessibility to observation and measurement (Bolker 1995, Robert 2004, see also Fagan 2013a, chap. 7). Cultured pluripotent stem cells exhibit all these features. Their body size, though flexible in principle, is in practice constrained by dimensions of cell culture dishes and flasks. Rapid, continuous growth is ensured, since cultured cells are selected for frequent division. Passaging occurs about once a fortnight, ensuring a rapid reproductive rate. And cultured pluripotent stem cells make cellular developmental processes visible and accessible—in striking contrast with normal mammalian development. Tractability is at present more goal than reality, as researchers work to streamline and standardize stem cell culture conditions. However, when any new stem cell line is established, the next tasks are to reduce variation among cells and manipulate culture conditions to yield reproducible experimental results. These efforts resemble earlier standardization practices for classic model organisms such as inbred mice and fruit fly strains. Cultured pluripotent stem cells also play the dual epistemic role of a model organism, being objects of study in their own right as well as tools for representing target phenomena. Whereas classic model organisms are altered to be tractable representatives of a wide variety of unmanipulated organisms, cultured pluripotent stem cells are created to be tractable representatives of a wide variety of developing multicellular organisms. So they are model organisms in a double sense—artificial biological entities designed to exhibit features of developing multicellular organisms. Their intermediate character, between cell and organism, is crucial to their scientific role.

7.6 Complementarity

The preceding discussion reveals an epistemic aspect to core questions about biological individuality. Focusing on the epistemic issue of uncertainty (section 7.4) also leads to a surprising cross-disciplinary parallel. To briefly recap: we cannot directly measure a single cell's capacity for self-renewal or differentiation, either separately or together. The analogy with uncertainty relations in quantum physics has not gone unremarked (e.g., Nadir 2006, Fagan 2013b). Heisenberg's uncertainty principle states that certain pairs of quantities cannot be simultaneously known with precision. The most familiar instance is position and momentum: we cannot simultaneously attribute a well-defined position and momentum to a single particle.[19] This reciprocal uncertainty has a theoretical basis, entailed by probabilistic principles of quantum mechanics involving noncommuting operators (see, e.g., Omnés 1999). But quantum uncertainty can also be understood in terms of mutually exclusive experimental arrangements: one can either measure a particle's direction of motion and so derive its position at a given time, or measure its wave characteristics and from these calculate its momentum. In the former, the particle's wave aspect is ignored (preventing calculation of its energy and momentum), while in the latter the wave's probabilistic "spread" ensures that its direction of motion cannot be known with precision. Uncertainty relations thus quantify the "reciprocal limitations" on what we can measure for a given physical system (Saunders 2005).

Obviously, the objects of study and properties of interest for stem cell research differ from those of quantum physics. Instead of a single particle with physical properties at a given time, stem cell experiments focus on a single cell with developmental properties relative to a lineage extended in time. However, mutually exclusive experimental measurements are implicated in both cases. In stem cell biology, experimental conditions for measuring self-renewal (even provisionally) rule out the possibility of measuring differentiation potential, and vice versa. And similarly for differentiation potential; experimental conditions for measuring whether a cell can give rise to, say, heart muscle, rule out the possibility of measuring that cell's ability to give to rise blood, bone, skin, neurons, and so on. Stem cell capacities are constrained by uncertainty relations grounded in the limitations of experimental measurement. What is the significance of the analogy to uncertainty in quantum physics? A full examination of this question, and its bearing on issues of realism and individuality, is beyond the scope of this chapter. A few preliminary points, however, seem clear.

First, there is a significant disanalogy between stem cell and quantum physical uncertainty. The former is not entailed by formal theory and involves no challenge to the ordinary ("classical") causal interpretation of experiments and their results, while the latter is emblematic of the challenge of articulating an intelligible interpretation of formal quantum theory. Stem cell experiments and their outcomes can be described in ordinary causal terms; the challenge is establishing evidential relations between these descriptions and hypotheses about single stem cells. Uncertainty in quantum physics challenges our ordinary causal understanding of physical events. Moreover, because stem cell biology lacks a formal, predictive theory, the analogy engages only the operational side of quantum physics. In consequence, much philosophical work on uncertainty in quantum physics does not extrapolate to the stem cell case.

However, the analogy cuts deeper if we consider quantum-mechanical uncertainty in the context of Bohr's principle of complementarity (1949).[20] This is because Bohr speculated that complementarity might apply to fundamental issues in biology, and was committed both to retaining classical concepts and to at least a modest operationalism in defining concepts for scientific use (Omnés 1999, Saunders 2005). So Bohr's account of quantum uncertainty highlights the role of experiments, strengthening the analogy with stem cell uncertainty. His operationalism restricted well-defined concepts to aspects of the world that can be directly measured by experiment. For cases involving uncertainty, then, the question of interpretation amounts to choice of experimental method. This is the core idea of complementarity: the experiments we choose to do determine which interpretation is correct for a given case. This prevents any formal contradiction from arising between wave and particle interpretations, for example; each applies to different experimental arrangements. So we can describe experiments and their outcomes in a way that respects quantum uncertainty relations, without at once employing incompatible notions to describe physical reality. But the price of such consistency is that experimental arrangements are implicated in the correct application of interpretive models of physical systems. In this sense, the experiment is involved in the phenomenon under investigation.

Similarly, experimental contexts and measurements are implicated in characterization of stem cells (see sections 7.3–7.4 above). So Bohr's "quantum postulate," that there is no absolute or fundamental separation between the object of inquiry and experimental context, applies to stem cells as well. Both physical and stem cell phenomena are "contextual" in the sense that the behavior of objects of inquiry cannot be sharply separated from interaction with measuring instruments ("the agencies of observation") that define the conditions under which phenomena appear. But this contextuality, or relativity to experiment, raises different challenges in the two cases, which accordingly require different solutions. In the quantum physics case, the challenge is to avoid contradictory descriptions of the objects of inquiry. Bohr's principle of complementarity offers a solution: "evidence obtained under different experimental conditions cannot be comprehended within a single picture, but must be regarded as complementary, in the sense that only the totality of the phenomena exhausts the possible information about the objects" (Bohr 1949, 209). Contradiction is avoided by separating experimental contexts that mutually exclude the operational definition of paired concepts such as position and momentum, particle and wave.

In the stem cell case, the challenge is not theoretical consistency but operationalization. A stem cell's defining capacities require multiple experimental contexts to demonstrate. So instead of separating experimental contexts to avoid contradiction, application of the stem cell concept requires a kind of unification across experimental contexts; inclusivity rather than exclusivity, so to speak. Multiple copies of a cell are needed to measure the extent to which that cell has both capacities and the full range of its developmental potential. So evidence of stem cell capacities depends on the assumption that the population of candidate stem cells is homogeneous, relative to the characters and environments of interest. An experiment demonstrating stem cell capacities involves a set of systematically varied experiments using replicates of the cell under investigation. So resolution (or rather, management) of uncertainty in the stem cell case takes a different form than in Bohr's quantum physics.

Finally, there is a sense in which Bohr's principle of complementarity holds for stem cell biology, though its scientific significance is not as he predicted. Saunders (2005) argues that Bohr's principle is best understood as a conjecture about different branches of science, proposing a "novel explanatory framework ... [that] applied in principle to any empirical domain in which concepts could only be applied under mutually exclusive experimental conditions" (435). Stem cell biology is such an empirical domain, and substantive definition of its core concept requires integrating mutually exclusive experimental conditions. But Bohr saw this integration as the task of new nonclassical laws, which the principle of complementarity provides theoretical space to discover (within a classical framework), but does not articulate. Stem cell biologists, however, do not seek laws and show scant interest in securing theoretical grounds for them. Instead, integration across experimental contexts is accomplished by provisional empirical assumptions (homogeneous cell populations with respect to characters of interest) and technical innovation. So Bohr's conjecture is both vindicated and subverted, in the case of stem cell biology.

7.7 Conclusion

A number of philosophers and scientists have argued that cells of multicellular organisms are biological individuals, or, more cautiously, that they exhibit some important characteristics of biological individuality. This chapter extends the question to stem cells: undifferentiated cells that self-renew and give rise to differentiated cells. This general definition suggests that a stem cell is a specialized type of cell, like neurons, red blood cells, muscle cells, and so on. If cells that compose multicellular organisms are biological individuals, then it follows that stem cells are biological individuals. However, counterintuitive as it may seem, we should resist this conclusion. I have argued that stem cells are not biological individuals in the way of cells of multicellular organisms. My argument uses a modeling approach, explicating the general working definition of "stem cell" as a simple abstract model: a structure of objects and relations. The modeling approach sheds light on central concepts and practices of stem cell biology, as well as the relation between cell and organismal individuality. This approach reveals two contrasts between stem cells and cells of multicellular organisms. First, the stem cell concept is more complex than that of a single cell bearing marks of biological individuality. Cell lineage relations, cell populations, and whole organisms are also implicated. Second, stem cell experiments cannot provide evidence that stem cells as defined in the model correspond to single cells in biological systems. So stem cells as currently defined and experimentally identified do not correspond to single cells. These contrasts show that stem cells are not biological individuals in the same way as cells of multicellular organisms (assuming that the latter are individuals).

This conclusion may sound paradoxical: a stem cell is not a cell. In fact, some scientists have argued exactly this, proposing instead the notion of "stemness," a state defined by wide developmental potential, which individual cells may enter or leave (e.g., Zipori 2004, Lander 2009).[21] The abstract stem cell model proposed above is compatible with the stemness view (Fagan 2013a). But we can learn more, I suggest, by pressing further on the question of stem cell individuality. At least some stem cells are, arguably, biological individuals in the way of multicellular organisms. More precisely, cultured pluripotent stem cells are analogous to multicellular organisms, the paradigmatic biological individuals. The key to understanding the biological individuality of stem cells is the concept of a model organism. Cultured pluripotent stem cells are designed to display paradigmatic features of biological individuality in simple, accessible, and exaggerated ways. In this sense, at least some stem cells are constructed to be *model individuals*. The stem cell case also exhibits a surprising parallel with physics. Bohr's principle of complementarity offers an intriguing perspective on the uncertainty inherent in measurements of individual stem cells.

Acknowledgments

Many thanks to Thomas Pradeu and Alexandre Guay for the opportunity to contribute to this volume, and for organizing the conference that inspired it ("Individuals Across the Sciences," Sorbonne, Paris, May 2012). This chapter has greatly benefited from questions and comments of participants at the Sorbonne event, particularly Matt Haber, Lucie Laplane, and Simon Saunders, as well as Jordi Cat, Hasok Chang, Allen Franklin, Elisabeth Lloyd, Kirstin Matthews, Sean Morrison, John Norton, and Irv Weissman. Special thanks to Marc Ereshefsky, Lucie Laplane, Makmiller Pedroso, and Thomas Pradeu for comments on an earlier version of this chapter. Support for this research was provided by Rice University's Faculty Innovation Fund, the Division of Humanities (Rice University), and the Mosle Foundation.

Notes

1. For philosophically inflected discussions of these various criteria, see Wilson and Barker 2013, Pradeu 2012, 228–232.

2. Historically, principles of cell theory have been expressed in many different ways (see references in section 7.1). Statements (1)–(4) express views that are very widely accepted in life sciences today. They do not capture tenets of cell theory across all relevant historical contexts.

3. There are two modes of cell division: mitosis and meiosis. In mitosis, the genome replicates once before the cell divides. In meiosis, the genome replicates once, but two rounds of cell division follow, yielding four offspring cells with half the complement of DNA. Because stem cell and somatic cell phenomena involve mitosis, the term "cell division" is used throughout this chapter to refer to that mode only.

4. The term "cell lineage," in what follows, refers to a biological entity composed of successive cell generations, organized by reproductive relations.

5. This coincidence cuts both ways; O'Malley and Dupré (2007) discuss interactive and multicellular aspects of microorganisms, which have traditionally been underemphasized. These insights do not affect the arguments in this chapter.

6. More precisely, single-cell organisms qualify as evolutionary individuals if we assume they are grouped into populations that vary with respect to properties that

influence reproductive success (e.g., rate of cell division). This assumption is satisfied in many (if not all) cases, as cell reproduction leads to formation of clonal populations of cells, which can then diversify. More nuanced differences among philosophical accounts of evolutionary individuals can be set aside for the purpose of this chapter. Thanks to T. Pradeu for pushing me to clarify this point.

7. Another familiar criterion for biological individuality is genetic homogeneity; i.e., that biological individuals are distinguished from one another by a unique chromosomal DNA sequence (Wilson 1999). Because single-cell organisms of the same lineage often have nearly identical DNA sequences ("clones"), they do not satisfy this genetic criterion. However, due to recent discoveries of microbial diversity in functioning human organisms (and those of many other species), the genetic criterion is itself subject to question, and currently de-emphasized by many theorists (see Wilson and Barker 2013, and references therein). An updated version, tying individuality to genetic regulatory states rather than DNA sequence, does apply to single-cell organisms (Klipp et al. 2009, chap. 6). A full discussion of this issue is beyond the scope of this chapter. Here I claim only that failure to satisfy the genetic criterion as traditionally conceived does not disqualify single-cell organisms as biological individuals (and, similarly, cells of multicellular organisms).

8. In fact, certain cells of the immune system participate in selective processes as a prerequisite for their specialized defensive functions (Paul 2003). Evolutionary theories of cancer posit a pathological counterpart to these processes (Nowak 2006).

9. Another aspect of differentiation is *diversification*. A population of cells diversifies over some time interval relative to a set of characters, if and only if variation in the values of those characters increases over that interval. For simplicity, I bracket this population-level aspect of differentiation here.

10. The term "correspondence" is used loosely, to refer to a mapping relation from components of the model to real biological systems. The argument of this chapter is neutral with respect to technical accounts of the correspondence relation.

11. Characters: chromosome number and appearance, telomerase activity, cell surface molecules. Character values: rapid division in culture, homogeneous appearance, lack of specialized traits, flat round shape, large nuclei surrounded by correlatively thin cytoplasm, and prominent nucleoli.

12. This is why ESC are described as "immortal."

13. This is why a thoroughgoing focus on cell populations cannot get around this evidential problem. Evidence for population-level models of stem cells also depends on the assumption of a homogeneous "founder" stem cell population. For details, see Fagan 2013a.

14. Pluripotent stem cell lines can also be produced from partly or fully differentiated cells extracted from adult organisms, which are then altered by manipulating genes, culture conditions, or both (yielding induced pluripotent stem cells; iPSC).

15. The selection process used in passaging resembles that of cell-level selection in plants: many cells could give rise to a new organism, but few actually do.

16. This has been demonstrated for mice, but (for ethical reasons) not in humans.

17. It is interesting to consider whether cultured stem cells with more limited differentiation potential, or cell lines that are not identified as stem cells (such as cancer cell lines), also qualify as model organisms. Though I do not rule this out,

I think a stronger case can be made for cultured pluripotent stem cell lines as minimally developing multicellular organisms than other kinds of cultured cell. So my claims here are restricted to cultured pluripotent stem cells, leaving the question open for other cases. Thanks to L. Laplane for pushing me to clarify this point.

18. It is not even the only standpoint relevant for evolution, as research in experimental evolution makes use of highly artificial bacterial populations. Thanks to T. Pradeu for raising this point. It is also worth noting that cultured pluripotent stem cells may count as "borderline Darwinian populations" on Godfrey-Smith's evolutionary account, exhibiting high B-, intermediate G-, and low I-values (2009, chap. 5 of this volume).

19. Another familiar example is description of light as *at once* an electromagnetic wave and a particle.

20. This insight is due to Simon Saunders (pers. comm., May 2012).

21. For philosophical treatments of this issue, see Leychkis, Munzer, and Richardson 2009, Laplane 2013.

References

Bohr, Niels. 1949. Discussion with Einstein on Epistemological Problems in Atomic Physics. In *Albert Einstein: Philosopher-Scientist*, ed. P. A. Schilpp, 199–242. La Salle, IL: Open Court.

Bolker, Jessica. 1995. Model Organisms in Developmental Biology. *BioEssays* 17: 451–455.

Conklin, Edwin G. 1939. Predecessors of Schleiden and Schwann. *American Naturalist* 73: 538–546.

Fagan, Melinda B. 2013a. *Philosophy of Stem Cell Biology*. London: Palgrave Macmillan.

Fagan, Melinda B. 2013b. The Stem Cell Uncertainty Principle. *Philosophy of Science* 80: 945–957.

Godfrey-Smith, Peter. 2009. *Darwinian Populations and Natural Selection*. Oxford: Oxford University Press.

Hall, T. S. 1969. *History of General Physiology, 600 B.C. to A.D. 1900*. Vol. 2. Chicago: University of Chicago Press.

Hull, David. 1992. Individuality. In *Keywords in Evolutionary Biology*, ed. Evelyn Fox Keller and Elisabeth A. Lloyd, 181–187. Cambridge, MA: Harvard University Press.

Klipp, E., W. Liebermeister, C. Wierling, A. Kowald, H. Lehrach, and R. Herwig. 2009. *Systems Biology: A Textbook*. Weinheim: Wiley-VCH.

Lander, Arthur. 2009. The "Stem Cell" Concept: Is It Holding Us Back? *Journal of Biology* 8: 70.1–70.6.

Laplane, Lucie. 2013. *Cancer Stem Cells: Ontology and Therapies*. Ph.D. dissertation, Paris Sorbonne University.

Leychkis, Y., S. Munzer, and J. Richardson. 2009. What Is Stemness? *Studies in History and Philosophy of Biological and Biomedical Sciences* 40: 312–320.

Melton, David, and Chad Cowan. 2009. Stemness: Definitions, Criteria, and Standards. In *Essentials of Stem Biology*, 2nd ed., ed. Robert Lanza et al., xxii–xxix. San Diego, CA: Academic Press.

Nadir, A. 2006. From the Atom to the Cell. *Stem Cells and Development* 15: 488–491.

Nicholson, Daniel. 2010. Biological Atomism and Cell Theory. *Studies in History and Philosophy of Biological and Biomedical Sciences* 41: 202–211.

O'Malley, Maureen, and John Dupré. 2007. Size Doesn't Matter: Towards a More Inclusive Philosophy of Biology. *Biology and Philosophy* 22: 155–191.

Omnés, Roland. 1999. *Quantum Philosophy*. Trans. Arturo Sangalli. Princeton, NJ: Princeton University Press.

Nowak, Martin. 2006. *Evolutionary Dynamics: Exploring the Equations of Life*. Cambridge, MA: Belknap Press of Harvard University Press.

Paul, William E., ed. 2003. *Fundamental Immunology*. 5th ed. Philadelphia: Lippincott, Williams, and Wilkins.

Pradeu, Thomas. 2012. *The Limits of the Self: Immunology and Biological Identity*. Trans. Elizabeth Vitanza. Oxford: Oxford University Press.

Ramalho-Santos, M., and H. Willenbring. 2007. On the Origin of the Term "Stem Cell." *Cell Stem Cell* 1: 35–38.

Reynolds, Andrew. 2007. The Theory of the Cell State and the Question of Cell Autonomy in Nineteenth and Early Twentieth-Century Biology. *Science in Context* 20: 71–95.

Robert, Jason S. 2004. Model Systems in Stem Cell Biology. *BioEssays* 26: 1005–1012.

Saunders, Simon. 2005. Complementarity and Scientific Rationality. *Foundations of Physics* 35: 417–447.

Thomson, J., J. Itskovitz-Eldor, S. Shapiro, M. Waknitz, J. Swiergel, V. Marshall, and J. Jones. 1998. Embryonic Stem Cell Lines Derived from Human Blastocysts. *Science* 282: 1145–1147.

Virchow, Rudolf. [1859] 1958. Atoms and Individuals. In *Disease, Life, and Man: Selected Essays by Rudolf Virchow*, trans. Lelland J. Rather, 120–141. Stanford, CA: Stanford University Press.

Weiss, Paul. 1940. The Problem of Cell Individuality in Development. *American Naturalist* 74: 34–46.

Wilson, Jack. 1999. *Biological Individuality*. New York: Cambridge University Press.

Wilson, Robert, and Matthew Barker. 2013. The Biological Notion of Individual. In *The Stanford Encyclopedia of Philosophy*, Spring 2013 ed., ed. Edward N. Zalta. http://plato.stanford.edu/archives/spr2013/entries/biology-individual/.

Zipori, Dov. 2004. The Nature of Stem Cells. *Nature Reviews Genetics* 5: 873–878.

CHAPTER 8 | Collective Individuals
Parallels Between Joint Action and Biological Individuality

CÉDRIC PATERNOTTE

8.1 Joint Action and Biological Individuals

When can a group of individuals be considered as an individual itself? When does the aggregation of individuals produce a collective akin to an individual? These general questions have been asked in very different fields and answered with variable success. When philosophers of action ask them, they consider sets of human individuals and ask why and how they could cooperate, and under which conditions this cooperation becomes a case of joint action, in which the interaction is so tight that the group itself behaves similarly to an individual agent. Such interrogations stem from countless observations of human beings coordinating and cooperating with ease to produce salient collective outcomes. One temporary answer is that just as individual agents have intentions, such cohesive groups typically have *joint intentions*. Joint intentions then need to be defined in terms of more elementary mental states. This task has occupied theorists for decades, and so far no definitive consensus has appeared.

When philosophers of biology ask our opening questions, they consider sets of biological individuals (such as organisms or evolutionary units) and ask two things: when may evolutionary processes lead them to aggregate so as to form higher-level individuals or units? What then defines biological individuality in general? The idea is that individuals can exist at various biological levels, and that a definition of biological individuality must rely on conditions that are general enough to abstract away from the specific characteristics of each level. Even if no consensus has been reached here either, lists of relevant criteria have been established and there are hints that their integration and unification is possible.

Despite the apparent similarity of these questions, intuitively their respective answers may not coincide or even hinge on one another. Human cooperation is the result of intertwined mental states and dedicated psychological mechanisms that can operate in interactions that occur only once. Biological individuality is the result of a process of natural selection that operates within populations, stems from countless interactions, and spans generations. Surely, these mechanisms are so different that there is no reason to presuppose that they would produce similar kinds of collective entities. Or is there?

Evolutionary theorists are familiar with the distinction between ultimate and proximate causes of a behavior (Mayr 1961), often taken to be independent. For instance, the evolutionary causes that led human altruistic behavior to spread are distinct from the psychological mechanisms behind it and do not constrain them. Even without considering recent doubts considering this distinction (e.g., Sterelny 2013 in the case of human cooperation), at best it establishes the *possible* independence of causes—actual independence still needs to be established on a case-to-case basis. Moreover, even if group formation can arise from different causal processes, it does not follow that the characteristics of intentional and biological groups are different. Indeed, their definition may well share a common structure. In this chapter, I show that both their definitions and explanations share strong common characteristics.

About a hundred years ago appeared the movement known as organicism. Although there had been a long history of analogies between human societies and biological organisms, organicists aimed to establish systematic parallels such that our biological knowledge would fertilize our understanding of how a society works.[1] Organicism fell into disrepute, both because of its ideological proximity with, and exploitation by, doctrines such as totalitarianism, fascism, and so on, and because of its dubious scientific dimension. In effect, it multiplied analogies without a clear idea of what they showed and where they should stop (Simiand 1987).

This unfortunate past notwithstanding, it now seems that organicism contained a hint of truth and could be resurrected in a more controlled, modest fashion so as to escape the flaws that plagued its ancestor. This chapter aims to do just that, by showing why and how the debate on joint intentionality could and should be informed by our knowledge about biological individuality. Why biological individuality rather than organismality? This is because theorists usually consider that an organism is merely a kind of biological individual. In principle, any entity that can be individuated in the biological world and that is of theoretical importance can be called a biological individual—a gene, an organ, a population—although we do not consider it as an organism. By contrast, organismality typically involves specific physiological or metabolic properties. Focusing on biological individuals rather organisms allows us to consider a larger number of possibly relevant criteria, the list of which should not be restricted from the get-go.

Section 8.2 briefly reviews the literature on joint intentionality, the main defining criteria it contains as well as its current shortcomings. Section 8.3 introduces the existing criteria for biological individuality, arguing that the related literature is considerably less problematic, though. Section 8.4 shows why biological individuality should be relevant for the joint intentionality debate and dispels several objections to this claim. Two helpful analogies between the two fields are then introduced: between criteria for joint intention and biological individuality in section 8.5; and between biological and psychological mechanisms in section 8.6, which provides a concrete example of the epistemological import of such analogies. Section 8.7 concludes.

8.2 Joint Action

Within the last three decades, philosophers have offered many definitions of human joint action. Their aim was to find out what relations exist between individuals who are said to cooperate, to act "together" or "jointly," and in some sense what makes a group of individuals similar to a (higher-level) individual itself. The idea is not that groups only appear to behave as individuals, but that they can possess some properties akin to that of individuals. This literature is often found under the label "joint intentionality": joint intentions are supposed to be to groups what intentions are to individuals, that is, what causes and explains their actions. The objective is then to define joint intentions so that they can play that role. Although some definitions of joint intentions are not analyses,[2] most of them are: they take joint intentions to be constituted by a set of ingredients, most of which are mental states of the individuals that compose the group. Consider an example of such a definition:

We intend to J if and only if

(1) (a) I intend that we J and (b) you intend that we J.
(2) (a) I intend that we J in accordance with and because of meshing subplans of (1)(a) and (1)(b); (b) you intend that we J in accordance with and because of meshing subplans of (1)(a) and (1)(b).
(3) (1) and (2) are common knowledge between us. (Bratman 1993)

Joint intention is here defined on the basis of intertwined individual mental states that are common knowledge among individual agents, that is, that are public among them. Although Bratman's definition embraces methodological individualism,[3] this is not a necessary condition for definitions in general: some of their ingredients may refer to collective properties, which is fine as long as they themselves receive an analysis. Still, most definitions share the structure of the above example: a set of conditions that are common knowledge among agents. Disagreement typically arises concerning the nature of these conditions. One striking feature of this literature is the multiplicity of definitions and their apparent irreducibility to one another. Indeed, most authors seem to have various intuitions as to the conceptual core of human joint action. Here is a summary of the most important ingredients, which will prove important for the rest of the chapter:

> Mutually consistent plans of action (Bratman 1992, 1993, 1999): In joint actions, agents should be disposed to adjust their actions to that of others so that they help, or at least do not hinder, one another.
> Normative commitments (Gilbert 1989): In joint action, agents are committed to doing their part, and only the whole group can release them of this commitment. In particular, if agents do not do their part, they see as legitimate that others would react, ask them for justifications, and possibly punish them.
> Collective goals (Miller 2001): In joint actions, agents must have an identical goal, that is, one realized by the same set of state of affairs.
> Collectively built reasons to act (Tuomela 2007): In joint actions, the group is the origin of the agents' reasons to think and to act.[4,5]

All these ingredients stem from intuitive real-life examples and are prima facie equally compelling but not readily reducible to one another. If intuition cannot adjudicate between them, what can? And should we adjudicate between them? Maybe human beings are such that whenever they are collectively committed, then they act in mutually consistent ways. In other words, some conditions may not logically follow from others but still be regularly caused by them. To clarify the relations between these accounts, we need some constraints on joint action that go beyond intuition. Several recent efforts have been made in this direction: they consist in trying to find minimal definitions of joint action, for instance that would fit cognitively unsophisticated agents such as human children (Tollefsen 2005) or situations in which the links between agents are very thin (Paternotte 2014).

One alternative is to formulate constraints on definitions of joint action that would allow us to exclude some of them. As more and more researchers accept the above diagnosis, this option is likely to become more and more popular (see, for instance, Butterfill, submitted). I suggest that any satisfying definition should meet an *efficiency constraint*.[6] Whatever defines human cooperation, it is by and large successful not by being automatic or wired but by being *reliable*.[7] We are routinely very good at cooperating, and examples of successful cooperation abound in our daily life; overall, cooperative behavior benefits us. Any component of joint action should contribute to explain this general reliability. As a consequence, ingredients of a definition have to be part of explanations of the success of cooperation. For instance, it is not enough to say that there is joint action whenever agents share a goal without an explanation of how agents usually manage to identify partners with whom they share a goal.

How could the efficiency constraint be met? One could look for rational or evolutionary explanations of joint action, that is, the benefits of cooperative behavior could be cashed out in terms of material payoffs or of fitness. To put it differently, joint action can be efficient because it regularly allows us to reap collective material benefits. Alternatively, it could be efficient because it is based on mechanisms that allowed our evolutionary ancestors—and possibly biological entities in general—to succeed in reaping collective evolutionary benefits. Some theorists have started to note conceptual similarities between definitions of joint action and its rational explanations (Hakli, Miller, and Tuomela 2010), although without using it to formulate constraints.[8] This chapter explores the biological alternative. Before developing the argument, I next introduce the debate on biological individuality.

8.3 Biological Individuality

What is a biological individual? Philosophers and biologists have put forward many answers to this question.[9] We have strong intuitions as to what counts as an organism—most animals and plants that surround us are considered to belong to this category. But our intuitions concerning biological individuals are less clear. Indeed, the concept of biological individuality has proved elusive, and finding a set of defining characteristics difficult.[10] The problem is not merely semantical. In evolutionary biology, one must be able to measure the change in gene frequencies between generations. Counting the sheer number of existing genes cannot suffice, though, lest we rule fatter individuals as fitter than others merely because they are composed of more cells and thus contain more DNA molecules. In other words, knowing what entities qualify as individuals allows us to determine the right headcount needed to measure reproductive success.

Just as for joint intention, there is a wealth of candidate defining criteria for biological individuality. In the recent literature, we can find 15 such criteria.[11] I cannot possibly describe all of them or discuss their relevance in this chapter—such a discussion, which can already be found in Clarke (2011), for instance, pertains to philosophy of biology.[12] I rather focus on those criteria from which links with the joint intentionality debate may be established.

One first criterion is that of *genetic identity/similarity*. Some biological individuals are made of parts that contain the same genotype (or similar enough genotypes). Accepting this criterion amounts to seeing cases of clonal reproduction (such as dandelions, fungi, aphids) as examples of individuals expanding rather than reproducing; a set of clones would be an individual, although a scattered one. Second, there is often a *division of labor* between the parts of a biological individual; parts have different functions that jointly contribute to the individual's survival and reproduction. One general instance of this is the division of reproductive labor, or the separation between germ cells (from which genetic material is transmitted to the progeny) and (sterile) somatic cells—the classical example is that of queens and workers in colonies of eusocial insects. Third, biological individuals often contain *policing mechanisms* that ensure continued cooperation between their parts and prevent them from behaving in ways detrimental to the whole. Fourth, parts of biological individuals typically enjoy a *shared evolutionary fate*: their reproductive output are perfectly aligned, or correlated, so that one cannot reproduce without the others reproducing as well (for instance, think of the organelles of eukaryotic cells).[13]

How can a list of disjoint and mutually inconsistent criteria (Clarke 2011) be of help to another field? Although consensus is absent from discussions both on joint intentionality and on biological individuality, our understanding of the latter concept is much better, for several reasons. First, it is informed empirical data; by contrast, joint intentionality is often assessed from intuitions and thus threatens to be a merely semantic matter. It is much easier to observe the behavior and functioning of biological entities than the precise ways in which human mental states combine and interact in order to reliably produce a behavior. Definitions of joint intentions often start from intuitive examples; variations in intuition allow for variations in definition. They are also based on the combination of many mental states, the behavioral consequences of which are hard to cash out. By contrast, the different mechanisms that produce and sustain biological individuality are easier to observe, and their biological consequences easier to assess.[14] Second, and relatedly, the interactions between possibly relevant biological factors are explored, both through observation and by way of modeling and mathematical models. For instance, Bourke (2011) is able to survey and determine which factors are crucial to the formation and maintenance of groups in many concrete cases. Similar analyses are scarcely available in the joint action literature. Third, there exist promising forays toward unification. For instance, Clarke (2013) suggests that the role of all existing criteria can be gathered in a more fundamental explanation (based on the relative strengths of within-group and between-group selection). So the apparent incompatibility of the criteria is not necessarily definitive. By contrast, joint intentionality is assessed from idiosyncratic perspectives that threaten to be irreducible.

Consider a particular example that will prove useful below. In an influential paper, Gardner and Grafen (2006) have tried to determine mathematically conditions under which a group can be considered as a "superorganism." They equate biological individuality with being a unit of adaptation (a controversial stance), and analyze formally when a group of individuals can become the bearer of group adaptations. The use of a population genetics model then allows them to show that clonal (genetically uniform) populations can always give rise to organismality, and populations with repressed competition slightly less so. Regardless of whether their conclusions are correct, this is one of many examples in which precise, assessable arguments lead to conclusions that can favor some criteria over others. More generally, even if there is little consensus on the definitions of biological individuality, there is concerning the set of candidate mechanisms and their explanatory scope, as witnessed by several recent clarificatory and unificatory works concerning cooperation in general (Lehmann and Keller 2006, West, Griffin, and Gardner 2007, Bourke 2011). Overall, the biological individuality debate is much more developed and informed by empirical data and conceptual arguments such as Gardner and Grafen's than the joint intentionality one. So the latter may in principle benefit from the former if bridges can be built between them.

8.4 Efficiency: Options and Objections

How can the evolutionary constraint help relate the joint intentionality and the biological individuality debates? According to the evolutionary version of the efficiency constraint, any definition of joint action should contain ingredients that explain or justify its efficiency, or reliable success, cashed out in terms of fitness (expected survival and reproductive value). There are reasons for preferring to tackle the evolutionary version of the efficiency constraint rather than the rational one. Rational efficiency is assessed in terms of material payoffs. As many collective actions happen in situations of one-shot interactions, it is often hard to determine whether one really is a joint action. Success in one particular case may be exceptional—other slightly different cases would have led to failure. Efficiency, or reliable success, can only be assessed by witnessing several collective actions accomplished in similar contexts. There is no such problem in an evolutionary setting: talk of fitness benefits already takes into account the entire consequences of an agents' behavior during its lifetime.

How can the evolutionary efficiency constraint be met? There are at least two possible approaches, respectively based on *explanations* and on *analogies*. The former looks for evolutionary explanations of psychological mechanisms involved in cooperative behavior. Consider the fact that a nonnegligible proportion of human individuals cooperate with strangers. Maybe such behavior is not efficient nowadays—we tend to trust others too easily. Nonetheless, it might have been efficient for our ancestors, for instance if encounters with strangers were less common or happened in different contexts. The explanation-based evolutionary approach identifies psychological mechanisms and then strives to show how they could have been adaptive in the past (regardless of their current efficiency). Note that although such explanations of social behavior abound in the literature, their consequences on definitions of joint action are hardly ever considered. Moreover, they are typically based on just-so stories and models that are difficult to scientifically confirm or disconfirm.

The second approach, which this chapter aims to introduce, is absent from the literature. It focuses on biological entities, although not in order to find the roots of our social behavior. Rather, it takes seriously the idea that joint action or joint intentions are ways by which groups become so tight that they resemble individuals, and so may be informed and constrained by definitions of individuality in other fields. Many biological individuals too are groups of lower-level entities; so their study may reveal fundamental abstract mechanisms of efficient grouping that also operate in other contexts—namely in human joint action. There may only be a small number of ways by which individuals manage to gather in stable groups. These are likely to be realized by very different causal processes in joint intentions and in biological individuals; however, they may still share some main properties. In virtue of their being kinds of groups, many biological individuals are possibly relevant sources of evidence for theories of cooperation in general.

Moreover, because the efficiency of biological factors, and of combinations thereof, can usually be cashed out directly in terms of fitness benefits, accounts of biological individuality naturally meet the efficiency constraint. A definition of joint action definition may be justified by its close similarity with one of biological individuality, as the latter will be constituted of factors that warrant fitness benefits.

Still, the idea that analogies between joint intentions and biological individuality can help shed light on the former may seem far-fetched. Before I develop these analogies, some objections to the proposal need to be addressed. There are at least three such criticisms: the *process* objection, the *transience* objection, and the *selection* objection.

According to the process objection, the strategies based on the efficiency constraint are unduly mixing distinct categories. When looking for *definitions* of joint intentions, considerations concerning their *explanations* should be irrelevant. What matters is the type of ingredients and of mental states that constitute a joint intention, not the process by which it appeared or was formed. There is a relatively straightforward reply though: even classical definitions of joint intentions mention causes of or reasons for the relevant intentions. Recall Bratman's definition: one clause states that one must intend to do her part "because of" the presence of meshing subplans. Intentions formed without considering subplans do not qualify. Moreover, the presence of such causes or reasons is a general feature of most definitions of joint intention. For them to act jointly, agents must not only intend to do their part; their intentions must have been formed in the right way, or for the right reasons. In other words, definitions have always referred to explanatory elements; nothing is new here.

The transience objection goes as follows. The focus on evolutionary or biological aspects implies a long-term perspective: interactions repeated over a lifetime, or individuals that are stable and maintain their unity during their life cycle. Whatever processes are responsible for these properties are unlikely to be implied in joint intentions, which are mostly transient: they are often formed shortly before, and disappear immediately after the action is accomplished. Considerations of maintenance or stability are thus irrelevant to joint intentions. This objection is misguided though. Of course, the respective time-scales of a joint action with immediate benefits and of a set of repeated interactions with fitness benefits are wildly different. However, even joint intentions enjoy a form of stability. For instance, individual intentions must be at least minimally mutually reinforcing: there could not be a joint action if I wanted to stop doing my part as soon as I am aware that you are doing yours. This is why in definitions, others' intentions are typically taken to provide reasons for one to form her own intention; this is another instance of the presence of causes or reasons to form intentions in definitions. In addition to such mutual reinforcement, the common knowledge clause also ensures that knowing others' intentions never provides a reason not to do one's part. Here again, stability considerations have always been part of definitions. Indeed, several philosophers have recently emphasized that joint intentions are equilibria and that their intertwined components render them fundamentally stable (Gold and Sugden 2007, Chant and Ernst 2007).

Lastly, the selection objection states that the background processes on which joint intentions and biological individuals depend are too different. In particular, adaptive explanations presuppose the action of natural selection; but no such selection is at work in joint intentions. So how could the explanations be similar? This objection mistakes a possibility claim with a necessity one. Nothing proves that joint intentionality and biological individuality involve similar mechanisms. Even if they do, nothing proves that we can learn about the former from studying the latter. So far, the analogy-based evolutionary approach is intended as a heuristic, based on the fact that grouping mechanisms may crosscut disciplinary boundaries and that our scientific understanding of biological individuality may inform our philosophical understanding of joint action, which has been much less empirically constrained so far. Of course, the proof of the pudding is in its eating. The following sections suggest ways in which the analogy between joint intentionality and biological individuality can be developed: the first while.

8.5 A First Analogy: Criteria

One aim of this chapter is to bring joint action and biological individuality closer. To the best of my knowledge, philosophers of joint action have never referred to the biological individuality debate. Moreover, quite understandably, biologists interested in individuality have never cared about joint intentionality. It is thus all the more surprising that many of the criteria with which they have come up in their respective fields are direct counterparts. In this section I will be content with listing a few of these analogies—one for each classical introduced in section 8.2.

First, recall Bratman's emphasis on the importance of meshing subplans for joint intentions. For agents to have meshing subplans they must be disposed to act in *complementary* ways. If I paint the wall that you have already (properly) painted, or if I hinder your action by trying to paint the very same area that you are painting, then our subplans of action do not mesh. This echoes the perceived importance of division of labor and even of functional integration for biological individuality. Despite its absence from some recent surveys, the idea that organisms should display some level of functional integration—that their parts should be differentiated and produce a joint effect—is widespread in the biological literature. One problem is that this criterion is hard to define precisely. Still, meshing subplans and division of labor or functional are two analogous conditions that appeared independently in both debates.[15]

A second striking analogous pair involves Gilbert's normative commitments. These imply agents' expectations that others would react if they did not do their part; in a full-fledged joint action, agents do not decide to stop doing their part unilaterally. There is a clear biological counterpart in the form of policing mechanisms—mechanisms whereby parts that do not fulfill their role are punished, or the consequences of their deviant acts are negated. The analogy may seem artificial, as it compares *expectations* of a reaction and *actual* reactions to a deviant behavior. However, as a matter of fact, most policing mechanisms do not actually result in constant punishment or compensation. Most biological individuals function well, and their parts only occasionally fail to fulfil their roles. When they do, policing mechanisms can act. Most of the time, they will not; their presence merely forces the rate of deviant behavior to stay low.[16] Biological mechanisms need not have a continuous effect to exist; reactions to deviation from joint action need only be expected.

At this stage, the following objection arises. Is the approach not circular, as it suggests that philosophers of joint intentionality should be inspired by the work of biologists that themselves already conceive of biological individuality in intentional and economic terms? Finding analogies between the two fields would thus be unsurprising and possibly uninformative. I think the objection fails for two reasons. In evolutionary biology, the intentional vocabulary is always used as a shortcut—it is "as if" individuals intended to act so as to maximize their fitness. First, the definitions of joint intentionality do not appeal to the vocabulary of rational choice theory. Whether the mental states that may be ingredients for a joint intention arise from a rational process is irrelevant to whether they are actual ingredients. Second, the suggested parallels are not between real and as-if intentions, but concern the relations between higher- and lower-level properties. These properties are respectively joint intentions and various mental states on the one hand, individuality and various nonintentional states on the other hand. But none of the candidate criteria for individuality presented in section 8.3 presuppose an intentional interpretation of the behavior of biological individuals. There is no threat of circularity: the analogies aim to gather different notions of collective individuals, realized by highly different mechanisms.

The two remaining analogous pairs are less easily established because of their conceptual proximity. First, the concept of a collective goal echoes that of a shared evolutionary fate. Second, collective reasons to act find a natural counterpart in genetic similarity. However, are goals not reasons to act? Does a shared evolutionary fate not imply a coincidence of genetic consequences? Both answers are positive. Goals underdetermine reasons to act. Let us consider goals as certain outcomes—certain states of the world.[17] A given outcome may be a goal for agents for different reasons. I may aim to acquire a car because I love to drive, because I need it to commute to work, because I do not want to be despised by my car-owning neighbors, and so on. We may aim to sing together because I enjoy your company and you like singing duets. So collective goals can be realized by multiple combinations of reasons to act. Briefly put, if agents have the same reasons to act, then their interests will coincide in all situations. If they have the same goal, their interests coincide only insofar as the realization of this goal is concerned. Collective reasons to act imply identical preferences or interests; collective goals merely imply a coincidence of interests.

The same relation holds between a shared evolutionary fate and genetic similarity. The latter induces identical genetic "interests," as a consequence of kin selection—the higher the genetic similarity between individuals, the more aligned their evolutionary interests are. A shared evolutionary fate is but a way whereby evolutionary interests come to be aligned—which can happen even among genetically dissimilar individuals. So the double analogy is justified: in one case agents' preferences or evolutionary interests are identical, in the other they are only partly correlated. More generally, this interpretation underlies more general analogies between genes and reasons to act, and between fitness and utility.

Overall, these four pairs of conceptual counterparts induce a partial translatability between the criteria for biological individuality and those for joint intentionality, which will be put to use below.

Interestingly, there are also parallels between the prevalence or neglect of similar criteria in their respective literatures. Consider spatial contiguity. As seen above, according to our folk concept of an individual, this is a necessary property for individuality. However, closer and better-informed looks have suggested that it is actually unnecessary—it may help us recognize some entities as biological individuals, but some other individuals are made of parts that are not spatially contiguous.[18] Interestingly, similar considerations have arisen with respect to joint intentionality. In biology, the spatial contiguity of parts ensures that many of them are in contact and directly exchange resources, signals, and so forth. Similarly, philosophers have long considered that joint action necessitates that agents have direct information about (some of) others' mental states, which is ensured by the common knowledge condition and the fact that intentions mutually cause or justify one another. Such common knowledge typically arises from witnessing public events. However, it has been suggested that collective actions such as mass actions and anonymous one-shot cooperation, in which agents assist in no such public event, can be genuine cases of joint actions (Kutz 2000, Paternotte 2014). This entails that direct epistemic sources are not necessary for joint action; and, as a consequence, that the common knowledge condition should be weakened or dropped. Regardless of the ways to amend or replace existing definitions, here we witness a parallel evolution in the literatures on joint action and biological individuality: a departure from a prima facie compelling condition that agents/parts should be copresent in a given situation and mutually perceivable. So not only do the criteria echo one another; some analogous pairs also enjoy a similar (ir)relevance.

8.6 A Second Analogy: Mechanisms

8.6.1 Fraternal and Egalitarian Evolutionary Transitions

I now explore a second analogy that concerns the explanation of cooperation rather than the definition of joint intentions. First, I need to introduce two particular strands of research—respectively on evolutionary transitions and on group identification—that provide the basis for this analogy. Its combination with the previous section will allow me to issue a verdict about some criteria for joint intentionality.

Bourke (2011) provides a systematic map of the ways by which biological individuals can form. Most of his analysis belongs to the field of so-called *evolutionary transitions* (Maynard Smith and Szathmáry 1995), which has triggered a renewed interest in biological individuality as it considers the history of life as a succession of episodes in which individuals aggregate to form new, higher-level individuals. Bourke decomposes the process that culminates in an evolutionary transition in three steps and describes the many mechanisms that are responsible for each stage in various contexts. These steps are social group *formation*, in which individuals gather and start cooperating; social group *maintenance*, in which their interactions are stabilized by resisting to exploitation from the outside (by recognizing the self and the nonself)[19] and from the inside (by self-limitation or coercion of defecting parts); and social group *transformation*, where division of labor intensifies and interactions become so tight that the group becomes an individual. Unsurprisingly, several criteria for biological individuality already feature in this extremely brief summary.

Following Queller (1997, 2000), Bourke emphasizes that there are two fundamental kinds of evolutionary transitions. *Fraternal* evolutionary transitions happen in populations of related individuals—in particular, they are genetically similar in virtue of their common descent. Examples include the transition from unicellular to multicellular individuals, and the emergence of eusocial societies. *Egalitarian* evolutionary transitions happen between unrelated individuals, either between species or within species. They include the emergence of interspecific symbiotic relationships, or the transition from prokaryotes to eukaryotes (nuclear cells).

These two kinds have distinct features and cannot receive the same explanation. For instance, altruistic behavior cannot evolve in egalitarian transitions, nor can parts lose their reproductive ability (Bourke 2011, 8). Still, crucially, they share a common general explanatory schema: in both cases, "a basic principle ... is that social groups grow more stable the more their members achieve a coincidence of fitness interests. In groups of relatives this occurs via shared genes, and in groups of non-relatives via shared reproductive fate" (Bourke 2011, 94). This view is uncontroversial in biology, as it amounts to saying that there are different ways to limit the relative strength of within-group selection (see, for instance, Wilson and Sober 1989).

8.6.2 Joint Action and Group Identification

Currently, there are only two main psychological explanations of human cooperation. The first one relies on social norms, and has been chiefly developed by Bicchieri (2006), although less psychologically inclined game-theoretic analyses of social norms also exist (Binmore 2006, 2010, Gintis 2010). The second explanation, on which I focus here, invokes group identification: the faculty by which individuals perceive themselves or conceive of their identity as members of groups before anything else.[20] Originally introduced by Tajfel (Tajfel et al. 1971, Tajfel 1973; see also Brewer and Gardner 1996), group identification has the striking characteristic that it can be triggered by many factors, even apparently irrelevant ones,[21] the result being an increased tendency to cooperate. The common point between these factors is that they all make salient a property that is shared by some agents;[22] as they perceive this similarity, their tendency to act for good of the group and to help their "fellow" members increases. Perceived similarity facilitates cooperation; in terms of biological criteria, this could be analogous to genetic similarity.

Surprisingly, Postmes and coauthors (2005) have provided empirical evidence that there is not one but two routes to social identification. The *inductive* one, just described, is based on group properties. However, the *deductive* one stems from individual differences and perceived complementarity between agents. Overall, agents' perception of themselves as part of a group can be triggered either when they share a characteristic with other members or if they are markedly individualized and complementary with others.

Group identification now offers a more fundamental analogy, for it reflects the fraternal/egalitarian distinction in evolutionary transitions. Recall that Bourke saw them as explained respectively by shared genes or shared evolutionary fate. As said above, genetic similarity is analogous to shared group properties; while individual differences and complementarity—which implies the presence of an outcome for which individuals are complementary—echo a shared evolutionary fate. One tentative implication would be that group identification is indeed a necessary mechanism for the formation of joint intentions. However, more precise and specific consequences can be drawn.

8.6.3 Knotting It All Together

It could be tempting to establish a parallel between Bourke's three stages for evolutionary transitions and joint intentions. Of course, as seen in section 8.4, because joint intentions can be transient, we cannot hope to find three similar temporal stages. But recall the structure shared by the definitions of joint intention, exemplified by Bratman's. The mention of reasons or causes for forming intentions echoes social group formation; just like social group maintenance, the common knowledge condition ensures that these intentions are stable; and the realization of a joint action would be analogous to social group transformation. However, this analogy is unhelpful: it is obvious that to become units, groups of any sort have to form, to stabilize, and to increase the tightness between parts. To find this structure in joint action is uninformative—it only shows that biological groups and groups that perform joint actions are instances of groups indeed.

The informative analogy is the following. Recall the parallels established in section 8.5 between genetic identity and collective reasons to act, and between shared evolutionary fate and collective goals. Also recall that evolutionary transitions arise through shared genes or shared fate, depending on whether they are fraternal or egalitarian; and that this distinction is present in explanations of cooperation by group identification. Finally, recall that shared genes are just a way to ensure a shared fate—to align reproductive interests; just as collective reasons to act are just ways for agents to have a collective goal. So in biology, it is a widespread view that the factor common to all evolutionary transitions is the similarity of reproductive interests rather than genetic identity. By analogy, the crucial ingredient for joint intentions would be that agents share goals, not reasons to act.

There are two consequences for definitions of joint intentions. Some philosophers have emphasized that in joint actions, collective goals must meet a *collectivity condition* (Tuomela 2007, 48): agents' goals must necessarily be realized by the same state of affairs—it is not enough that they differ but happen to coincide this time. In essence, this amounts to asking that agents have the same reasons to act, so that their goals will always coincide. As a result of the previous analogies, the collectivity condition is too strong: mere goal coincidence is enough for joint action. The agents' interests need to be actually—rather than necessarily—aligned.[23] Here, we see how insights from evolutionary biology can put pressure on an apparently compelling intuition, according to which agents cannot really act together without having exactly the same goals. Even if this is not a knockdown argument, it is still valuable because such constraints on conditions for joint intentionality have long been lacking.

A more general consequence is that adequate conditions for joint action should be stated in terms of coincidence of interests. More specific conditions, such as the collectivity conditions, are too strong and unduly exclude genuine cases of joint action; less specific conditions that are entailed by such coincidence will at best be necessary, but not sufficient for joint action. For instance, Bratman's meshing subplans *stem from* a coincidence of interests (we want to paint the house); so they describe a subspecies of joint action. In general, definitions of joint intentions should be *minimally teleological*: based on goal coincidence rather than goal identity.[24] In particular, agents may have a variety of reasons to act.

8.7 Conclusion

There are strong analogies, wholly unnoticed until now, between joint intentionality and biological individuality: between their defining criteria and between the mechanisms that lead to them. These can be used to infer characteristics of joint intentionality from those of biological individuality, as the latter have been more thoroughly investigated. Taking these analogies seriously entails that, for instance, joint intentions should be defined on the basis of the coincidence rather than the identity of individual goals.

The analogies I favored may be viewed as obvious. If they seem so, then the mission is partly accomplished—partly, because these are only the beginning of the story: they should be more thoroughly investigated, and the parallels between the particular mechanisms for biological individuality should be systematically compared with the processes, such as collective reasoning and deliberation, leading to the formation of joint intentions. If the parallels can be made precise, then the tentative conclusions reached in section 8.6.3 would be confirmed. Analogies are hints of systematic structural resemblances: the latter are the reasons why we pursue the former. Overall, analogies can bring to the table constraints for definitions of joint intentions; an increased role of empirical evidence in the joint intentionality debate, so far a mostly semantic enterprise; and a general sense of interdisciplinary unification between studies of groups of various sorts.

Acknowledgments

I thank Thomas Pradeu, Marc Ereshefsky and Julian Reiss for their very helpful comments and suggestions. This work was supported by the Alexander von Humboldt Foundation.

Notes

1. See, for instance, Worms 1895.
2. See, for instance, Searle 1990, 1995, who takes joint intentions to be fundamental and irreducible to individual properties. As my aim is to shed light on analyses of joint intentions, I will not discuss such options here.
3. But note that the definition refers to individual intentions "that we J"—individual intentions with a collective content. Nothing in what follows hangs on this.
4. Tuomela's full characterization of joint action depends on what he calls the we-mode, which in addition to collective reasons also includes collective commitments and collective goals.
5. Some authors have tackled the similarity between groups and individual agents without reference to joint action, arguing, for instance, that one key to this similarity is rational consistency (List and Pettit 2010). However, consistent behavior provides *evidence* for agency but does not *constitute* it; it is arguably the consequence of some more fundamental properties of a group. Accordingly, I do not consider such options in this chapter.
6. See Paternotte 2014 for an exploration of various possible constraints, as well as a full defense of the efficiency constraint.
7. Joint action must not be mistaken with cooperation. While the latter is often observable and can be defined as the obtention of a collective benefit, the former is not, as it involves mental states. Crucially, cooperation does not have to be reliable: people may cooperate out of sheer luck or in fragile ways. By contrast, joint action is supposed to capture the part of human cooperation that is reliable. Cooperation is more robust, or less sensitive to variation in external conditions when realized as a joint action than when not.
8. Also see Paternotte 2015 for a justification of the common knowledge clause based on the (rational) efficiency constraint.
9. Ironically, this debate started not long after (but independently from) that on joint intentionality. Buss 1987 is usually considered as the seminal work in this field.
10. Biological individuals include organisms, evolutionary individuals, but also other entities such as developmental individuals and interactors. Strictly speaking, most of what follows concerns not biological but evolutionary individuality. This is because the set of possibly relevant criteria has been mainly established and discussed in the evolutionary individuality literature. However, the cases in which biological individuals are not evolutionary individuals often involve physiological and metabolic properties that do not have clear counterparts in the joint intentionality literature. For this reason, and because this is not a philosophy-of-biology chapter, I treat biological and evolutionary individuality equivalently in the text, and use exclusively the former term in order to avoid confusion.
11. Clarke's (2011) list contains 13 criteria. She mentions but rejects two more, namely functional integration and autonomy, because of their vagueness. I include them because they are fairly present in the literature.
12. For other discussions, see, for instance, Buss 1987, Michod 1999, Godfrey-Smith 2009.

13. I neglect other criteria such as sexual reproduction, life cycle, bottlenecks, histocompatibility, and unity of adaptation because they concern properties that seem exclusive of the living realm—that is, they mainly concern organism and so are not general enough for my purposes, which concern biological individuals in general. I also neglect criteria such as the degree of cooperation and conflict (Queller and Strassmann 2009) because it is arguably explanatorily reducible to other criteria. Some other criteria, such as spatial contiguity and fitness maximization, are evoked in later sections.

14. I do not mean to imply that works on joint intentionality cannot be informed by empirical data. I fully agree with Guala (2007) that, as parts of the philosophy of social sciences concerned with social ontology, they can and should be. However, for most authors of such works, "ontological issues are to be resolved by means of *linguistic analysis* and *intuition*" (Guala 2007, 966; his emphasis). My focus on the efficiency constraint and on mechanisms for biological individuality can be seen as a way to bring empirical aspects to bear on definitions of joint intentionality.

15. A possible objection is that functional integration is by and large a physiological property, which concerns organisms but not biological individuals in general; so it is too specific to be relevant to our discussion. However, note that division of labor, which is commonly seen as a necessary step toward the emergence of higher-level individuals during evolutionary transitions, is not (see, for instance, Michod et al. 2006 on the transition toward multicellularity). One may even talk of functional integration in such cases, as cells come to play different individually necessary and jointly sufficient roles. So division of labor and functional integration are relevant to biological individuality in general.

16. One may be tempted to establish a firmer parallel between policing in biological individuals and punishment in human societies—the effect of which on cooperation is formally shown through classical theorems in game-theoretic models of repeated interactions (Osborne and Rubinstein 1994). However, in explanations of cooperation (rather than definitions) such punishment has most of its effects in virtue of being expected. For instance, in Bicchieri's (2006) definition of social norms as an explanation of cooperative behavior (among others), even expectations of punishment are no more than an optional clause (11).

17. Strictly speaking, goals are themselves underdetermined by outcomes: one can imagine a given outcome that simultaneously realizes distinct goals. If a prince aims to inherit a kingdom, while rival kings want to get rid of the current king, this king's death simultaneously fulfills these goals. However, this does not jeopardize the underdetermination of reasons to act by goals. The prince may aim to govern may because he craves power or because he has the country's interests at heart.

18. See, for instance, Janzen 1977 on evolutionary individuals.

19. I resort to this traditional vocabulary for convenience; but note that it is debatable and has been criticized by Pradeu (2012).

20. Although sometimes considered as an explanatory rival to social norms (for instance, by Bicchieri), the two may actually be compatible, in which case group identification would be ubiquitous in all situations of cooperative behavior. Space constraints prevent me from defending this view here.

21. Such as allocation to subgroups supposedly on the basis of shared characteristics, but actually arbitrary.

22. For instance, relevant factors can be common fate, unanimous promises, interdependence, the presence of a competing group ... (Bacharach 2006).

23. Note that with respect to evolutionary biology situations, Okasha and Paternotte (2012) have reached a similar conclusion, although based on wildly different arguments, when critically assessing Gardner and Grafen's (2006) claims as to the role of genetic identity condition; namely, that the crucial condition for biological individuality and even organismality is repression of competition, that is, alignment of reproductive interests; genetic identity is but one way to realize it. The conclusion itself is far from being new; but its mathematical treatment and the fact that it is derived from the hypothesis that individuals are units of adaptation is. (By contrast, Gardner and Grafen's mathematical treatment suggested that genetic identity was the royal path toward superorganismality.)

24. This does not mean that we should distinguish between coincidental and identical evolutionary fates—a distinction that evolutionary biologists do not make. Recall that the analogy is between the goals / reasons to act and shared fate / genetic identity couples. So the coincidence/necessity distinction is on the intentional side of the analogy.

References

Bicchieri, Cristina. 2006. *The Grammar of Society: The Nature and Dynamics of Social Norms*. Cambridge: Cambridge University Press.
Binmore, Ken. 2006. Why Do People Cooperate? *Politics, Philosophy and Economics* 5: 81–96.
Binmore, Ken. 2010. Social Norms or Social Preferences? *Mind and Society* 9: 139–57.
Bourke, Andrew F. G. 2011. *Principles of Social Evolution*. New York: Oxford University Press.
Bratman, Michael. 1992. Shared cooperate activity. *The Philosophical Review* 101(2): 327–341.
Bratman, Michael. 1993. Shared Intention. *Ethics* 104: 97–113.
Bratman, Michael. 1999. *Faces of Intention: Selected Essays on Intention and Agency*. New York: Cambridge University Press.
Brewer, Marilynn B., and Wendi Gardner. 1996. Who Is This "We"? Levels of Collective Identity and Self Identification. *Journal of Personality and Social Psychology* 71: 83–93.
Buss, Leo W. 1987. *The Evolution of Individuality*. Princeton, NJ: Princeton University Press.
Butterfill, Stephen A. Submitted. What Is Joint Action? A Modestly Deflationary Approach.
Chant, Sara R., and Zachary Ernst. 2007. Group Intentions as Equilibria. *Philosophical Studies* 133: 95–109.
Clarke, Ellen. 2010. The Problem of Biological Individuality. *Biological Theory* 5(4): 312–325.

Clarke, Ellen. 2011. Plant Individuality and Multilevel Selection Theory. In *The Major Transitions in Evolution Revisited*, ed. Kim Sterelny and Brett Calcott, 227–250. Cambridge, MA: MIT Press.

Clarke, Ellen. 2013. The Multiple Realizability of Biological individuals. *Journal of Philosophy* 110(8): 413–435.

Gardner, Andy, and Alan Grafen. 2006. Capturing the Superorganism: A Formal Theory of Group Adaptation. *Journal of Evolutionary Biology* 22: 659–671.

Gilbert, Margaret. 1989. *On Social Facts*. Princeton, NJ: Princeton University Press.

Gintis, Herbert. 2010. Social Norms as Choreographer. *Politics, Philosophy, and Economics* 9: 251–264.

Godfrey-Smith, Peter. 2009. *Darwinian Populations and Natural Selection*. New York: Oxford University Press.

Gold, Natalie, and Robert Sugden. 2007. Collective Intentions and Team Agency. *Journal of Philosophy* 104(3): 109–137.

Guala, Francesco. 2007. The Philosophy of Social Science: Metaphysical *and* Empirical. *Philosophy Compass* 2(6): 954–970.

Hakli, Raul, Kaarlo Miller, and Raimo Tuomela. 2010. Two Kinds of We-Reasoning. *Economics and Philosophy* 26(3): 291–320.

Janzen, Daniel H. 1977. What Are Dandelions and Aphids? *American Naturalist* 111: 586–589.

Kutz, Christopher. 2000. Acting Together. *Philosophy and Phenomenological Research* 61(1): 1–31.

Lehmann, Laurent, and Laurent Keller. 2006. The Evolution of Cooperation and Altruism: A General Framework and a Classification of Models. *Journal of Evolutionary Biology* 19(5): 1365–1376.

Maynard Smith, John M., and Eors Szathmáry. 1995. *The Major Transitions in Evolution*. New York: Oxford University Press.

Mayr, Ernst. 1961. Cause and Effect in Biology. *Science* 131: 1501–1506.

Michod, Richard E. 1999. *Darwinian Dynamics: Evolutionary Transitions in Fitness and Individuality*. Princeton, NJ: Princeton University Press.

Michod, Richard E., Yannick Viossat, Cristian A. Solari, Mathilde Hurand, and Aurora M. Nedelcu. 2006. Life-History Evolution and the History of Multicellularity. *Journal of Theoretical Biology* 239: 257–272.

Miller, Seumas. 2001. *Social Action: A Teleological Account*. New York: Cambridge University Press.

Okasha, Samir, and Cédric Paternotte. 2012. Group Adaptation, Formal Darwinism and Contextual Analysis. *Journal of Evolutionary Biology* 25(6): 1127–1139.

Osborne, Martin J., and Ariel Rubinstein. 1994. *A Course in Game Theory*. Cambridge, MA: MIT Press.

Paternotte, Cédric. 2014. Minimal Cooperation. *Philosophy of the Social Sciences* 44(1): 45–73.

Paternotte, Cédric. 2014. Constraints on Joint Action. In *Perspectives on Social Ontology and Social Cognition*, ed. Mattia Gallotti and John Michael, 103–123. New York: Springer.

Paternotte, Cédric. 2015. The Epistemic Core of Weak Joint Action. *Philosophical Psychology* 28(1): 70–93.

Postmes, Tom, Russell Spears, Antonia T. Lee, and Rosemary J. Novak. 2005. Individuality and Social Influence in Groups: Inductive and Deductive Routes to Group Identity. *Journal of Personality and Social Psychology* 89(5): 747–763.

Pradeu, Thomas. 2012. *The Limits of the Self.* New York: Oxford University Press.

Queller, David C. 1997. Cooperators since Life Began. *Quarterly Review of Biology* 72: 184–188.

Queller, David C. 2000. Relatedness and the Fraternal Major Transitions. *Philosophical Transactions of the Royal Society of London B* 355: 1647–1655.

Queller, David C., and Joan E. Strassmann. 2009. Beyond Society: The Evolution of Organismality. *Philosophical Transactions of the Royal Society of London B* 364: 3143–3155.

Searle, John. 1990. Collective Intentions and Actions. In *Intentions in Communication*, ed. Philip Cohen, Jerry Morgan, and Martha E. Pollack, 401–415. Cambridge, MA: MIT Press.

Searle, John. 1995. *The Construction of Social Reality*. New York: Free Press.

Simiand, François. 1987. Organisme et société. In *Méthode historique et sciences sociales*, 47–54. Paris: Éditions des archives contemporaines. Reprinted from the 1897 original edition.

Sterelny, Kim. 2013. Cooperation in a Complex World: The Role of Proximate Factors in Ultimate Explanations. *Biological Theory* 7(4): 358–367.

Tajfel, Henri. 1973. The Roots of Prejudice: Cognitive Aspects. In *Psychology and Race*, ed. Peter Watson, 76–95. Chicago: Aldine.

Tajfel, Henri, Michael Billig, R. P. Bundy, and Claude Flament. 1971. Social Categorization in Group Behavior. *European Journal of Social Psychology* 1: 149–178.

Tollefsen, Deborah. 2005. Let's Pretend! Children and Joint Action. *Philosophy of the Social Sciences* 35(1): 75–97.

Tuomela, Raimo. 2000. *Cooperation: A Philosophical Study*. Dordrecht: Kluwer.

Tuomela, Raimo. 2007. *The Philosophy of Sociality: The Shared Point of View*. New York: Oxford University Press.

West, Stuart A., Ashleigh S. Griffin, and Andy Gardner. 2007. Social Semantics: Altruism, Cooperation, Mutualism, Strong Reciprocity and Group Selection. *Journal of Evolutionary Biology* 20(2): 415–432.

Wilson, David S., and Elliott Sober. 1989. Reviving the Superorganism. *Journal of Theoretical Biology* 136: 337–356.

Worms, René. 1895. *Organisme et société* [Organism and Society]. Paris: Giard & Brière.

CHAPTER 9 | On the Emergence of Individuals in Physics

SIMON SAUNDERS

I TAKE "INDIVIDUAL" TO mean an object that (i) persists, somehow, in time, and (ii) can be uniquely identified throughout the time that it persists. I take "object" (and interchangeably, "thing") to mean anything that can stand in predicate position, typically the value of a bound variable; in this I follow Quine.

The main task of this chapter is to show how individuals arise from classical and quantum indistinguishable particles—from objects that are *permutable*, meaning, whose descriptions are invariant under permutations. There is a prima facie difficulty with (ii): how can something be uniquely identified—at any time—if when interchanged with other things the overall description is unchanged? The problem arises, I shall argue, as much in classical physics as in quantum physics, although it takes a slightly different form in the two cases. In both cases (in the quantum case only in certain circumstances) we can identify something else as not being subject to interchange; for example, we can pass from talk of particles that have states to talk of the states themselves—to points of phase space or one-particle states in Hilbert space (one-dimensional subspaces). But as we shall see, this option has no real connection to the way we ordinarily refer to individuals in the laboratory, or the use of names in defining the state spaces of individuals as distinguishable things.

In this chapter I am interested in the question: what is the metaphysics appropriate to the way individuating reference actually goes, in the laboratory, consistent with the requirement of indistinguishability? This amounts to the question (or I shall take it as the question) of how permutation symmetry can be broken, at one level of description, whilst remaining intact at a more fundamental level of description.

9.1 Logic and Ontology

My starting point is the notion of object in first-order logic in simple declarative sentences. This would seem the most secure ground for relating representations in physics (phase space, Hilbert space) to things, via referring expressions in ordinary language. I see the detour through language as reflecting the fact that physical theories are primarily about *quantities*, rather than things, so we cannot simply consult our best physical theories to discover what there is. It may even be that the world is at bottom a mathematical structure, or "has" a mathematical structure; but in trying to be more precise as to what that involves, I see no safer way than to put questions of ontology into words, using simple declarative sentences and the standard apparatus of first-order quantifiers. So given the mathematical structure of a physical theory, if there are puzzles about what aspects of it are real, or correspond to reality, or what has purely mathematical as opposed to physical significance, we should first see what can be said about things in simple declarative terms—in terms of objects, identity, properties, and relations. Objectual structure in this sense I see as a coarse-graining of the mathematical structure of the world: the pegs and poles that gather its materials and most reliably tie them together.[1]

Evidently what is needed for this to work is a close association of predicates with physical quantities, on the one hand, and with the domain of quantification, on the other. My suggestion is that the "allowed" predicates be those that can be constructed from values (and changes in values) of physical quantities, and specifically quantities that are *invariant* under the exact symmetries of a theory. This plausibly includes all quantities that are actually measurable. And further, that allowed predicates be tied to the domain of quantification by the requirement that no more is admitted than are required by Leibniz's law.[2]

This needs some explanation. Let $\mathcal{L}$ be a first-order language with a finite lexicon, including identity (what I shall call primitive identity). Suppose further that only certain (perhaps complex) predicates in a subset $\mathcal{P} \subset \mathcal{L}$ are allowed (those corresponding to physically real properties, defined by invariant quantities). The principle is then: physical objects are values of variables (in the logical sense) that can be discerned by the allowed predicates. By this I mean, if s and t are terms for physical objects:

$$s = t \leftrightarrow \bigwedge_{F \in \mathcal{P}} F...s... \leftrightarrow F...t.... \tag{9.1}$$

The implication from left to right follows from Leibniz's law (so the language is extensional). It is the implication from right to left that is controversial, enforcing, for physical objects, a version of the Principle of the

Identity of Indiscernibles (PII).[3] Of course, if primitive identity is itself an allowed predicate, (9.1) is a tautology.

Generalizing on free variables, one obtains from among the conjuncts of (9.1) not only sentences like

$$\forall y(Fsy) \leftrightarrow \forall y(Fty) \tag{9.2}$$

but also like

$$\forall y(Fsy \leftrightarrow Fty). \tag{9.3}$$

It is sentences of the form (9.3) that are false of the familiar supposed counterexamples to the PII (so by (9.1), $s = t$ is false as well, whereupon they *cease* to be counterexamples); those of the form (9.2) are all true. Thus consider Black's two iron spheres, exactly alike, one mile apart, in an otherwise empty Euclidean space. Suppose allowed predicates are those invariant under the symmetries of this space (translations and rotations). Every monadic predicate true of one sphere is true of the other, including complex monadic predicates with embedded quantifiers, as in (9.2). But taking F as the symmetric, irreflexive, and invariant (so clearly allowed) relation "is one mile apart from," (9.3) is false, so $s \neq t$.

As Quine (1976) showed (but using the terminology "discriminables"), identity construed in this way yields the following exhaustive classification: objects s and t are *absolutely discernible* if for some monadic predicate, Ps but not Pt; *relatively discernible* if for some binary predicate, Fst but not Fts; and *weakly discernible* if for some binary predicate, Fst and Fts, but not Ftt and not Fss. When I say an individual is *identifiable* (at a time or throughout a period of time), I mean it is absolutely discernible (at a time or throughout a period of time); thus individuals, in my sense, are always absolutely discernible. By "indistinguishables" I mean things that are at most weakly discernible, if discernible at all.[4]

Should the negation of identity itself be an allowed predicate? It is not, or not obviously, a relation definable in terms of the invariant values of any particular physical quantity (although, e.g., as suggested by causal set theory, a speculative program in quantum gravity, that might yet change). Rather, and more plausibly if physics is at bottom about quantities rather than things, it should be implicitly defined by them all, by the PII in the sense (9.1). So my proposed methodology counts against it. If it is allowed anyway, with identity taken as primitive, the PII in the form (9.1) is trivialized, and as an account of identity of physical things it loses its philosophical interest. But Quine's classification remains: it is just that among the things that are only weakly discerned, are those that are only discerned by the negation of identity. The essential notions for our purposes are absolute and weak discernibility,[5] categories that are defined whether or not primitive identity is allowed.

In summary: individuals are absolutely discernible. Indistinguishables, assuming they can be discerned at all, are at most weakly discernible. Since a permutation leaves their state-description unchanged, it should leave their predicative description—in terms of allowed predicates—unchanged as well. But that means reference to only one of a number of indistinguishables, by an allowed monadic predicate, however complex, is impossible: if it applies only to one indistinguishable, and not to any other, it would absolutely discern it, contrary to supposition. If it applies to one indistinguishable, it must apply to every; if it is a binary predicate and applies to one pair of indistinguishables, it must apply to every pair; and so on. Call such predicates *permutation invariant* (or invariant for short). Any talk of indistinguishables, if it is to respect permutation symmetry (if it is to be allowed, if it is to be invariant under permutation symmetry) must be conducted in terms of invariant predicates. I shall also talk of indistinguishable as *permutables*.

The contrast, evidently, is with predicates and function symbols (including names) that do not respect this symmetry—as are used in descriptions of laboratory systems and everyday things. And now there is an obvious difficulty. The descriptions (states) of electrons, protons, and neutrons are invariant under permutations; electrons, protons, and neutrons are therefore at most weakly discernible. In short, they are indistinguishable. Yet ordinary objects are constituted by electrons, protons, and neutrons, so reference to ordinary objects must break permutation symmetry. Or to turn the problem around: what, from the point of view of a symmetry-preserving description in terms of indistinguishables, is being described by a symmetry-breaking description in terms of individuals?

Call it the *paradox of constitution*: descriptions of macroscopic things may be singular; but specification of the electrons, protons, and neutrons of which they are composed is impossible without breaking permutation symmetry.

9.2 Particles and Trajectories

The paradox seems stronger in the case of classical particles, where we think we know what we are talking about. But here there is an easy response: simply deny that classical particles *should* be treated as indistinguishable; insist that permutability of classical particles is simply unintelligible.

Since weak discernibles are permutable, and weak discernibility (as we have just seen) is a perfectly well-defined logical category of objects, this claim is hardly obvious. But it is supported by a simple argument: let the state-description pick out the exact motions of particles (so the state is not just a probability distribution); then the particles can be identified by their trajectories, *and are, therefore, not indistinguishable.*[6]

There is something right about this argument,[7] but there is something wrong with it too, for in certain circumstances the same can be said in quantum mechanics. In place of points in phase space, we have rays in Hilbert space. Let Π_N be the permutation group for N elements, and let c be a normalization constant. Consider a state of N particles of the form

$$|\Phi\rangle = c \sum_{\pi \in \Pi_N} |\phi_{\pi(a)}\rangle \otimes |\phi_{\pi(b)}\rangle \otimes \ldots \otimes |\phi_{\pi(c)}\rangle \otimes \ldots \otimes |\phi_{\pi(d)}\rangle, \qquad (9.4)$$

where each summand is the tensor product of N pairwise-orthogonal one-particle vectors $|\phi\rangle$, elements of the one-particle Hilbert space $\mathcal{H}$ labeled by the symbols "*a*," "*b*," etc. (so there are no repetitions of these symbols). Such a state is totally symmetrized; as such it is invariant under permutations. It describes indistinguishable particles, specifically, bosons. If we allow for superpositions of states of the form (9.4), we obtain the entire Hilbert space of N bosons.[8] But restricting to states like (9.4), we can speak of one-particle states instead; and we may take it that each of these states, as orthogonal to any other, is absolutely discernible. It is true that in this case we do not have *trajectories* as such, but there is something just as good under a further restriction. Thus, let the unitary dynamics **U** factorize, so it is of the form

$$\mathbf{U}|\Phi\rangle = c \sum_{\pi \in \Pi_N} U|\phi_{\pi(a)}\rangle \otimes U|\phi_{\pi(b)}\rangle \otimes \ldots \otimes U|\phi_{\pi(c)}\rangle \otimes \ldots \otimes U|\phi_{\pi(d)}\rangle. \qquad (9.5)$$

Then each one-particle state has a unique trajectory, namely, its orbit under U. The analogy with the classical case is complete. But now, if classically we can simply identify particles with trajectories—so that particles are *not* indistinguishable—then why not in quantum mechanics simply identify particles with the orbits of one-particle states? And if we can: does it follow that *quantum* particles are distinguishable after all?

Of course symmetrized N-particle states are *not* in general of the form (9.4),[9] whereupon it is no longer possible to identify (or replace) particles by one-particle states. And there is another difficulty in the case of fermions, in the analogous state to (9.4) (but antisymmetrized rather than symmetrized): in that case the state can also be rewritten in terms of *other* collections of N orthogonal one-particle states, as the singlet state of spin makes clear. Which is the right collection? Both difficulties show that something else is going on in quantum mechanics.

There surely is, but it seems to have very little to do with indistinguishability. The same problems arise for *distinguishable* quantum particles. The state space for N distinguishable particles is spanned by product states

$$|\Phi\rangle = |\phi_a\rangle \otimes |\phi_b\rangle \otimes \ldots \otimes |\phi_c\rangle \otimes \ldots \otimes |\phi_d\rangle \qquad (9.6)$$

rather than by states of the form (9.4). For any state (9.6), each particle can be assigned a unique one-particle state (the kth in the sequence as specified by the tensor product)—a unique pairing of particles with states. But no such assignments of one-particle states to particles is possible for superpositions of states (9.6).[10] And as for the second problem, the ambiguity of the one-particle states in the case of fermions: it arises for the singlet state of spin, regardless of whether or not the two fermions are indistinguishable. (I shall come back to this question in section 9.6.)

The correct conclusion to draw, surely, is not that particle indistinguishability makes no sense in either classical *or* quantum mechanics;[11] it is that at least in some circumstances in quantum mechanics, and nearly always in classical mechanics, one can shift from a description in terms of indistinguishables (particles that have trajectories or that are in one-particle states) to a description in terms of distinguishables (the trajectories, the one-particle states). And notice, in this shift, we pass from a description in terms of all indistinguishable particles (in (9.4) and (9.5), of all N particles), to a description of a particular trajectory, a particular one-particle state, apparently without any need to make reference to any other trajectory or any other one-particle state.

9.3 Indistinguishability in Ordinary Language

The same shift in ontology can be mimicked in ordinary language. Consider:

(i) Buckbeak the hippogriff flies higher than Pegasus the winged horse.

Permuting the expressions "Buckbeak the hippogriff" with "Pegasus the winged horse" would give an entirely different sentence, one that contradicts (i). But suppose we omit proper names and make do with descriptive predicates instead, for example, "is Buckbeak-shaped" and "is Pegasus-shaped," where being Buckbeak-shaped includes being a hippogriff and so on, as descriptivists about proper names recommend. Assume for convenience that these are the only two mythical creatures (the only two things in our domain of discourse), so we do not have to worry about uniqueness. In that case (i) is equivalent to (dropping quantifiers)

(ii) there is x and there is y such that x is Buckbeak-shaped and y is Pegasus-shaped and x flies higher than y.

Now consider the complex predicate

> (iii) x is Buckbeak-shaped and y is Pegasus-shaped and x flies higher than y, or y is Buckbeak-shaped and x is Pegasus-shaped and y flies higher than x.

Evidently (iii) is invariant under permutation of x and y. Taken as a single complex predicate, and assuming it is the only allowed predicate (or that there are others but they are all equally permutation-invariant), then x and y are only weakly discernible (note that (iii) is irreflexive). Under this constraint, it is impossible to make reference to x or y singly; yet (iii) under existential quantifiers conveys the same information as (ii).

In terms of properties, we have a way of understanding properties of permutables as disjunctive properties. It carries over to quantum mechanics, where properties are represented by projection operators. Sums of orthogonal projectors correspond to disjunctions of the corresponding properties. Since they all sum to one, to obtain the negation of a property subtract it from one. Then for N indistinguishable quantum particles, the projector corresponding to the fact that there is exactly one particle that is an A, with corresponding 1-particle projector P, is

> (iv) $P \otimes (I-P) \otimes \ldots \otimes (I-P) + (I-P) \otimes P \otimes (I-P)$
> $\ldots \otimes (I-P) + \ldots + (I-P) \otimes \ldots \otimes (I-P) \otimes P.$

where there are N factors in each term of the summation, and $\binom{N}{1} = N$ summands (it is clear how this generalizes to properties shared by $k \leq N$ particles). (iv) is not a property that one of the N particles has, and none of the others: it is a property of the collective. It is the property that exactly one of N indistinguishable things is an A, or has the corresponding property P.

In the predicate calculus the parallel construction is

> (v) $(Ax_1 \wedge \neg Ax_2 \wedge \ldots \wedge \neg Ax_N) \vee (\neg Ax_1 \wedge Ax_2 \wedge \neg Ax_3 \ldots \wedge \neg Ax_N \vee$
> $\ldots \vee (\neg Ax_1 \wedge \ldots \wedge \neg Ax_{N-1} \wedge Ax_N).$

As with (iii), (v) is permutation-invariant, and as with (iv), it is a complex N-ary predicate that is true of all N particles. Of course there are permutation-invariant predicates with arity $n < N$, but they only report what is true of every sub-collection of n particles out of N; of what is true of every particle ($n = 1$), of every pair of particles ($n = 2$), and so on. Thus, "there is an x that is a Buckbeak shape or a Pegasus shape, that does not fly higher than itself" exhausts what can be said of one of the two (hence of both). The more informative predications are those that include all their relations. In this sense, permutability forces a kind of structuralism: it is the global ascriptions of properties and relations that are the most informative.

Are they informative enough? It seems so, at least to the extent that sentences in the predicate calculus about collections of N things are informative enough *without* the restriction to permutation symmetry. For it is a theorem that any categorical first-order sentence (hence a sentence whose models all have the same finite cardinality) describing N objects is logically equivalent to one of the form

$$\exists x_1 \exists x_2 \ldots \exists_N F x_1 \ldots x_N \tag{9.7}$$

where F is totally symmetric (Saunders 2006a). And here I am relaxed about the restriction to finitely many things, because states on classical phase space and in Hilbert space (supposing it is separable) are likewise restricted to descriptions of finite numbers of particles.

Having understood how to go from ordinary descriptive sentences to descriptions invariant under permutations, it is obvious how to go back again. From (iii), use the "that which" construction instead of quantifiers and variables in each disjunct, to obtain

(vi) That which is Buckbeak-shaped flies higher than that which is Pegasus-shaped, or that which is Buckbeak-shaped flies higher than that which is Pegasus-shaped.

The disjunction is then redundant; at the same time, there is no question of interchanging "that which is Buckbeak-shaped" with "that which is Pegasus-shaped" *salva veritate*, for "flies higher" is asymmetric. And from (vi), pass to "the Buckbeak shape," "the Pegasus shape," and then to "Buckbeak" and "Pegasus," like passing from "that which is butter" to "butter." We obtain the sentence (9.1) that we started with. The predicates "is Buckbeak shaped" and "is Pegasus shaped" function as what I shall call *individuating predicates*.

Evidently there are two steps involved: first, find an individuating predicate; second, make use of it as a mass noun, in object position in predicates, without any requirement of permutability. But to be serviceable—to be available in a wide variety of states of affairs—the individuating predicate should not include too much. It should be stable in time, if it is to serve as criterion that can actually be used, whilst allowing for plenty of change. "Animal shape" is reasonably robust in this sense, but only when understood in terms of general anatomy, not in terms of the precise shape that an animal has at a particular moment in time. It should not be too generic, either, if it is to absolutely discern one out of a collection of things at one time. It is a Goldilocks property, that is just right as a referring expression for a general context of use. Considerations like these are familiar to descriptivist theories of names.

9.4 Reasons for Permutation Invariance

Agreed that we can see how the trick is done; what is the point of it? The lengthy disjunctive properties corresponding to permutation-invariant predicates seem rather contrived. Why restrict to properties like this, or equivalently, to predicates like this?

One answer is that in quantum statistical mechanics and field theory, there are important *empirical* consequences of permutability. But while there are advantages to symmetrizing in classical statistical mechanics, in that case it leads to no directly testable consequence. It is unlikely to be of practical use in formal logic, either, so let me put the pragmatic answer to one side. The question remains: why symmetrize?

An obvious line of thought is that the symmetry arises because it doesn't matter which particle has which trajectory (or one-particle state), because the particles involved are "simples"; they all have exactly the same intrinsic properties (the same mass, charge, and spin). The various permutations of these particles (with everything else unchanged) yield observationally indistinguishable states of affairs, so they should not be conceptually distinguished, either. But it is not obvious how this is to work at the level of everyday language. In the case of Buckbeak and Pegasus, it invites us to picture a realm of things, all with the same intrinsic properties, each of which can have one or other of a number of animal-shapes. What are these things, exactly?

It is the wrong picture. It may be the wrong picture in quantum mechanics too, where—for example in string theory—the intrinsic (state-independent) properties of simples are in danger of disappearing altogether. If none of the intrinsic properties of elementary particles turn out to be state-independent (not mass, charge, or spin),[12] what, precisely, remains?

An alternative picture is that the redundancy attaches not to a choice among physical particulars that are intrinsically the same, but to something else; to a referential device, for example, or perhaps to something more metaphysical.[13] In formal logic: to values of variables as elements of some class $\mathcal{D}$, over and above their function as relative pronouns and the expression of pluralities. In phase space and Hilbert space: to values of particle labels, over and above their function of keeping track of sequence position (in terms of ordered sequences of 6-tuples or coordinates, for phase space points, and tensor products of one-particle states, for rays in

Hilbert space). The redundancy in each case is eliminated by passing to permutation-invariant descriptions. Let us take each case in turn.

9.4.1 Logic and Model Theory

The suggestion is that permutation symmetry may function in logic and model theory in roughly the way that it functions in a physical theory: it corresponds to a certain kind of redundancy of representation that can be eliminated by passing to an invariant description. How might this work?

Recall that an interpretation $\mathfrak{U}$ of a first-order language $\mathfrak{L}$ consists of a universe of discourse $\mathcal{D}$, an assignment of relations $P^{\mathfrak{U}}$ on $\mathcal{D}$ for each $\mathfrak{L}$-predicate P, and of functions $f^{\mathfrak{U}}: \mathcal{D} \times \ldots \times \mathcal{D} \to \mathcal{D}$ (n factors) to function symbols f (n arguments) in $\mathfrak{L}$. Proper names are 0-ary function symbols, assigned designated elements of $\mathcal{D}$. A valuation σ on $\mathfrak{L}$ is a mapping of variables $x \to x^{\sigma} \in \mathcal{D}$, inducing a mapping to truth values as: Px is true if and only if $x^{\sigma} \in P^{\mathfrak{U}}$, with obvious extensions to quantified variables. A model of an $\mathfrak{L}$-sentence (with no free variables) is an interpretation under which it is true for every valuation.

A model, therefore, comes equipped with relations on $\mathcal{D}$. How are they defined? Proper names are assigned designated elements of $\mathcal{D}$. Designated how? Predicates are interpreted by their extensions, for monadic predicates, by subclasses of $\mathcal{D}$. What are those elements, and how are they specified? The primary role of variables in syntax, apart from generality, is that the same variable may be repeated in a sentence: this their function as relative pronouns (the "that which" construction). But under a valuation of $\mathfrak{U}$, they are also assigned elements of $\mathcal{D}$. These elements are given in advance. They are, perhaps, abstract particulars. How are they related to physical particulars?

There are the usual suspects: by way of proper names; by the intentions of the users of the language; by indexicals; by way of identity (the model just is the world); by an antecedent understanding of the referents of proper names and of the extensions of predicates—take your pick. Or, returning us to our topic, it doesn't matter which element of $\mathcal{D}$ is associated with which physical particular. It is the structure of the model as a whole that represents the world.

Indeed, there is a puzzle here that has long been an embarrassment to philosophers. It was first stated by Quine, as one of a number of arguments for his doctrine of "ontological relativity."[14] According to Quine, at bottom, values of variables are no more than "neutral nodes" that can be shuffled among one another without any linguistic effect. They look very much like permutables.

Quine argued as follows. Let $\lambda: \mathcal{D} \to \mathcal{D}$ be a bijection, what Quine called a "proxy function" on the universe of discourse, and consider, for any interpretation U, the interpretation U^* under which P is true of λ of what P was true of under U (with the obvious action of λ on sequences); and similarly for function symbols. It then follows that any $\mathfrak{L}$-sentence true under U will be true under U^* as well, even though it talks about quite different things (elements in D). In this way λ induces a new and "unintended" interpretation $\mathfrak{U}^*$ of $\mathfrak{L}$ (to put it in Putnam's terms). Quine viewed the matter as an extension of his doctrine of underdetermination of meaning: reference was "relative to a manual of translation." But he also put the matter like this:

> Reference and ontology recede thus to the status of mere auxiliaries. True sentences, observational and theoretical, are the alpha and omega of the scientific enterprise. They are related by structure, and objects figure as mere nodes of the structure. What particular objects there may be is indifferent to the truth of observation sentences, indifferent to the support they lend to the theoretical sentences, indifferent to the success of the theory in its predications. (Quine 1990, 31)

In the case of finite models, the parallel with permutability is hard to ignore. But it cannot be the same: Quine's argument applies to any first-order language, any sentence, any interpretation, whether or not its allowed predicates are permutation invariant. It is also clearly paradoxical—and was seen as such by Putnam.[15] On the other hand, the method of section 9.3 *also* applies to any first-order language, any sentence, any interpretation, so long as it describes a finite number of things. And indeed, permutation-invariant sentences like (iii) are indifferent between U and U^* as their intended model; for them there is no paradox. Permutability is not the same as ontological relativity; it is the cure for it. If we mimic the procedure used in physics, the problem is solved.

There remains, however, the peculiarity that on passing to a description in terms of permutation-invariant predicates, singular reference to any of these "neutral nodes"—to anything less than the entire state of affairs—is impossible. Quine, on the reading I am giving, was half-right in his diagnosis: in classical and quantum mechanics, true sentences, the alpha and omega of ontology, are related by structure, and values of variables serve as mere nodes of the structure, tying it together, but not tying it to anything: it is the structure as a whole that is instantiated in the world. Only if this structure is sufficiently variegated is there a passage to singular reference, and that proceeds quite differently: it is reference to qualitative features of this structure, whether using proper names, Fregean senses, Russellian descriptions, causal chains, or ostension.

If, further, these qualitative features are robust (they admit variation) and are stable (they persist in time), then they are individuating properties. They are the values of variables and referents of proper names, and relations on them are the extensions of predicates, where now there is no requirement of permutation symmetry. Indeed, permuting one qualitative feature into another, altogether different, with all else unchanged, not only yields a distinct state of affairs, but is likely to take us out of the space of physically possible states of affairs altogether.

But can't we just run Quine's argument all over again, in talk of these qualitative features? Of course we can, but then there are other responses to the puzzle, one being that we have an antecedent understanding of what the elements of the universe of discourse are, of which of them are the referents of proper names, and of which of them lie in the extension of one predicate, rather than another—in short, of what are the "perfectly natural properties." All of that is expressed, or represented, in the ground-level representation, in terms of permutables and invariant predicates of the structure as a whole.

9.4.2 Phase Space and Hilbert Space

A point $\langle \vec{q}, \vec{p} \rangle \in \Gamma^N$ is specified by $2N$ triples of numbers, where each triple is indexed by a particle label $k = 1, \ldots, N$, thus:

$$\langle \vec{q}, \vec{p} \rangle = \langle q_1, p_1; \ldots; q_k, p_k; \ldots; q_N, p_N \rangle \in \Gamma^N. \tag{9.8}$$

Given that the Hamiltonian is a symmetric function of the N particles—the sequence position of the arguments of the Hamiltonian does not matter—the labels become irrelevant to the dynamics. The permutations are thus symmetries. The phase space point $\langle \vec{q}, \vec{p} \rangle$ yields the same set of N trajectories in μ-space as the initial data

$$\pi \langle \vec{q}, \vec{p} \rangle = \langle q_{\pi(1)}, p_{\pi(1)}; \ldots; q_{\pi(k)}, p_{\pi(k)}; \ldots; q_{\pi(N)}, p_{\pi(N)} \rangle \tag{9.9}$$

for $\pi \in \Pi_N$. If there are no repetitions of phase space coordinates, there will be $N!$ distinct sequences of the form (9.8), (9.9), each corresponding to a set of N one-particle trajectories in μ-space, differing only in which trajectory is assigned which particle label (passive view), or which is assigned which particle (active view).

The distinction is as real, no more and no less, as the distinction between which element in $\mathcal{D}$ is assigned to which extension $P^\mathfrak{U}$ of each predicate P. Just as permutation-invariant predicates are oblivious to such distinctions, invariant functions on phase space (and in particular the Hamiltonian)—functions invariant under permutations of particle labels—are blind to them too.

Similar remarks apply to particle labels in quantum mechanics, and property ascriptions (in terms of totally symmetrized projectors, of the form (iv)) that are indifferent to distinctions as to which particle is in which one-particle state, for states of the form (9.6). And as in quantum mechanics, it is clear how to proceed to a new universe of discourse, in which the distinctions, previously ignored, no longer arise at all. Classically, points of Γ^N related by permutations can be simply identified: that is, Γ^N can be replaced by the quotient space Γ^N/Π_N of Γ^N under Π_N. This is *reduced phase space*. In place of $N!$ equivalent points in Γ^N, there is a single point, denote $\widetilde{\langle \boldsymbol{q}, \boldsymbol{p} \rangle} \in \Gamma^N/\Pi_N$. Here $\widetilde{\langle \boldsymbol{q}, \boldsymbol{p} \rangle}$ is the same set of N pairs of configuration space and momentum space coordinates as in (9.8) and (9.9), yielding the same N trajectories in μ-space as all the $\overrightarrow{\langle \boldsymbol{q}, \boldsymbol{p} \rangle}$s, but expressed as an unordered set:

$$\widetilde{\langle \boldsymbol{q}, \boldsymbol{p} \rangle} = \{\langle q_a; p_a \rangle, \langle q_b; p_b \rangle, \ldots, \langle q_c; p_c \rangle, \ldots, \langle q_d; p_d \rangle\} \in \widetilde{\Gamma}^N. \tag{9.10}$$

Assuming there are no repetitions,[16] then just as with (9.4), we may speak directly of the one-particle values of position and momenta (or phase-space coordinates) $\langle q_a; p_a \rangle, \langle q_b; p_b \rangle$, and so on, as points or point-like regions of μ-space, rather than of particles that have those coordinates or those trajectories; or equivalently, just use the word "particle" to denote such coordinates, or values of position and momentum, or point-like regions of μ-space, or properties.

Isn't this just to revert to particle labels, and won't the same considerations apply as before? No: particle labels were defined by sequence-position in $\overrightarrow{\langle \boldsymbol{q}, \boldsymbol{p} \rangle}$, but $\widetilde{\langle \boldsymbol{q}, \boldsymbol{p} \rangle}$ is not a sequence. In place of $\langle q_a; p_a \rangle, \langle q_b; p_b \rangle$, etc. we could just as well have written $\langle q; p \rangle, \langle q'; p' \rangle$ and so on. Likewise in (9.4): in $|\phi_a\rangle, |\phi_b\rangle$, etc. "$a$" and "$b$" are not labeling particles, but orthogonal one-particle states: they are distinguishable.

With no use of the machinery of sequences and particle labels, there is no reference to values of labels; so no redundancy either. Predications true of one trajectory, or one one-particle state, will no longer be true if that trajectory or one-particle state is substituted for another—indeed, will in general be an out-right mathematical impossibility. With no restriction to permutation-invariant predicates, singular reference to particulars—so long as there are no repetitions—is straightforward, or as straightforward as it ever is in the use of coordinates to define positions and momenta (or velocities) of particles. And notice that in passing to reduced state space, and making no use or mention of particle labels or names (but only of coordinates $\langle q_a; p_a \rangle, \langle q_b; p_b \rangle$ and thereby of places a, b in one-particle phase space), we are implementing the reductionist ploy recommended by Quine:

> Those results [in quantum statistics] seem to show that there is no difference even in principle between saying of two elementary particles of a given kind that they are in the respective places a and b and that they are oppositely placed, in b and a. It would seem then not merely that elementary particles are unlike bodies; it would seem that there are no

such denizens of space-time at all, and that we should speak of places *a* and *b* merely as being in certain states, indeed the same state, rather than as being occupied by two things. (Quine 1990, 35).

Places, Quine is recommending, should take object position, and predications should be made of them. The state they are both in is the one-particle state. Note, however, that in making no mention of place in *momentum* space, Quine makes it seem easier than it is to pass to an ontology of places. It is places in μ-space (one-particle phase space) that replace particles, not physical space. It is not a reduction of material bodies to regions of space: it is a reduction of permutable particles to point-like regions of μ-space, and in quantum mechanics, to one-particle states in Hilbert space.

Can we think of either of these as properties? Classically, if strictly point-like, this is a property represented by a delta-function (a distribution) on μ-space, whereas properties are ordinarily associated with functions (characteristic functions). But in quantum mechanics it is purely a matter of terminology: there states are rays, and rays are "regions" of $\mathcal{H}$—meaning subspaces—and subspaces are properties, denoted by the associated projectors $P_{|\phi\rangle}$. The reduction, then, in the quantum case, consists in passing from a global description of N indistinguishables, in terms of the totally symmetrized projection

$$P_{|\phi_a\rangle} \otimes P_{|\phi_b\rangle} \otimes .. \otimes P_{|\phi_c\rangle} \otimes .. \otimes P_{|\phi_d\rangle} + \text{all permutations} \tag{9.11}$$

acting on $\mathcal{H} \otimes ... \otimes \mathcal{H}$ (N factors), to talk of distinguishable properties, the projections $P_{|\phi_a\rangle}, P_{|\phi_b\rangle}$, and so on, each acting on $\mathcal{H}-$ or equally, to talk of one-particle states. And when the unitary evolution factorizes (as in (9.5)), we can talk of the orbits of these one-particle states as well, or sequences of properties; just as we speak of trajectories in classical mechanics, or sequences of positions in μ-space.

But whether point-like regions of phase space, or one-particle states, can really function as individuating properties is another question. Given the dynamics, they may imply a kind of persistence in time (pass to the trajectories, where available); but the property itself, of having such-and-such position and momentum or such-and-such a one-particle state, does not persist, and cannot function as a state-independent property, even (or rather especially) when such properties are carefully identified in the laboratory (say by a sufficiently precise state-preparation device, whereby a particle is produced in a definite state)—the investigation, experimental and theoretical, of particles in such a state consists exactly in seeing how the particle evolves in time, how its state changes in time. Whatever characterizes it as a state cannot be a state-independent property. We must look elsewhere for individuating properties that can be used as names to define a state-space of distinguishable particles.

We have explained the need for permutation symmetry; we can see, roughly, why it requires global descriptions, and how, contra Quine, they are descriptions of definite things nonetheless. We have an answer of sorts to the paradox of constitution in the classical case, at least at one instant of time. But not even that is available, in general, in quantum mechanics, and the answer is in any case unsatisfactory. So what are the state-independent properties that can function as names for distinguishable things?

9.5 Demarcating Properties

I return to the notion of stable and robust properties that are fine-grained enough to stand in for single objects, but not so fine-grained as to have no interesting state space; Goldilocks properties, properties that are just right. But by now it should be evident that to solve the paradox of constitution we do not need uniqueness: it is enough to speak of relatively small numbers of particles, entangled or otherwise, as apart from all particles. Permutation symmetry only has to be broken enough to get down to a definite collective, stable in time; the description of the collective itself may be in terms of permutables. Or for another way of putting it: it is enough to show how a total symmetrized state-space can be replaced by a Cartesian product of two state-spaces, each totally symmetrized.

In fact, it is enough to show how this goes in a sufficiently good approximation, in some regime of energy and scale of interest in the laboratory (or for that matter that of everyday things). By our general ansatz, we should look for properties shared by some but not all permutables, properties that are robust and stable in time for the dynamical regime in question. Call them *demarcating* properties; individuating properties arise as the limiting case, where the "small number" is unity. Thus, in place of properties like (9.11) that specify everything, or, like (iv), the property of there being exactly one thing with some property, we need a more coarse-grained projector—say a demarcating property P that applies to n out of N particles. In that case the collective has the totally symmetrized property:

$$\underbrace{P \otimes ... \otimes P}_{n \text{ factors}} \otimes \underbrace{(I-P) \otimes ... \otimes (I-P)}_{N-n \text{ factors}} + \text{permutations}. \qquad (9.12)$$

Because demarcating properties will only be available in a certain dynamical regime, and for limited periods of time, we shall say they are *emergent* properties, rather than fundamental ones. We then use the method of section 9.3, and use P in object position—but as a mass term, like "butter," or better, a natural kind term, like "metal." Thus, if the demarcating property is the projection onto bound states of the electron and proton, it is "hydrogen." We can equally construe P as the name of a plurality, and speak of Ps, or of instances of P, or of particles of kind P. There will now be a range of dynamical behavior available to Ps (consistent with the fact that they are Ps, that it is hydrogen), ensured by the fact that the demarcating property is stable and robust. P can thus function as a label in state space, hosting a non-trivial dynamics, that applies to P particles, as opposed to not-P particles. And evidently the interchange of Ps with not-Ps will not be a symmetry; typically a transformation like that will take us out of the state-space altogether. But the Ps are still permutable among themselves, and the not-Ps are still permutable among themselves.

The idea, then, to return to the classical case, is to refer to more coarse-grained regions of μ-space than the point-like places of section 9.4.2: in effect, to find regions of phase space that can function like natural kinds. Let us see how this works in the simplest case of two particles. Because there are just two, the demarcating property will in fact be an individuating property, but the example extends easily to the N-body case.

Let μ-space be coarse-grained with respect to two independent degrees of freedom h, v ("horizontal" and "vertical" respectively), each into two regions, as in figure 9.1. Thus each particle can be either in A or B, and, independently, in U or D ("up" or "down"). There are four cells in all, denote AU, AD, BU and BD. This induces a coarse-graining of the two-particle phase space. If the particles are distinguishable, this is the space Γ^2: it is partitioned into 16 regions, as illustrated in figure 9.2a. If the particles are indistinguishable, it is the space Γ^2/Π_2. It is partitioned into 10 regions, as shown in figure 9.2b. (There are just 10 distinct descriptions of the two particles in terms of permutation-invariant predicates; 10 ways of distributing two indistinguishable particles over the four cells of figure 9.1.)

The Quine redux is to pass from talk of things being in regions AU, AD, BU and BD, to talk of regions being in the occupied or unoccupied states—in the 0-particle state, the 1-particle state, and the 2-particle state. However, there are correlations among these states: if AU is in the 2-particle state, then AD, BU, and BD are in the 0-particle state; and so on. The constraint is conservation of total particle number.

The similarity with the occupation-number formalism of nonrelativistic quantum field theory is obvious:[17] the dynamics, as in that theory, consists in variations among integer-valued states of distinguishable things (modes of the field, labeled, typically, by wavelengths or frequencies), subject to conservation of total particle number. But now in addition each individual has a smooth "internal" degree of freedom—thus, in the case of AU,

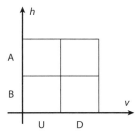

FIGURE 9.1 μ-space with degrees of freedom h and v, coarse-grained into regions A and B, and U and D, respectively

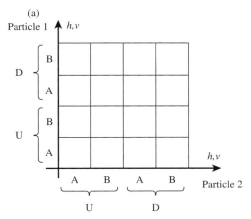

FIGURE 9.2A Γ-space for two distinguishable particles each with two degrees of freedom h and v. Each axis represents both degrees of freedom, coarse-grained into region A, B and U, D respectively

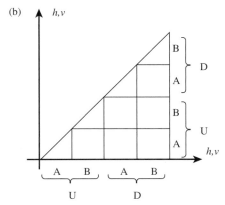

FIGURE 9.2B Reduced phase space for two indistinguishable particles, with the same coarse-graining of μ-space as figure 9.1

variation with respect to values of h restricted to A, denote h_A, and of v restricted to U, denote v_U; and similarly for AD, BU, and BD.

But we are looking for a correspondence with a classical Hamiltonian evolution for distinguishable particles, not with quantum theory. To that end, let us just treat regions defined by the coarse-graining of the horizontal degree of freedom as things, that is, call them A and B respectively; leave the vertical degrees of freedom as common to each. In addition, A and B have the internal degrees of freedom h_A and h_B respectively. As before, A and B are to be found in one of the three states, the 0-particle, the 1-particle, and the 2-particle state, subject to constraints. In general, as A and B change in time, this description has the further peculiarity than the *numbers* of degrees of freedom of A and B change as well. Thus when one of them (say A) is in the 0-particle state, it no longer has any degrees of freedom, whereas B has four: two horizontal (in this case, both h_B) and two degrees of freedom v. But still, peculiar as they are, with these rules we can pass from the coarse-grained description of two indistinguishable particles to the coarse-grained description of two distinguishable regions of phase space.

Notice now that the peculiarities disappear in the special case that A and B are each in the 1-particle state, *and the dynamics is such as to keep them there*. In that case, effectively, certain regions in phase space will not be accessible, the regions shaded in figure 9.3a. In this regime, particle number is frozen as a degree of freedom of A and B, and it can be left out of the effective description. So long as the dynamics acts in this way, A and B are always in the one-particle state; they can each be called "particles." The only degrees of freedom remaining are the two internal ones and the vertical degree of freedom. The accessible (unshaded) region of Γ^2/Π_2 has the structure of a phase space Γ^2 for two *distinguishable* particles, figure 9.3b, where the effective Hamiltonian lives. It is an effective dynamics—it gives a good approximation to the underlying dynamics of the permutables—only so long as the underlying dynamics keeps the shaded regions inaccessible; so long as the properties A and B are stable and robust over time.

How does our toy model generalize? It is obvious how to add additional degrees of freedom; what about additional particles? Evidently, if we have unique demarcating properties for each extra particle, then we have individuating properties, whereupon we pass to a description of N distinguishable particles, completely breaking permutation symmetry.[18] But more typical, when N is large and the particles are microscopic, the dynamics only freezes out a small number of coarse-grained properties. Thus, for two such properties (as in our toy model), the N particles may be divided into N_A permutables confined to region A, and N_B permutables confined to region B, by properties of the form (9.12), where $n = N_A$ (and expressed in words as disjunctions of predicates related by permutations,

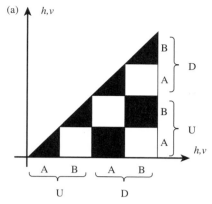

FIGURE 9.3A Shaded areas represent dynamically inaccessible regions of reduced phase space

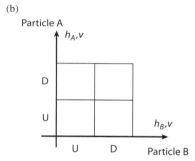

FIGURE 9.3B Γ-space for distinguishable particles, labeled by individuating properties A and B, with degree of freedom v, coarse-grained into regions U and D. Particles A and B also have internal degrees of freedom h_A and h_B respectively

as in (v)). We can then pass to an effective phase-space structure, but now using "A" and "B" as mass terms rather than proper names. The result is a phase space for N_A indistinguishable A-particles interacting with N_B indistinguishable B-particles, each permutable only among themselves.[19]

In real physics, of course, these demarcating properties are often only statable in a technical language. They do not line up with ordinary words. But the border is porous; talk of air easily goes over to talk of nitrogen, oxygen, and carbon dioxide. It may be that to talk of "spin" as a mass term seems odd, but less so with "charge" (and "mass," after all, is a mass-term par excellence). Mass, various kinds of charge, and spin are expected to suffice in a grand unified theory. Specify these (the combination of these demarcating properties) and you specify quarks of one flavor rather than another, or of quarks rather than leptons, and so on. In the Standard Model, demarcating properties are bound states of quarks: one bound state is a proton, another is a neutron, and so on through the hadrons. In all these cases we can see what I have been calling "internal" degrees of freedom clearly do arise, and can be experimentally probed (for example in the case of longer-lived hadrons, in elastic scattering experiments). In nuclear physics demarcating properties are bound states of protons and neutrons, yielding the 118 known nuclei. Internal degrees of freedom concern variations in nuclear structure that keep the nucleus intact. But typically, in effective theories, these "internal" degrees of freedom are neglected altogether: the whole point of an effective theory is to provide a simplified description at the appropriate regime of energy and scale.

However demarcating properties like these do not on their own provide an answer to the paradox of constitution; nor by themselves can they yield individuating properties—this time, not through lack of robustness, but through lack of uniqueness. However complex a bound state of a large number of molecules, or protons and neutrons, if it is stable, it is unlikely to be instantiated just the once in the universe. This much was right about our earlier solution to the constitution puzzle in terms of points in of μ-space: what is needed in addition are *spatial* demarcating properties, of being of definite spatial extent at a given time, in spatiotemporal relation to various other things, likewise localized in space and time. How, exactly, do these arise? This has nothing to do with questions about the physical meaning of coordinate systems, or inertial or freely falling frames. I mean, *given* all that: how do localized systems arise, in a way that it is stable over time?

That they arise, in this way, is not in doubt. And of course spatial localization is routinely studied—and assumed—in the physical sciences, whether in terms of boundary conditions and confining potentials, rigid bodies, crystals and lattices, or in terms of dynamical models (as for example in plasma physics and cosmology). There are countless questions about all of these uses and studies—and answers. But the burden of proof, at this point, is unclear. The original question was about permutation symmetry; the provenance of properties of spatial localization might be thought a larger question.

The paradox of constitution can therefore be solved: in terms of quantum demarcating properties, on the one hand (in terms of the formation of stable bound states, robust over long periods of time); and in terms of spatial demarcating properties, on the other (as used throughout the physical sciences).

9.6 The Problem of Measurement

You who are content with this answer, and are relaxed about the various options for solving foundational problems in quantum mechanics, can stop reading here. The puzzle as we have posed it, in terms of permutation symmetry, is solved. But others not so sure, or who are persuaded that the foundational problems in quantum mechanics exactly concern questions of ontology, read on.

A first point is that the dynamics, relevant to the formation of bound states and questions of stability over time, is the unitary dynamics; it is Schrödinger's equation. The definition and investigation of dynamics like this is more or less what quantum mechanics at low energy scales is about. But the fact that there are localized events at all, in quantum mechanics, is not without difficulty.

Why not make do with properties as defined by projectors—and specifically, projectors on localized regions Δ of configuration space or phase space? In place of the much too fine-grained properties $P_{|\phi\rangle}$ (projectors onto rays in $\mathcal{H}$), use coarser-grained projectors P_Δ instead. Thus, given (as before) a minimally entangled state (9.4), suppose that n of the N one-particle states $|\phi_a\rangle, |\phi_b\rangle, \ldots$ are localized in Δ, some small spatial region, with the remainder localized in the complement of Δ, and that the supports of these states (in configuration space) are changing slowly in time. In that case the N particles have the property (9.12), with $P = P_\Delta$, over some appreciable period of time. In that case too, the expectation value of any local dynamical quantity is the same, whether the state is totally symmetrized over all N particles, or the product of symmetrized states for n and $N-n$ particles.

This is an important consistency check. It also has the supreme virtue that it works for superpositions of such states, too, as long as in all the states superposed, n of N permutables are located in Δ. This follows from linearity: any superposition of states each with property P itself has property P, meaning, is an eigenstate (with eigenvalue 1) of the associated projector. For the same reason, the problem of nonuniqueness of fermion one-particle states, briefly remarked on in section 9.2, is solved. That problem, recall, was that for fermions, even when minimally entangled, there is no unique set of N orthogonal one-particle states entering into the entanglement. But whether or not a state is an eigenstate of an operator is independent of the basis in which the state is written.

This point is perfectly familiar in special cases. I have already mentioned an example: the spherically symmetric singlet state of spin, which can be written with respect to z-component of spin, $|\psi_+^z\rangle$ and $|\psi_-^z\rangle$, as

$$|\Psi_0\rangle = \frac{1}{\sqrt{2}}(|\psi_+^z\rangle \otimes |\psi_-^z\rangle - |\psi_-^z\rangle \otimes |\psi_+^z\rangle).$$

It can equally well be written with respect to any other components of spin, say $|\psi_\pm^y\rangle$, as

$$|\Psi_0\rangle = \frac{1}{\sqrt{2}}(|\psi_+^y\rangle \otimes |\psi_-^y\rangle - |\psi_-^y\rangle \otimes |\psi_+^y\rangle).$$

It is the same state nonetheless. But then a similar non-uniqueness of composition applies to any state of this form. Thus for any orthogonal one-particle states $|\varphi_a\rangle, |\varphi_b\rangle$, the state

$$|\Phi\rangle = \frac{1}{\sqrt{2}}(|\phi_a\rangle \otimes |\phi_b\rangle - |\phi_b\rangle \otimes |\phi_a\rangle) \tag{9.13}$$

can be written as

$$\frac{1}{\sqrt{2}}(|\phi_+\rangle \otimes |\phi_-\rangle - |\phi_-\rangle \otimes |\phi_+\rangle),$$

where

$$|\phi_+\rangle = \frac{1}{\sqrt{2}}(|\phi_a\rangle + i|\phi_b\rangle), \quad |\phi_-\rangle = \frac{1}{\sqrt{2}}(i|\phi_a\rangle + |\phi_b\rangle)$$

or, indeed in terms of $|\phi_1\rangle, |\phi_2\rangle$, where

$$|\phi_1\rangle = \frac{1}{\sqrt{2}}(|\phi_a\rangle + |\phi_b\rangle), \quad |\phi_2\rangle = \frac{1}{\sqrt{2}}(|\phi_a\rangle - |\phi_b\rangle).$$

In the case of the singlet case of spin, this indeterminateness is explained in terms of rotational symmetry: no one direction in space is preferred. But the indeterminateness afflicts any two fermions in a minimally entangled state[20]—in fact, it afflicts any *N* fermions, in a minimally entangled fermion state. So it afflicts fermions, period, whatever their state. This method for passing to individuals, for special states of the form (9.4), as one-particle states, was pointless anyway, but now we see that in the case of fermons, failing a notion of preferred basis, it was completely illusory.

But the underdetermination of fermion one-particle states is irrelevant to the dynamical emergence of any demarcating property, if defined by the spectrum of a self-adjoint operator and written as a projection operator. The basis in which a state is written makes no difference to the subspace in which it lies. Quite generally, so long as the dynamics is unitary, dynamical considerations are indifferent to the basis used to represent the state: the choice of basis is like the choice of coordinates in classical spacetime theories. Of course, like a coordinate system in classical spacetime theories, a basis can be better or worse "adapted" to the dynamics and initial state; it may be more or less convenient (from the point of view of explicit equations and calculations) to use one basis rather than another.

Thus, if the initial state is spherically symmetric (and this is a symmetry respected by the Hamiltonian), use spherically symmetric coordinates.

The only obstacle, then, to defining spatial demarcating properties in quantum mechanics in a dynamically stable way is in understanding how states satisfying projectors like (9.12) for $P = P_\Delta$ arise in the first place. Such states describe systems of particles well-localized in Δ over appreciable periods of time.[21] How do they arise?

But *this* problem, we now see, is not at all trivial. For in quantum mechanics states like these do *not* readily arise under the purely unitary evolution. There is no point in simply positing them at much earlier times, either; quantum mechanical states, initially localized, tend to spread over time. Admittedly, if they are states of large numbers of bound particles (so of relatively large mass), this dispersion is slow, especially if they are left to themselves. Thus, a rock of mass 1 kg, prepared in a state localized to within a micron, freely evolving in time, will remain localized to within a few microns for the entire lifetime of the universe. But if it is subject to complex external forces, if, say, it is tumbling chaotically, in the way that Hyperon tumbles in its orbit about Saturn, it will become delocalized much sooner—much, much sooner. Hyperion has a mass of about 10^{20} kg, and diameter 10^5 m; according to the estimate of Zurek and Paz (1995), if initially well-localized, it will become completely delocalized over its entire orbit about Saturn in less than 10 years.

Of course if we are willing to apply quantum mechanics to macroscopic bodies, there are any number of ways in which initially localized quantum states become wildly delocalized on much smaller timescales: virtually any quantum experiment can be arranged to have this result. Rather than Schrödinger's cat being alive or dead, depending on the outcome of the experiment, have it end up in one corner of the laboratory rather than the other, if possible alive either way. Evidently we are up against *the quantum-mechanical problem of measurement*.

We may even be at the heart of it. For the three worked-out solutions to the measurement problem (worked out, at least, for nonrelativistic quantum mechanics) all take the quantum state as representative of something physically real, and they all engineer a process by which the state is localized, whether effectively (according to effective equations), or fundamentally (by tampering with the unitary equations). They all involve "wave-packet collapse," over length and timescales that are under theoretical control. In Everettian (many-worlds) quantum theory and in pilot-wave theory, according to which the universal state propagates unitarily at the fundamental level, the collapse is effective, as defined in decoherence theory (taking slightly different forms in the two cases). In dynamical collapse theories like the GRWP theory (due to Ghirardi, Rimini, Weber, and Pearle), the collapse is fundamental, and the equations at the fundamental level are no longer unitary. In all cases the collapse, effective or real, is associated with probabilistic events, so it is also inseparable from that other dimension to the problem of measurement, the role of the quantum state in determining probabilities.

But in all these theories, collapse, however it is achieved, is onto states of mesoscopic bodies that are sharply localized in position and momentum space. Any account of ontology based on the structure of the quantum state, subject to collapse onto well-localized states at a rate subject to theoretically control, stable in time, will ipso facto provide an account of how spatiotemporal demarcating properties can be dynamically defined. If defined by fundamental equations, as fundamental properties; if by effective equations, as emergent properties.

I plump for effective collapse and its explanation in terms of decoherence theory, preserving the unitary dynamics unchanged, without any additional hidden variables. The spatial localization of quantum states of ordinary things, as branches decohere, is then dynamically emergent. The states best adapted to this dynamics, for reasonably massive particles, are Gaussians, localized in position and momentum space. This is the "decoherence basis"—the right basis for resolving the ambiguity in (9.13), when $|\Phi\rangle$ has a decohering structure, consistent with the unitary dynamics.

In whatever way decoherence theory solves the measurement problem,[22] it solves the problem of spatiotemporal localization as well, the final piece of the jigsaw of how to break permutation symmetry at the effective level of description, consistent with permutation symmetry at the fundamental level. Decoherence theory defines spatial demarcating properties for molecules and larger structures, from decoherence scales and times on up, as dynamically emergent properties. Given a sufficiency of other demarcating properties, we obtain individuating properties too. And thereby we arrive at an account of how individuals arise at an emergent level in physics, and thence across the special sciences.[23]

Acknowledgments

Thanks to Adam Caulton and David Wallace for helpful discussions.

Notes

1. What more, precisely, of the mathematical structure of physical theories might be understandable as physical ontology (or the structure of physical ontology) is an open question. In the tradition of Russell, Carnap, and Quine, presumably, *all of it*—when reformulated in formal logic. But the ontology then is essentially made up of the members of sets in an elaborate construction in set theory, with no distinction

between mathematical and physical particulars. I am sceptical of this tradition, especially if cashed out in tems of the Ramsey sentence (as it so often is by structural realists): see Saunders and McKenzie 2014.

2. For an indication of the scope of this method in application to space-time points and classical particle and field theories, see my 2003a, 2003b, 2013a and, for mirror-symmetry, 2007.

3. This definition of identity was first proposed by Hilbert and Bernays (1934), based on the axiom schema for identity introduced by Gödel in 1930 in his proof of the completeness of the predicate calculus. It played a largely implicit role in Quine's early writings on identity (e.g. in his 1950), but it was explicit in Quine 1960.

4. Are elementary particles in quantum discernible at all? I have argued that fermions are, in my 2003a, 2006b, 2013b, and in Muller and Saunders 2008, but elementary bosons (as opposed to bosons that are multiplets of fermions) pose a special difficulty. It may this can be solved (see, for example, Muller and Seevink 2009), but elementary bosons, with the sole exception of the Higgs, are all gauge bosons, like photons, which we might well do better to talk of as excitations of modes of quantum fields. The Higgs, meanwhile, is exceptional for a number of reasons.

5. Why no particular role for *relative* discernibility? But there could be; for example, if we were to systematize talk of causal relations, or the before-after relation.

6. Assuming of course, their trajectories are absolutely discernible. That need not be the case (for example, the trajectories of Black's iron spheres, unchanging in time in an otherwise empty universe, are only weakly discernible).

7. It was explicit in Bach (1997 7), but it was surely implicit in a number of other criticisms of the notion of classical indistinguishables; see e.g. Van Kampen 1984.

8. Since quantum states are rays, rather than vectors, there is another possibility: the ray will still be invariant if permutations produce a change of overall phase. In fact the only consistent assignment of phase changes of this kind is alteration of sign for odd permutations, i.e. antisymmetrization (fermionic states, which therefore satisfy the Pauli exclusion principle). (Important as the distinction between states [rays] and vectors is, I shall not belabor it in what follows.)

9. Such minimal entanglements have been called "trivial" by Ghirardi et al. 2002, Ghirardi and Marinatto 2004 (see also Penrose 2004, 598).

10. As before, for "minimal" entanglements of the form (9.4), there is just enough entanglement to obliterate the correspondence between place in the tensor product and one-particle state, while still picking out a set of N one-particle states. This does not yield a determinate assignment of one-particle states to particles, however, when the latter are distinguishable.

11. Although that has been suggested; see Dieks and Lubberdink 2011.

12. There is a physically deep way of making, e.g., spin state-dependent (supersymmetry), but there is also a shallow way (as shown by Goldstein et al. 2005a, 2005b), which applies to any supposedly intrinsic property of elementary particles. The shallow method suffices for our purposes.

13. "Bare particulars," for example, or perhaps "haecceities." Huggett (1999) suggests that haecceitism is the culprit, a position endorsed by Albert (2000, 47–48) and elsewhere by me (Saunders 2013b, 356).

14. Quine 1968. Another concerned the Lowenheim-Skolem theorem, which I shall not consider here (restricted as we are to finitary theories).

15. It is otherwise known as "Putnam's paradox" (Lewis 1984). Lewis's solution, in terms of "perfectly natural properties," bears some relation to the one that I offer; see below.

16. Cases of repetitions are important to the observable differences between classical and quantum statistics. For a detailed discussion, see Saunders 2006a, 2013b.

17. Otherwise known as the Fock space representation: see any introductory text in quantum field theory; see also Teller (1995). I single out the nonrelativistic case because in relativistic quantum theory only conservation of total energy, not particle number, is ensured.

18. As can sometimes arise in statistical mechanics: in the case of colloids, for example (see Swendson 2002, 2006), or, say, stars in stellar nebulae.

19. This connects fairly directly to the Gibbs paradox, and the extensivity of the entropy, both in classical and quantum theory. See my 2006a and 2013b.

20. There is something to the explanation from spherical symmetry. The triple of operators $\{P_{|\phi_a\rangle} - P_{|\phi_b\rangle}, P_{|\phi_+\rangle} - P_{|\phi_-\rangle}, P_{|\phi_1\rangle} - P_{|\phi_2\rangle}\}$ satisfy the same commutation relations as the spin operators $\{S_x, S_y, S_z\}$, for orthogonal directions x, y, z in space. In this sense any state of the form (9.13) also has a kind of spherical symmetry.

21. It follows that they are states well-localized in momentum-space as well (subject of course to the constraints imposed by noncommutativity). They pick out—or rather define, assuming phase space is an emergent structure—small patches of phase space. There are no strictly point-like classical objects (points of μ-space), if quantum mechanics is true.

22. See Saunders 2010, Wallace 2012, chaps. 1–3 for more on emergence, decoherence theory, and many worlds. I do not believe decoherence theory in a one-world setting can solve the measurement problem, if the unitary dynamics is fundamental, for their still remains a multiplicity of decohering states, i.e., there are many worlds.

23. Notice that neither the trajectories of pilot-wave theory, nor the "flashes" or mass-densities of collapse theories, have any role to play in the emergent notion of individuals here in play. What need, then, for "primitive ontology" in these theories?

References

Albert, D. 2000. *Time and Chance*. Cambridge, MA: Harvard University Press.

Bach, A. 1997. *Indistinguishable Classical Particles*. Berlin: Springer.

Dieks, D., and A. Lubberdink. 2011. How Classical Particles Emerge from the Quantum World. *Foundations of Physics 41*: 1051–1064. Available at http://arxiv.org/abs/1002.2544.

Ghirardi, G., and L. Marinatto. 2004. General Criterion for the Entanglement of Two Indistinguishable States. *Physical Review A 70*(1): 012109.

Ghirardi, G., L. Marinatto, and Y. Weber. 2002. Entanglement and Properties of Composite Quantum Systems: A Conceptual and Mathematical Analysis. *Journal of Statistical Physics 108*(1–2): 49–122.

Goldstein, S., J. Taylor, R. Tumulka, and N. Zanghi. 2005a. Are All Particles Identical? *Journal of Physics A 38*: 1567–1576.

Goldstein, S., J. Taylor, R. Tumulka, and N. Zanghi. 2005b. Are All Particles Real? *Studies in the History and Philosophy of Modern Physics 36*: 103–112.

Hilbert, D., and P. Bernays. 1934. *Grundlagen der Mathematik*. Vol. *1*. Berlin: Springer.

Huggett, N. 1999. Atomic metaphysics. *Journal of Philosophy 96*: 5–24.

Lewis, D. 1984. Putnam's Paradox. *Australasian Journal of Philosophy 62*: 221–236.

Muller, F., and S. Saunders. 2008. Distinguishing Fermions. *British Journal for the Philosophy of Science 59*: 499–548.

Muller, F., and M. Seevink. 2009. Discerning Elementary Particles. *Philosophy of Science 76*: 179–200. Available at http://arxiv.org/PS_cache/arxiv/pdf/0905/0905.3273v1.pdf.

Penrose, R. 2004. *The Road to Reality*. New York: Vintage.

Quine, W. van. 1950. Identity, Ostension, and Hypostasis. *Journal of Philosophy 47*: 621–633. Reprinted in *From a Logical Point of View* (Cambridge, MA: Harvard University Press, 1953).

Quine, W. van. 1960. *Word and Object*. Cambridge, MA: Harvard University Press.

Quine, W. van. 1968. Ontological Relativity. *Journal of Philosophy 65*: 185–212. Reprinted in *Ontological Relativity and Other Essays* (New York: Columbia University Press, 1969).

Quine, W. van. 1976. Grades of Discriminability. *Journal of Philosophy 73*(5): 113–116.

Quine, W. van. 1990. *Pursuit of Truth*. Cambridge, MA: Harvard University Press.

Saunders, S. 2003a. Physics and Leibniz's Principles. In *Symmetries in Physics: New Reflections*, ed. K. Brading and E. Castellani, 289–307. Cambridge: Cambridge University Press.

Saunders, S. 2003b. Indiscernibles, General Covariance, and Other Symmetries: The Case for Non-reductive Relationalism. In *Revisiting the Foundations of Relativistic Physics: Festschrift in Honour of John Stachel*, ed. A. Ashtekar, D. Howard, J. Renn, S. Sarkar, and A. Shimony, 151–174. Dordrecht: Springer.

Saunders, S. 2006a. On the Explanation of Quantum Statistics. *Studies in the History and Philosophy of Modern Physics 37*: 192–211. Available at http://arXiv.org/quantph/0511136.

Saunders, S. 2006b. Are Quantum Particles Objects? *Analysis 66*: 52–63. Available at http://philsci-archive.pitt.edu/2623/.

Saunders, S. 2007. Mirroring as an A Priori Symmetry. *Philosophy of Science 74*: 452–480.

Saunders, S. 2010. Many Worlds: An Introduction. In *Many Worlds? Everett, Quantum Theory, and Reality*, ed. S. Saunders, J. Barrett, A. Kent, and D. Wallace, 1–52. Oxford: Oxford University Press.

Saunders, S. 2013a. Rethinking Newton's Principia. *Philosophy of Science 80*: 22–48.

Saunders, S. 2013b. Indistinguishability. In *Oxford Handbook of Philosophy of Physics*, ed. R. Batterman, 340–380. Oxford: Oxford University Press.

Saunders, S., and K. McKenzie. 2015. Structure and Logic. In *Physical Theory: Method and Interpretation*, ed. L. Sklar, 127–162. Oxford: Oxford University Press.

Swendson, R. 2002. Statistical Mechanics of Classical Systems with Distinguishable Particles. *Journal of Statistical Physics 107*: 1143–1166.

Swendson, R. 2006. Statistical Mechanics of Colloids and Boltzmann's Definition of the Entropy. *American Journal of Physics 74*: 187–190.

Teller, P. 1995. *An Interpretative Introduction to Quantum Field Theory*. Princeton, NJ: Princeton University Press.

van Kampen, N. 1984. The Gibbs Paradox. In *Essays in Theoretical Physics in Honour of Dirk ter Haar*, ed. W. E. Parry, 303–312. Oxford: Pergamon Press.

Wallace, D. 2012. *The Emergent Multiverse: Quantum Theory according to the Everett Interpretation*. Oxford: Oxford University Press.

Zurek, W. H., and J. Paz. 2005. Quantum Chaos: A Decoherent Definition. *Physica D 83*: 300–308.

CHAPTER 10 | Are There Individuals in Physics, and If So, What Are They?

JAMES LADYMAN

10.1 Introduction

It is often said that if there are individuals there must be a principle of individuation for them, perhaps a different one for each class of individuals. Principles of individuation are supposed to say in virtue of what it is that a particular individual is the one it is and not any other. For example, according to a popular, if not dominant, view in biology, species are individuals, and the principle of individuation says that a species is the particular one it is in virtue of its speciation and extinction events (Hull 1978). Accordingly the identity and diversity of species, and consequently the criteria for whether or not individual organisms are members (which is to say parts) of the same species, are completely determined by whether or not they share those events. In the philosophy of personal identity candidate principles of individuation advert to bodies, lives, memories, and psychological continuity. Other celebrated debates concern the principles of individuation for events, for numbers and for everyday material objects (see, e.g., Lowe 1998). In the case of the latter, as with the case of persons, questions of individuality are usually focused on identity over time in the face of change of parts and change of intrinsic properties.

It is usually supposed that individuals are a subclass of objects, where the latter may be taken to form a subclass of entities even more broadly understood. There is no consensus among philosophers, however, about what individuals as opposed to objects are. Standardly, following Quine, to be an object is to be the value of a first-order variable, so the issue can be put more precisely as follows: what else if anything is required of something in the domain of first-order quantification for it to be an individual?

Steven French and Michael Redhead (1988) take it to be necessary and possibly sufficient for an object to be an individual that it persists. In doing so they follow a venerable line of argument, adopted by, for example, Peter Strawson (1959), according to which individuation is always by means of space and time. Similarly, as mentioned above, in many philosophical discussions about specific classes of individuals it is assumed that they must persist and the issue is what the criteria for their doing so are. However, in recent debates about individuals in physics identity over time has not been to the fore. The controversy has largely concerned intraworld and transworld not transtemporal identity. This is because the putative individuals are quantum particles and space-time points. Quantum particles have been argued to lack even the most minimal synchronic individuality (Cassirer 1956), and identity over time is of course out of the question without that. Space-time points clearly do not persist over time, but it is important for the interpretation of General Relativity whether permutations of them necessarily constitute distinct worlds. It is sometimes assumed that if the objects in a domain are individuals, then permuting them must give rise to a different state of affairs. Since such permutations do not always seem to give rise to different physical models of General Relativity, some conclude that space-time points are not individuals (Smolin 2000). Likewise permutations of quantum particles are usually taken not to give rise to different physical states, and this is often taken to be due to a lack of individuality (see French and Krause 2006, and Saunders 2006a for a critique of them and others).

Whether or not identity over time and transworld identity are completely independent of each other, or logically and/or metaphysically related, is not clear, nor is it obvious how these matters relate to two other core ideas of individuality, namely determinate self-identity and numerical distinctness from other things. In recent discussion of the individuality or not of objects in physics there has been little explicit discussion of how these different notions of individuality relate; rather the focus has been on whether the relevant physical objects violate the Principle of the Identity of Indiscernibles (PII), and the implications of whether they do or don't for the latter issues of determinate self-identity and numerical distinctness. French has consistently argued for the "received view" that they do not (French and Redhead 1988, French and Krause 2006). On the other hand, Simon Saunders argues that quantum particles satisfy PII (Saunders 2006b). However, they seem to agree that either way quantum particles are not individuals. For example, French and Décio Krause talk of quantum particles as nonindividual objects, and Saunders and Fred Muller (2008) argue that they are "'relationals," not individuals but still objects. However, while the above philosophers therefore seem to agree about what is required for objects to be individuals (namely absolute discernibility as explained below), it is not clear that this criterion should be taken to have any metaphysical significance. Given that more generally there are different ideas associated with individuality over and above objecthood, it may be better to think in terms of various categories depending on what is at issue rather than attaching great significance to a single concept of individuality.

Hence, it is worth questioning (*a*) whether there is a single notion of individuality that is metaphysically significant; and even if there is (*b*) whether a substantive account of how objects are individuated is required. In the next section, the individuation of particles is reviewed. In section 10.3, the idea of reducing individuality in the sense of numerical identity to qualitative features of the world via discernibility relations is considered and it is argued that if the definability of identity is at stake, then the cut does not come between individuals and nonindividuals as usually defined, but rather between weakly discernible and completely indiscernible entities. On the other hand, if what is at stake is the principle that identity must be grounded in qualitative properties, even things that are absolutely discernible, in the sense of there being at least one property that one has and not the other, may only be so in virtue of relational properties. If it is allowed that individuality be grounded in nonqualitative features of the world, or if a logically thin notion of individuality is adopted, then quantum particles may be individuals after all. In section 10.4, the discussion is briefly broadened and a different approach to these issues is sketched according to which the metaphysically significant notion of individuality apt for quantum particles is that of emergent "real patterns" (in the sense of Ladyman and Ross 2007 following Dennett 1991).

10.2 The Individuation of Particles

As stated above, to the extent that a substantive account of individuality is given, it takes the form of a principle of individuation that says in virtue of what a thing is what it is and not anything else. Insofar as empiricists accept that there are individuals in need of principles of individuation, they have been inclined to adopt the bundle theory according to which a particular thing is to be somehow be identified with all its properties. It has usually been thought that this requires that PII hold for the entities in question. PII is best stated in contrapositive form (the discernibility of the distinct) as follows: there are no two individuals that have all the same properties. If we allow the property of being identical with some particular individual (call this the individual's haecceity) to be within the scope of PII, then it is satisfied because every object is discernible from every other by such properties.[1] The bundle theory of individuation is motivated by the rejection of principles of individuation that appeal to haecceities or similar notions such as primitive thisness (Adams 1979), transcendental individuality (Post 1963, French and Redhead 1988), bare particulars (Allaire 1963), individual substances, substratum (Robinson 2014), or self-individuating elements (Lowe 2003). It is hard to see how we could have empirical knowledge of such properties, and so if individuation requires them, then we could not have epistemic access to it. This is taken to require that PII be true when it is restricted to "qualitative" properties.

A metaphysically deflationary way of looking at the issue is as follows: there are facts about the identity and diversity of individual objects in a particular domain, and there are facts about indiscernibility and discernibility. The question is whether the identity and diversity relations are primitive, or somehow reducible to or supervenient on the facts about what is discernible or indiscernible from what. While the precise relationship between individuation and qualitative features is some kind of ontological dependence, grounding, determination, in virtue of relation, or supervenience, to the extent that it is supposed that PII is necessary for any of these to obtain they can be modeled by derivability in a formal language. Such languages may have different resources; for example, they may or may not include names. The formal correlate of the metaphysical content of the identity of indiscernibles can be construed as the reduction of identity to predication involving qualitative properties and relations. Saunders (2006b) and following him others (Saunders and Muller 2008, Muller and Seevinck 2009), have argued that "weak discernibility" is sufficient for such a reduction. Ladyman, Øystein Linnebo, and Richard Pettigrew (2012) show that weak discernibility is the weakest nontrivial form of discernibility.[2] By focusing on definability or entailment of facts about identity and diversity from other facts as explained below, they shed new light on the relationship between different kinds of discernibility that have been discussed in the literature and on their metaphysical status.

Given names for every object in the domain, in a language with identity we can construct predicates for haecceities as follows: $Pa(x)$ iff $x = a$. As mentioned above, to avoid triviality PII must be restricted to "'qualitative" properties in the properties over which it ranges to exclude haecceity. Qualitative properties are those that are not "identity involving," but this notion is ambiguous between not involving the identity relation and not involving the identity of a particular object. Having a constant for every object does not amount to having haecceities, while in a language with identity but without constants objects may fail to be discernible from each other. A language that has neither constants nor the identity relation is certainly not going to be able to express haecceities.

Consider a language without identity or constants but with sufficient resources to express all the facts that can be stated without them. Further restrictions can be made. Much of the current literature concerns whether or not quantum particles can be discerned in various technical senses. Saunders uses a distinction between absolute, relative, and weak discernibility (similar ideas were discussed in Quine 1976). Two objects are absolutely discernible (by some theory of them in some language) iff there is a formula free in one variable that is satisfied by one object and not the other. Assuming the relevant comprehension principle, this is equivalent to there being a one-place predicate satisfied by one object and not the other.

Leibniz seems to have believed a very strong form of PII to be true: No two individuals can have all the same intrinsic qualitative properties (Look 2014). If this is false, PII may still be true when allowed to range over extrinsic or relational properties so that the formula/predicate above can express a spatiotemporal property, where spatiotemporal properties are not construed as intrinsic. There is universal agreement that neither classical nor quantum particles of the same kind are intrinsically discernible, but that classical particles are absolutely discernible by their spatiotemporal trajectories, which cannot cross by the principle of impenetrability.[3] Hence they satisfy PII when it is not restricted to intrinsic properties. Notoriously, according to orthodoxy, quantum particles do not have determinate spatiotemporal trajectories. More importantly the statistics they obey suggest that if they are permuted within a state, this does not amount to a different physical state of affairs; this is the "Indistinguishability Postulate" that is obeyed by all quantum particles.

The fundamental physical distinction between bosons and fermions is that many (of a given kind) of the former, for example photons, may exist in the same state, but only one of the latter, for example electrons, can exist in a particular state. This makes it seem as if fermions ought to be discernible. If one considers the naive form of the Pauli exclusion principle, which states that no two fermions can have the same set of quantum numbers, it seems any two fermions can be discerned by their quantum numbers and so satisfy PII. The complication that makes this inference erroneous is with the way composite systems are described in quantum mechanics. Any pair of fermions must have a full state that is antisymmetric, that is of the form |a>|b> − |b>|a>. The minus sign means that when the particles are permuted, the state differs by a phase factor of minus one, though since individual phases have no physical significance, it is the same physical state, and so as with bosons permutation of fermions does not produce a difference that counts statistically. If |a> and |b> are states that describe every physical property, then bosons can be in states of the form |a>|a> but fermions cannot. However, neither bosons nor fermions can exist in states of the form |a>|b> (or |b>|a>); rather, fermions are always in states like the one above and bosons in symmetric states of the form |a>|b> + |b>|a>, or in states of the form |a>|a>.

The antisymmetric states of fermions (and the corresponding symmetric states without the minus sign for bosons) are not in the product space of their individual state spaces, and so according to orthodoxy separate states cannot be attributed to the particles in such states—they exist, rather, in entangled states.[4] Hence, it seems that there are kinds of states that quantum particles can be in that do not attribute to them any properties that are sufficient to discern them one from another, and so the orthodoxy is that they violate PII and hence are not individuals. This obviously presupposes that genuine individuals must satisfy PII, and, as discussed above, the reasoning relies upon rejecting transcendent forms of individuation.

However, the requirement of discernibility in a language without constants or identity does not entail absolute discernibility. Here is how weak discernibility is usually defined following Saunders:

> Two objects a and b are weakly discernible iff there is a relation R such that Rab and ~Raa.

A slightly weaker version is as follows:

> Two objects a and b are weakly discernible iff there is a formula free in two variables that is true when a and b are substituted into it but not true either when a or b is substituted into it twice.

Again the reason that this is weaker is because to infer the former definition from it we need an axiom of comprehension that ensures that every such formula corresponds to a relation. Either way it is important to note contrary to what is repeatedly said erroneously in the literature that the relation need not be an irreflexive relation but only irreflexive in a and b. The confusion is due to a quantifier shift fallacy. We require only that for any pair of objects there is a relation that obtains between them but not between one or the other of them and itself. We do not require that there is a single relation that discerns all pairs of objects.[5]

The basic intuition that Saunders articulated is that fermions in states such as the singlet state (which gives rise to the anticorrelations in measurement results that violate the Bell inequalities) of spin in an arbitrary direction, namely the state |up>|down> superposed with |down>|up>, are weakly discerned by having opposite spin. Muller and Saunders 2006 elaborated the argument to argue that quantum particles are weakly discernible by relations that are describable in the language of quantum mechanics. Although Nick Huggett and Josh Norton (2014) (and perhaps Tom Imbo) agree with the basic intuition above, they argue that quantum particles are not in fact discernible in all states. Whether, and if so exactly how, quantum particles can be weakly discerned is not discussed below. Rather the next section is about what if any significance to accord to weak discernibility.

10.3 PII and the Reduction of Identity to Qualitative Features of the World

Saunders, and following him Muller and Michael Seevinck (2009), argue that, for quantum particles to be objects, it is necessary and sufficient for them to satisfy the identity of indiscernibles understood in terms of weak discernibility. They define individuals to be objects that are absolutely discernible, and those that are only weakly and not absolutely discernible are "relationals." Adam Caulton and Jeremy Butterfield (2007) also take absolute discernibility as a necessary (and sufficient) condition for individuality. French and Kraus seem to agree (though they deny that weak discernibility is appropriate for a formulation of PII and regard things that are not absolutely discernible as "nonindividual objects"). However, being weakly discernible and not absolutely discernible in the senses defined above, has no obvious relation to whether the entities concerned are individuals. It is usually taken to have to do with the fact that weak discernibility is by relations, but as explained below absolute discernibility can also be by relations.

In order to address the issue of whether identity is reducible to qualitative features of the world, Ladyman, Linnebo, and Pettigrew investigated what can be done with the different discernibility relations in different languages describing a domain of entities.[6] The key issue is whether or not identity can be defined in a language that describes all and only the facts about the domain other than the facts about identity and diversity. (Matters become complicated when infinite domains are considered, so in what follows only the finite case is discussed.) In the case where every object is absolutely discernible from every other it is easy to see that identity can be defined. First consider a domain consisting of objects a and b; then since they are absolutely discernible there is a property P that a has and b lacks, and so being identical with a is equivalent to having property P, and being identical with b is equivalent to not having the property (or vice versa). If we add more objects to the domain, then a can be defined as the object having the conjunction of the properties that it has and lacks, that collectively discern it from all the other objects in the domain.

Investigating what can be done in formal languages is only relevant if we stipulate that the language has all and only the resources to describe the genuine properties of the objects in the domain. For example, if predicates for haecceities are included, then as argued above, PII is trivial because every object can be discerned from every other because it uniquely possesses its haecceity; yet it may be argued that these predicates do not correspond to genuine properties in the world. On the other hand, if the language lacks terms for genuine properties, then it fails to express all the qualitative facts, and so failure of discernibility in such a language is not metaphysically significant. All discernibility relations collapse to distinctness in a language with identity and constants for every object, but not in a language that lacks one or other of these components. Certainly in a language without either identity or constants weak discernibility is nontrivial because there are domains (such as the edgeless graph with two vertices) in which not every object is even weakly discernible from every other. In general, then absolute discernibility is not as discerning as weak discernibility, which is not as discerning as numerical distinctness.

Adding constants to the language for every object in the domain does not discern any more objects than were already weakly discernible, but it does make them absolutely discernible. (The converse, that two objects that are absolutely discernible in a language with constants are weakly discernible in a language without them, is also true, so weak discernibility is equivalent to absolute discernibility in a language with constants for every object.) In the light of these facts it is not surprising that weak discernibility is so powerful. Weak discernibility in a purely qualitative language is equivalent to weak discernibility in the enhanced language that enjoys singular reference to all the objects in its domain. Furthermore, weak discernibility is the most discerning nontrivial discernibility relation in the sense that, if a and b are not weakly discernible, they are not discernible by any means.

Weak but not absolute discernibility does entail individuation by relations, and this leads to the objection that weak discernibility cannot be taken as any kind of grounding of individuality analogous to the grounding of individuality in properties in the standard bundle theory. Hawley (2006) and French and Krause (2006) argue that it is circular to use weak discernibility in a principle of individuation. As Hawley (2009) puts it, weak discernibility is "grounded in the fact that the objects in question are distinct; weak discernibility cannot itself be the ground of distinctness" (110). Similarly, many philosophers since Russell have argued that there could not be individuals lacking intrinsic properties, whose individuality is conferred by their relations to other individuals. Russell famously made this objection to Dedekind's conception of the ordinal numbers as nothing more than terms in a progression, insisting that they each must have some intrinsic nature (1903, 249). However, this requirement is met by quantum particles, at least as modeled by nonrelativistic many-particle quantum mechanics, since they each possess their state-independent properties such as mass and charge intrinsically. The problem is that particles of a given kind all have the same such properties, so they are not sufficient to individuate them, but they do give them natures over and above the relations into which they enter with each other.

The argument against the relevance of weak discernibility must therefore be that not only must the individuals have intrinsic natures, but they also must be distinct individuals that are metaphysically prior to the relations into which they enter, or else there would be nothing to stand in the relations that are supposed to confer individuality on them. However, no explanation is given as to why relations are any more problematic than properties in this regard, since one might just as well insist that in order for a property to be instantiated there must be a metaphysically prior individual, or there would be nothing to bear the property.

Ladyman (2007) argues there can be purely relational individuation because the reduction of identity to qualitative relations is no more problematic than the reduction of identity to qualitative nonrelational properties. In any case it is wrong to equate grounding identity or individuality in relations with weak discernibility. Ladyman (2007) also shows that there are cases of absolute discernibility where relations account for the identity and diversity of the objects, so purely relational individuation need not imply lack of absolute discernibility, and equivalently absolute discernibility can be by relations only. Recall that two objects are absolutely discernible if there is a formula free in one variable that applies to one and not the other. It follows that entities can be absolutely discernible because they sit in a unique position in a nexus of relations. Hence, absolutely discernibility does not entail that the entities in question are individuated by properties rather than by relations. The cut between absolute discernibility and weak discernibility is not the cut between individuation by qualitative properties and by qualitative relations. Relational individuation, in the sense of the bundle theory of individuality where relations rather than intrinsic properties form the bundle, is implicitly accepted by all those who take absolute discernibility to be sufficient for individuation.

Note also that the issue with which Saunders at least is concerned is not that of grounding but of definability, and it is undeniably the case that identity is definable in a language that lacks it, provided that every object is at least weakly discernible from every other. Metaphysicians may object that the relations supposed to obtain in the domain could not really do so unless the facts about the identity and diversity of the objects were independently fixed, but no good argument has been given that does not carry over for nonrelational properties too (see Ladyman, forthcoming).

Structuralism is often construed as a claim about individuation to the effect that it is somehow dependent on relations. If weak discernibility is appealed to in support of the claim that quantum particles are individuals, then this is a structuralist view of the ontology of quantum mechanics. To repeat, it has not been shown that there is reason to deny that relations can individuate if properties can. However, it is contestable whether entities being "discernible" in some particular technical sense entails that they are distinguishable in the sense of it being possible to tell them apart (see Ladyman and Bigaj 2010), and so it is not clear that the original spirit of PII is adhered to in the move to weak discernibility.

However, in any case, even an empiricist could argue that individuals do not need to satisfy PII, and it is not clear that naturalism requires the rejection of primitive identity and individuality. The physics of both quantum particles and space-time seems to deny that permutations of objects necessarily give rise to distinct states of affairs, since quantum states in which particles are permuted are regarded as the same physical states and since models related by diffeomorphisms are regarded as representing the same physical space-time. The forms of primitive identity considered above are all intrinsic in that they involve ascription of an intrinsic property to objects that are individuals. This suggests that the permutation of individuals gives rise to qualitatively identical but nonetheless distinct possible worlds (haecceitism) since which individual has which set of qualitative features can be varied. However, there is no reason why primitive identity cannot be adopted for a structure as a whole so that brute permutations of indiscernible individuals are not defined at all since they cannot be reidentified outside of the structure to which they belong. Ladyman (2007) argues accordingly that there is a coherent structuralist idea of primitive identity that does not imply haecceitism.

10.4 Emergent Individuals

Of course, nonrelativistic many-particle quantum mechanics is not our best physical theory, and particles should be understood in terms of quantum field theory. It is worth broadening the discussion to consider the fact that quantum particles are not fundamental entities according to our best physics. Particles such as electrons are elementary in the sense that they are thought not to decay, but they are not elementary in the metaphysical sense, being excitation states of quantum fields. Doreen Fraser (2008) argues that strictly speaking there are no particles in quantum field theory. On the other hand, David Wallace (2001) argues that particles are emergent entities that exist when there are effective degrees of freedom in an effective description that behave like particles.

Lots of standardly recognized particles such as protons are not elementary, and there are also many so-called quasi-particles in physics. For example, phonons are vibrational modes in a crystal lattice that are associated with shifts in atomic states, and which propagate and behave as if they were particles. All quasi-particles have finite lifetimes, and therefore their existence can be considered as relative to a timescale. For example, according to Goldstone's Theorem, if a continuous symmetry is spontaneously broken at lower energy, then there are massless bosons associated with the lower energy regime. Pions were once thought to be fundamental particles and to be the carrier particles for the strong force. The pion has about 15% of the mass of the proton. In the effective quantum field theory of pion interaction this mass is set to zero (even though it is of the order of 139MeV, where the energy/nucleon released in nuclear fission is about 10MeV). Pions are then treated as the Nambu-Goldstone bosons of the spontaneously broken chiral symmetry of the strong force. Actually, not only are pions not massless but the strong force is not really chirally symmetric. So this is a doubly effective theory. Nonetheless, a Lagrangian can be written down in terms of pion degrees of freedom, and it provides an empirically adequate description of some phenomena.

The Renormalization Group is a formal approach to the limiting relationships between theories at different scales that justifies the elimination of degrees of freedom. Theories that are renormalizable are such that some physical quantities are independent of the exact length scale cutoff that is made. There is what Ladyman and Ross (2007) call "the scale relativity of ontology" because the phenomena associated with the intermediate asymptotic regime exist only at the scales associated with the limit. The scale relativity of ontology is with respect to both time and space because composition is diachronic and dynamical; for example, the characteristic properties of water depend on the complicated interactions among its parts.

Ladyman and Ross argue for a view of individuals that can be put in the form of negative and positive theses as follows:

Negative theses: every "thing" must go insofar as the world is not made of little things, in the sense of little material objects as modeled by intuition and as perceived in the manifest image; particles are not intrinsically individuated individuals, nor is there intrinsic individuation by properties.

Positive theses: the scale relativity of ontology, the real-patterns account of ontology, realism about modality, the unity of science, there may not be a fundamental level, and composition is a real feature of the world and there is higher-order ontology, but composition is diachronic not synchronic, and domain specific since dependent on the relevant kinds of interaction.

In this way the matter of whether elementary, nonelementary, and quasi-particles in physics are individuals is metaphysically deflated. The criterion of ontological commitment is as follows: real patterns (genuine individuals) must figure in projectable generalizations / causal laws that allow us to predict and explain the behavior of the world. It makes no sense to talk about real things that do not figure in generalizations that support prediction and counterfactuals, and explanations. The sciences (and common sense) posit such objects, properties, relations, and processes that allow the formulation of such projections. Hence, there is no further question to be asked about the existence of objects once we settle the matter of whether there are law-like/causal regularities in which they feature (nonredundantly). On such a naturalistic view, if quantum particles and/or space-time points are individuals, it is not in virtue of weak discernibility or any other general principle of individuation, but because the practice of science treats them as being so in effective descriptions applicable to certain regimes. In this context what makes the difference between objects and individuals would seem to be largely a matter of stipulation, though it may be appropriate to reserve the term *individual* for entities that persist over time after all.

Acknowledgments

Arts and Humanities Research Council "Foundations of Structuralism" project: www.bristol.ac.uk/structuralism.

Special thanks are due to Richard Pettigrew and Øystein Linnebo and to the organizers and participants in the "Individuals in Science" conference. I am also extremely grateful to Alexandre Guay, Kerry McKenzie, and Simon Saunders for very helpful comments on a draft of this chapter.

Notes

1. PII is sometimes thought to be a necessary truth. Furthermore, haecceities might be defined as necessary properties of individuals.

2. What follows is a summary of some of the results of their paper and a consideration of some of their implications for individuality in physics.

3. In a sufficiently symmetric universe particles' trajectories are not absolutely discernible from each other. In a fairly messy world like ours particles can be absolutely discerned by their positions relative to arbitrary background features. Hence, perhaps PII is only contingently true of classical particles.

4. It is important to note, however, that while technically all states of particles are entangled in this sense, it is only when there is entanglement in the state that represents a particular physical degree of freedom that the characteristic phenomena associated with entanglement, such as violation of Bell inequalities, occurs (Ladyman, Linnebo, and Bigaj 2013).

5. The standard examples of weakly discerning relations, such as being a mile apart, are all irreflexive, and there may be good reasons to insist on irreflexivity. Simon Saunders in comments on a draft of this chapter made this point. I do not explore them here as nothing in what follows hangs on the issue. At least from a purely logical point of view identity can be defined provided only that for every pair of objects a relation between them can be found that one of them does not bear to itself.

6. The language in question is taken to represent all the qualitative features of the world with predicates, but it is left open whether or not it includes constants and/or identity. Clearly where the identity relation is present, objects can be discerned by its negation.

References

Adams, Robert. 1979. Primitive Thisness and Primitive Identity. *Journal of Philosophy* 76: 5–26.
Allaire, Edwin. 1963. Bare Particulars. *Philosophical Studies* 14: 1–8.
Cassirer, Ernst. 1956. *Determinism and Indeterminism in Modern Physics*. New Haven: Yale University Press.
Caulton, Adam, and Jeremy Butterfield. 2012. On Kinds of Indiscernibility in Logic and Metaphysics. *British Journal for the Philosophy of Science* 63: 27–84.
Dennett, Daniel C. 1991. Real Patterns. *Journal of Philosophy* 88: 27–51.
Fraser, Doreen. 2008. The Fate of "Particles" in Quantum Field Theories with Interactions. *Studies in History and Philosophy of Modern Physics* 39: 841–859.
French, Steven, and Décio Krause. 2006. *Identity in Physics: A Formal, Historical and Philosophical Approach*. Oxford: Oxford University Press.
French, Steven, and Michael Redhead. 1988. Quantum Physics and the Identity of Indiscernibles. *British Journal for the Philosophy of Science* 39: 233–246.
Hawley, Katherine. 2006. Weak Discernibility. *Analysis* 66: 300–303.
Hawley, Katherine. 2009. Identity and Indiscernibility. *Mind* 118: 101–119.
Huggett, Nick, and Josh Norton. 2014. Weak Discernibility for Quanta, the Right Way. *British Journal for the Philosophy of Science* 65: 1–20.
Hull, David. 1978. A Matter of Individuality. *Philosophy of Science* 45: 335–360.
Ladyman, James. 2007. Scientific Structuralism: On the Identity and Diversity of Objects in a Structure. *Proceedings of the Aristotelian Society* Supplementary Volume 81: 23–43.
Ladyman, James. 2015. The Foundations of Structuralism and the Metaphysics of Relations. In *The Metaphysics of Relations*, ed. Anna Marmadoro. Oxford: Oxford University Press, forthcoming.
Ladyman, James, and Tomasz Bigaj. 2010. The Principle of the Identity of Indiscernibles and Quantum Mechanics. *Philosophy of Science* 77: 117–136.
Ladyman, James, Øystein Linnebo, and Tomasz Bigaj. 2013. Entanglement and Non-factorizability. *Studies in History and Philosophy of Modern Physics* 44: 215–221.
Ladyman, James, Øystein Linnebo, and Richard Pettigrew. 2012. Identity and Discernibility in Philosophy and Logic. *Review of Symbolic Logic* 5: 162–186.
Ladyman, James, and Don Ross. 2007. *Every Thing Must Go: Metaphysics Naturalized*. Oxford: Oxford University Press.
Look, Brandon C. 2014. Gottfried Wilhelm Leibniz. In *Stanford Encyclopedia of Philosophy*, Spring 2014 ed., ed. Edward N. Zalta. http://plato.stanford.edu/archives/spr2014/entries/leibniz/.

Lowe, E. J. 1998. *The Possibility of Metaphysics: Substance, Identity, and Time*. Oxford: Clarendon Press.

Lowe, E. J. 2003. Individuation. In *The Oxford Handbook of Metaphysics*, ed. Michael J. Loux and Dean W. Zimmerman, 75–95. Oxford: Oxford University Press.

Muller, Fred A., and Michael P. Seevinck, 2009. Discerning Elementary Particles. *Philosophy of Science* 76: 179–200.

Post, H. 1963. Individuality and Physics. *Listener* 70: 534–537. Reprinted in *Vedanta for East and West* 32: 14–22.

Quine, W. V. O. 1976. Grades of Discriminability. *Journal of Philosophy* 73: 113–116.

Robinson, Howard. Substance. 2014. In *Stanford Encyclopedia of Philosophy*, Spring 2014 ed., ed. Edward N. Zalta. http://plato.stanford.edu/archives/spr2014/entries/substance/.

Russell, Bertrand. 1903. *The Principles of Mathematics*. Cambridge: Cambridge University Press.

Saunders, Simon. 2006a. On the Explanation for Quantum Statistics. *Studies in History and Philosophy of Modern Physics* 37: 192–211.

Saunders, Simon. 2006b. Are Quantum Particles Objects? *Analysis* 66: 52–63.

Saunders, Simon, and Fred A. Muller. 2008. Discerning Fermions. *British Journal for the Philosophy of Science* 59: 499–548.

Smolin, Lee. 2000. *Three Roads to Quantum Gravity*. London: Weidenfeld.

Strawson, Peter. 1959. *Individuals*. London: Methuen.

Wallace, David. 2001. Emergence of Particles from Bosonic Quantum Field Theory. *Philpapers*. http://arxiv.org/abs/quant-ph/0112149.

CHAPTER 11 | Minimal Structural Essentialism
Why Physics Doesn't Care Which Is Which

DAVID GLICK

11.1 Introduction

In his (2002) paper, John Stachel claims that general relativity (GR) and quantum mechanics (QM) share a curious feature: they don't seem to care which objects possess which properties. More precisely, he argues that both theories contain models that are *Generally Permutable*.

> General Permutability (GP): "every permutation P of the **a** entities in **S**, **R**(*a*), **R**(P*a*) represent the *same* possible state of the world" (2002, 242),

where **S** is a set of *n* entities, $a_1, a_2, ..., a_n$, among which there is an ensemble **R** of M n-place relations, $R_1, R_2, ..., R_M$. **R**(*a*) is the ensemble of relations with their places filled by the sequence, $a = (a_1, a_2, ..., a_n)$ and P*a* is the permutation of the sequence *a*.

GP is an interpretative principle. It claims that *representations* or *models* that differ only by a permutation of the entities they contain represent the same physical state of affairs. In the case of GR, this amounts to the claim that "swapping" points in space-time models results in redundant representations of the same space-time. Thus, GP is committed to a *passive* reading of the permutations of model elements: such permutations reveal a formal symmetry of the theory rather than a symmetry of the (physical) world.

The purpose of this chapter is to explore the implications of GP for a metaphysics of objects in fundamental physics. In particular, it has been argued—by Stachel himself and more recently by Adam Caulton and Jeremy Butterfield (2012)—that GP recommends a "structuralist" metaphysics. I agree with these authors that structuralism is recommended, but I find their particular metaphysical proposals insufficient. Instead, I argue for a version of structural essentialism according to which an object occupies its place in a structure essentially. This view, I contend, offers a promising structuralist metaphysics that accounts for GP.

This view also has surprising implications for the question of individuality in physics. There is an appealing argument that GP (or something like it) renders elementary particles or space-time points nonindividuals by making it indeterminate which object is which. On my view, this argument is mistaken. The best explanation for GP implies that these basic physical objects are *structural* individuals, that is, objects that are individuated by their place in a relational structure.

At the onset, it is worth clarifying what is meant by a "structuralist metaphysics." "Structuralism" is a term of art in contemporary discussions, with both advocates and critics using the term differently. My own usage is informed by the debate in the general philosophy of science concerning structural realism. Structural realism—as I understand it in that context—is the view that we should be realists only about the *structure*[1] of our best scientific theories. Structural realism further divides (following Ladyman 1998) into epistemic and ontic varieties. The former holds that the limitation to structural knowledge posited by structural realism reflects our limitations as knowers ("all we can *know* is structure"), while the latter claims that our knowledge of the world is in principle complete ("all there *is* is structure"). Thus, according to ontic structural realism (OSR), there is nothing more to fundamental reality than structure. This implies the metaphysical thesis I intend by "structuralism": the correct metaphysics of the world is one that is fundamentally structural.

One can be a structuralist without being a structural realist. Structuralism is a claim about fundamental ontology that one may come to by many paths (in the present case, by taking aspects of physical theories seriously). The point is that the structuralism I am concerned with is the same one proponents of OSR are committed to. Saying more about what is meant by the world being "fundamentally structural" is difficult because of divergent views regarding the nature of structure, fundamentality, and objects. One thing on which there seems to be close to a consensus among structuralists, however, is that the traditional "atomistic" picture must be rejected. The view that there are atomic building blocks with intrinsic properties out of which everything is composed must be replaced by a view in which relational structures play an ineliminable role. As Esfeld and Lam put it, "Structural realism in the metaphysics of science is a sort of a holism in contrast to an atomism" (2010, 10) and later, "Structural realism as a metaphysical position is the claim that there are no fundamental intrinsic properties underlying the relations that we can know. That is to say, all there is to the fundamental physical objects are the relations in which they stand" (12).

This chapter will proceed as follows. Section 11.2 briefly explains GP and how it arises in the context of GR and QM. It is not my aim here to conclusively establish that GP holds in these theories, but rather to illustrate and lend some plausibility to the claim that it does. Section 11.3 will turn to metaphysical proposals aimed to account for GP. In section 11.3.1, I will contend that the proposals of Stachel and Caulton and Butterfield, while promising, are insufficient to the task of accounting for GP. In section 11.3.2, I return to GR and draw on the views of Tim Maudlin to motivate *structural essentialism*. In section 11.3.4, I offer my own proposal, *minimal structural essentialism*, which aims to provide a structuralist metaphysics adequate to the task of explaining GP. The view is clarified and defended from objections in section 11.4. Finally, in section 11.5, I make explicit the implications of the view for the question of individuality in physics.

11.2 The Case for General Permutability in Physics

The motivation for GP is provided by central topics in the interpretation of GR and QM: the hole argument and quantum statistics, respectively.

11.2.1 GP in GR: The Hole Argument

Models in GR are given by the triple $\mathcal{M} = \langle M, g, T \rangle$, where M is a manifold of points, g is a metric field defined on M, and T is a matter field defined on M. The manifold of points M represents all of the points in space-time (or "events"), the metric field g specifies the spatiotemporal distance relations between space-time points, and the matter field T specifies the distribution of matter and energy across space-time.

The hole argument concerns a particular transformation of the metric and matter fields (g and T) of a model $\mathcal{M}^2$. A *hole diffeomorphism* is a smooth transformation of the metric and matter fields that leaves them unchanged outside of some arbitrary region (the "hole"), but changes them within the hole region. The resulting model $d\mathcal{M} = \langle M, d*g, d*T \rangle^3$ *apparently* represents a world exactly like ours outside of the hole, but different within it. For example, on an active reading of the hole diffeomorphism, the worldline of a particular galaxy that $\mathcal{M}$ describes as passing through a space-time point p located within the hole no longer does so in the world represented by $d\mathcal{M}$.

Taking the result of a hole diffeomorphism to represent a distinct world is problematic, however. Such diffeomorphic models are not only empirically equivalent to the original model, but also indistinguishable from the perspective of the laws of GR. From the perspective of GR, there seems to be no reason to prefer M over dM, yet they seem to differ in their metaphysical implications. One way to see the problem concerns indeterminism. If M and dM are taken to represent distinct worlds, then specifying the entirety of the region outside of the hole underdetermines the facts inside the hole. If we let the hole be a future region of space-time, then a complete specification of the past fails to determine the future regardless of our dynamical laws. This is problematic not because of the unpalatability of indeterminism as such, but because in this case it results solely from an interpretative choice.

A popular way to resolve this representational underdetermination is to regard the distinct *models* M and dM as representing the same *world W*. Thus, this position—known as "Leibniz equivalence"—resolves the underdetermination of representations by "collapsing" the two distinct worlds apparently represented by M and dM into a single world, redundantly described.

It is a short step from here to GP. What the hole diffeomorphism does is to effectively "swap" points within the hole: manifold point α is now where manifold point β was and vice versa. As Stachel shows (2002, 235), if we abstract away the differentiability and continuity of space-time models, we are left with a bare manifold of points, and diffeomorphisms become permutations of the points. Thus, if Leibniz equivalence is the right solution to the hole argument, then GP—its discrete analog—is part of the correct understanding of GR.

11.2.2 GP in QM: Quantum Statistics

Stachel also claims that GP is part of QM in virtue of offering the best explanation for the statistics of indistinguishable particles. While the argument for this claim is not made explicit, there is a popular argument that connects quantum statistics with GP: the case of the "bosonic quantum coins."

Consider two bosons each of which can be in one of two states a or b. Our "classical" intuition is that there are four possible configurations: *aa, ab, ba, bb*. Assuming all configurations are equiprobable, then the probability of *aa* would seem to be 1/4. However, the Bose-Einstein statistics appropriate for bosons tell us that Prob(*aa*) = 1/3, and this is what is found experimentally. What accounts for the failure of the "classical" intuition?

The answer, so the argument goes, is that the configurations *ab* and *ba* "collapse" into a single state in QM.[4] We would get the statistics wrong if we take the bosonic states *ab* and *ba* to be distinct, so we must view those two possibilities as combining (in some sense) into a single possibility. Thus, there are three rather than four configurations and hence the probability for each of them is 1/3. The reason why they "collapse" is the same as in the hole argument: *representations* related by permutation describe the same physical state of affairs. In that case we reduced the possibilities by taking $d\mathcal{M}$ and $\mathcal{M}$ to represent the *same* world, and here we analogously take ab and ba to combine into a single state of affairs $\left(\frac{1}{\sqrt{2}}(ab+ba)\right)$. In other words, taking on GP justifies the move from classical to quantum statistics.

There is an important disanalogy with the situation in GR though.[5] The hole argument proceeds by presenting a case of representational underdetermination—between $\mathcal{M}$ and $d\mathcal{M}$—which compels us to "collapse" the two models into one by invoking GP. Yet, in the case of QM, there is no such representational underdetermination. The states ab and ba that are allegedly "collapsed" are in fact *impossible* for bosons to occupy. Instead there is a single mathematical representation of the quantum state of the system—a ray in Hilbert space—that corresponds to the superposition $\frac{1}{\sqrt{2}}(ab+ba)$, which cannot easily be interpreted as "either ab or ba."

Yet, all hope is not lost for establishing GP in QM. Adam Caulton and Jeremy Butterfield (2012) (henceforth, CB) argue that the representational underdetermination present in GR can be found in quantum statistics if a broader understanding of the latter is adopted. Rather than limiting quantum statistics to purely symmetric (for bosons) or antisymmetric (for fermions) states, CB propose that we also allow for "parastatistical" states that are neither symmetric nor antisymmetric. If we allow for the possibility of such parastatistics, then states that were formerly represented by a single ray now correspond to an entire equivalence class of such rays that are each related by permutation. Thus, CB argue that GP must be invoked in QM to remove representational underdetermination just as in GR.

Another possible approach would be to view quantum statistics itself as such a reduction. In the case of the bosonic coins, QM replaces the four "classical" possibilities with the three given by Bose-Einstein statistics. In so doing, one may regard QM *itself* as reducing the two "classical" possibilities ab and ba to the single quantum state $\frac{1}{\sqrt{2}}(ab + ba)$. For this reason, perhaps QM (at least *with* the symmetrization postulate) can be viewed as incorporating GP as part of the statistics it allows.

Far more would need to be said to defend either of these approaches to GP in QM. CB's approach rests on taking parastatistics more seriously than some may wish, while the second approach requires taking seriously the four "classical" possibilities despite their inapplicability to bosons.

There may be yet other approaches that find a place for GP in QM as well.[6] The claim I wish to defend here is only that the prospects for GP as a feature of QM are good. Permutation invariance, which lies at the heart of these considerations, is a central aspect of QM and strongly suggests GP or a principle very much like it.[7]

11.3 Explaining GP: Structuralism, Essentialism, and Structural Essentialism

In what follows I will leave behind debates over the status of GP in GR and QM and proceed on the assumption that it is a genuine feature of these theories. These are theories in which representations that differ only with respect to which entities have which properties should be viewed as physically equivalent. The question now is: what accounts for this feature of these theories?

11.3.1 Structuralism

Stachel and CB each argue that GP suggests an increased role for *structure* in one way or another. There are several different projects one may be engaged with in this vicinity. My own goal, as mentioned above, is to find a structuralist metaphysics that grounds GP. I take this to mean answering the following question: what is the nature of the world described by GR and QM such that GP follows from it?

My central criticism of the two proposals considered here is that they are inadequate to the task of providing such a metaphysical grounding of GP. Later I offer my own proposal, *minimal structural essentialism*, as an attempt to provide such a grounding.

11.3.1.1 Stachel's Structural Individuation
Stachel claims that GP shows that

> The basic building blocks of any model of the universe (for example, the elementary particles and the points of spacetime) are individuated entirely in terms of the relational structures in which they are embedded. (Stachel 2002, 249)

One difficulty in understanding this proposal comes from Stachel's use of the term "model." If this is a thesis only about *models*—that is, that which is used to represent the world—then it is hard to see how it could be supported by GP, a thesis about the relation between models and the world. If, however, we take the proposal to concern particles and space-time points *themselves*, the proposal suggests that such objects are structurally individuated. The intuitive idea here seems to be that models related by permutation attribute the same structure to the world, and hence, if objects are picked out by that structure, these models attribute the same properties to the same objects.

Yet if we take Stachel's proposal to be that certain physical objects are individuated structurally, further questions remain. For example, what is meant by "individuation" in this context? To *metaphysically* individuate an object x—following Lowe (2007)—is to (metaphysically) *determine* which object x is among members of a certain class of objects. For example, the set $A = \{1, 2, 3\}$ is individuated by its members 1, 2, 3 in this sense.

Alternatively, Stachel may intend a weaker, epistemic, sense of individuation where to individuate objects of the same kind is to render them *distinguishable*. On this reading, to say elementary particles are structurally individuated is to claim that one can only tell them apart by appeal to the structure they enter into. This reading, however, fails to provide a grounding for GP in the sense articulated above. In particular, it leaves open the question of what it is about these objects that makes them only distinguishable by appeal to the structures into which they enter.

The most promising version of Stachel's proposal, for our purposes, construes "structural individuation" metaphysically: structure determines which object an object is. However, while this certainly constitutes a claim about metaphysics, Stachel does not provide a specific account of the sense in which structures individuate these physical objects. In lieu of such an account, the view is at best incomplete; it is possible that the correct metaphysical account of GP involves structural individuation, but until such a notion can be spelled out, our task is not complete. Indeed, without further specificity, this view looks paradoxical: structural individuation entails that structure determines which object is which, but GP seems to deny there is a fact about such matters.

11.3.1.2 Caulton and Butterfield's Structuralism

Caulton and Butterfield argue that GP supports a view they call "structuralism" and characterize as follows.

> [Structuralism's] central claim is that individuality is grounded, if at all, only on *qualitative* properties and relations. (Caulton and Butterfield 2012, 236)

CB's structuralism amounts to a prohibition on nonqualitative accounts of individuality. That is, one cannot explain what makes something the individual object it is by appeal to features that would not be shared by a duplicate. What CB aim to rule out are views like that of Adams (1979), who holds that what grounds the individuality of an object a is its "primitive thisness," the nonqualitative property of being identical to a. Yet it is misleading to call such a view "structuralism" because, while structuralists would surely reject nonqualitative accounts of individuality, they would also reject certain views based on *qualitative* properties as well.

As mentioned in section 11.1, structuralism holds that structures are ontologically basic or fundamental. This leads structuralists to reject the atomistic picture of the world assumed by many traditional metaphysicians. One particularly clear version of atomistic metaphysics is David Lewis's "Humean supervenience" account (Lewis 1986, ix–xvi). On this picture, there are only local matters of fact and from these (and the spatiotemporal relations between them) all else emerges. Thus, on this view, relational structures emerge from individual objects with intrinsic properties bearing spatiotemporal relations. This is a view that structuralists deny. They claim that there are relations that are ontologically basic; there is nothing prior to them on which they depend.

My claim here is not that GP requires that Humean supervenience is false, or even that it is false. My claim is simply that structuralists—in virtue of taking the world to be fundamentally structural—must reject Lewis's thesis. CB's structuralism, however, is not sufficiently strong to rule out positions like Lewis's. Their version of structuralism leaves open these possibilities by giving the following ontological options: (1) individuals whose individuality is grounded in qualitative intrinsic properties, (2) individuals whose individuality is grounded in qualitative extrinsic properties, (3) nonindividuals. The problem with calling this view "structuralism" is that it leaves open option (1), which most structuralists strongly deny.[8]

Putting to one side whether or not CB's proposal counts as structuralism, how does the proposal fare as an explanation of GP? On one hand, it follows from CB's structuralism that *worlds* related by permutation/diffeomorphism—which agree on all the qualitative facts—are identical ($W = dW$). But on the other hand, their proposal amounts to little more than a bare modal claim[9] rather than a genuine explanation of GP. For individuality to come out as purely qualitative is surely a desideratum for a metaphysical grounding of GP, but it does not constitute an adequate explanation in itself. Each of these proposals aims to explain GP in terms of what grounds individuality: structure in the case of Stachel, and purely qualitative facts in the case of CB. My criticism of these views is not that they are wrong—I agree that qualitative, structural facts should ground the individuality of objects in physics—but rather, that they are insufficient qua *explanations* of GP. They do not provide an account of the nature of reality according to these physical theories from which GP follows. In order to find such an account, we must return to GR to consider another perspective on the hole argument.

11.3.2 Essentialism

There is one well-known approach that aims to account for the hole argument by appeal to the nature of space-time points: Tim Maudlin's metric essentialism (Maudlin 1988, 1990). On this view, space-time points bear their metrical relations essentially, so that it is impossible for the very same point p to enter into different metrical relations with other points. Yet, as an explanation of GP, there is an immediate problem: unlike most commentators, Maudlin *denies* Leibniz equivalence.

Recall that Leibniz equivalence is the view that diffeomorphic *models* in GR represent the very same world. Maudlin denies this and takes diffeomorphic models to represent distinct worlds W and dW, but he avoids the problematic consequences of the hole argument by taking dW to be *impossible* in light of the essential properties of space-time points. This approach certainly won't do as an account of GP, but it reveals an interesting relationship illustrated by the following inconsistent triad.

1. **Essentialism**: Space-time points have certain essential properties (or relations) that make swapping them impossible ($\neg[(W$ and $\Diamond dW)$ and $(W \neq dW)]$).
2. **Distinctness:** Diffeomorphic models represent distinct physical states of affairs ($W \neq dW$).
3. **Possibility:** Diffeomorphic models represent possible physical states of affairs ($\Diamond W$ and $\Diamond dW$).

Maudlin chooses to reject Possibility on the basis of Essentialism, but we may elect to reject Distinctness instead. If we do so, then Leibniz equivalence can be combined with essentialism. In fact, this constitutes the beginnings of a metaphysical grounding of GP: the essential properties of the objects in question make permuting *them* impossible, which is why *models* that appear to describe just that difference should be viewed as physically equivalent. That is to say, the *active* reading of the hole diffeomorphism is blocked by the impossibility of a distinct world represented by $d\mathcal{M}$, hence we must adopt the *passive* reading (i.e., GP).

11.3.3 Structural Essentialism

Structural essentialism is a generalization of the view just outlined. It holds that objects (in particular, points and particles) occupy their places in a relational structure essentially. In other words, it is part of the nature of being that particular object that it enters into certain relations with other objects.

In the case of space-time points, this comes very close to metric essentialism. The geometrical relations between space-time points *in the world* form a relational structure, and to be a given point p is to have a particular location in that structure. However, structural essentialism differs from metric essentialism in two ways. First, it endorses rather than rejects Leibniz Equivalence, and, second, it applies to particles in QM as well as space-time points in GR.

In the case of particles in QM, structural essentialism claims that particles occupy their place in structure essentially, but what are the relevant structures? A hint comes from the emphasis many structuralists place on the phenomenon of entanglement. In particular, the spin-singlet state of two fermions has become the standard example of the sort of irreducible relations taken to support a structuralist metaphysics for QM. Such a state may be written schematically as follows.

$$\psi = \frac{1}{\sqrt{2}}\left(|\uparrow\downarrow\rangle - |\downarrow\uparrow\rangle\right) \tag{11.1}$$

According to most interpretations of QM, ψ is a state in which neither particle has a determinate spin value in any direction, yet we know with certainty that a spin measurement of both particles along an arbitrary axis will find them to have opposite spin values along that axis. This leads Simon Saunders (2003, 2006a, 2006b) to argue that there is a relation of *having opposite spin from* that obtains in this case, despite the fact that one cannot attribute a spin value to the particles taken individually.

Generalizing from this particular case, the structural essentialist takes the relational structures in QM to be comprised of "state-relations" between particles. Entanglement relations are paradigmatic examples of such state-relations, but other relations are possible as well.[10]

11.3.4 Minimal Structural Essentialism

Even from this beginning sketch of structural essentialism, a problem becomes apparent in the cases of both points and particles. Consider space-time points. The structural essentialist (and the metric essentialist) claim that a point p must bear certain geometrical relations to other space-time points essentially. Yet, in GR such metrical relations are not independent of the material contents of space-time (as represented by the matter field T). Thus, if the material contents of space-time had been (significantly) different, then it follows that p would not have existed. This is a counterintuitive result. Typically we think that our space-time could have been such that there was a fly at p even if there was no such fly there in the actual world.

Maudlin suggests overcoming this challenge by appealing to Lewisian counterpart theory (see also Butterfield 1989). According to this reply, while it's true that according to metric essentialism p could not have been occupied by (significantly) different matter than it is in the actual world W, there may be a distinct point p' in a possible world W' that grounds the counterfactual claim that there could have been something different at that same point in space-time. This is not the place to evaluate Lewisian counterpart theory, but it does seem preferable to preserve the intuition that p itself could have been differently occupied, other things being equal.

In the quantum case things are worse. State-relations typically change as time passes (assuming the Schrödinger picture); an electron e may become entangled with other particles, or one of the particles it is entangled with may be detected or absorbed. According to structural essentialism, e would not survive such a change. It no longer enters into the same state-relations in the same way and hence it can no longer be the same object. Again, this is a counterintuitive result. While the reidentification of particles in QM is problematic, surely the same particle can change its state or which particles it is entangled with.

In each case the problem is that structural essentialism prohibits objects from occurring in distinct structures. But recall that GP only concerns cases in which different representations have the *same* structure. Thus, to ground GP, all that is needed is fixing an object's place in a structure *when that structure is present*. This suggests minimal structural essentialism.[11]

> *Minimal structural essentialism (MSE)*: for any relational structure S and any object a embedded in S, a has its place in S essentially whenever S obtains.

MSE claims that objects must occupy the place they do *in a given structure*, but allows that they may also be embedded in *different* structures. In the case of space-time points in GR, MSE says that p must have its place in S in any space-time with that geometrical structure. In the case of particles in QM, MSE says that the electron e must have its place in R in any world with that state-relation structure. In cases where the structure of the world is different, the object in question may or may not exist. It is often difficult to say under which conditions the same point or particle is present, and MSE does nothing to resolve this difficulty; it simply allows for the *possibility* of such transstructural identifications. In so doing, it avoids the counterintuitive consequences of structural essentialism mentioned above.

11.4 Essentialism and Sufficiency

In order to further clarify MSE, it is worth saying a bit more about the notion of essentialism at work here. A traditional understanding is that in any world in which a exists, it must have property P. We may express this, roughly, as follows:

Essentialism: $\Box \forall x (x = a \rightarrow Pa)$

This is perhaps enough to avoid the standard hole argument[12] in the manner endorsed above: the space-time point p has its location properties essentially (i.e., it enters into the geometrical relations between other points and material bodies essentially) and so a world (dW) in which p has different location properties is impossible, and therefore, we should view diffeomorphic models as redundant representations. But there are variants of the hole argument that cause problems for such a view.

Consider, for example, a model in GR in which the manifold point a is replaced by an "alien" point or "mole" μ.[13] This model—call it $m\mathcal{M}$—will also be empirically and nomically equivalent to $\mathcal{M}$, and hence GP would recommend regarding it too as merely a redescription of W. But the argument just sketched fails to deliver this result.[14]

On an active reading of the "mole transformation," one arrives at two distinct possible worlds, W and mW, that differ only in that space-time point p has been replaced by a distinct space-time point m. But the essential properties of p don't have anything to say about worlds in which p does not exist, and hence Essentialism is powerless to rule out the active reading. Thus, it seems that a different condition is needed to rule out worlds like mW.

Sufficiency: $\Box \forall x (Px \rightarrow x = a)$

Sufficiency prevents the "mole argument" by stipulating that anything with the particular location properties of p is (identical to) p in all possible worlds. This way the mole world mW cannot be a distinct possible world and hence we can run the argument just as we did in response to the ordinary hole argument. It is also evident that Sufficiency is strong enough to block the original hole argument by ruling out the world dW as well.

Yet Sufficiency gives rise to its own problems. In certain symmetrical models of space-time, distinct points may have all the same location properties. But, according to Sufficiency, any point with the properties of p is identical to p, so it follows that such distinct points are impossible. This point is especially pressing in the case of QM. Even if we limit our application to fermions like those in the spin-singlet state discussed above, such particles will often share all of their properties and relations (again, on most interpretations). It would be an unfortunate consequence if MSE required identifying particles sharing all of their properties.

Fortunately, MSE isn't committed to either Essentialism or Sufficiency. Instead, it employs a weaker claim.

Weak Sufficiency: $\Box (\exists x (Px) \rightarrow Pa)$

Weak Sufficiency says that if anything has p's location properties, then p does. This blocks the hole argument by denying that points with different location properties can swap places, and it also blocks the mole argument by denying that a mole can take the place of *p*. MSE endorses Weak Sufficiency by claiming that objects must have certain properties *whenever a given structure is present*. The claim that a structure *S* obtains entails the existential antecedent of Weak Sufficiency. If the world instantiates a relational structure *S*, then there exist relata—at least in the minimal sense of subjects of predication—that enter into those relations. Hence, MSE is simply Weak Sufficiency applied to a number of entities entering into a structure.

11.4.1 A Remaining Problem: Symmetric Structures

There is one aspect of the problem of symmetry mentioned above that remains for MSE. Unlike Sufficiency, Weak Sufficiency doesn't entail that points with the same properties should be identified, but it does seem to allow swapping them to create a distinct possible world. This may not be too troubling in the case of space-time, where perhaps perfectly symmetric models can be dismissed as unrealistic for our world,[15] but in the case of QM there appear to be cases in which we attribute the same properties and relations to indistinguishable particles. If MSE is to adequately account for QM, it must have something to say about such situations.

If we return to the spin-singlet state of two electrons considered above (equation 11.1), most interpretations provide no unique properties or relations for either electron. Hence, MSE's requirement that each electron have the properties it does in the actual world doesn't rule out a distinct possible world in which the two electrons are swapped. Initially, it may seem that one could reply in the same manner as with space-time points: if we consider the *global* structure of state-relations—the relations among (the states of) all quantum systems—then perhaps the symmetry will vanish. MSE claims that quantum systems enter into their state-relations essentially, but these state-relations are never as simple as the two-particle spin singlet state used as an illustration. The proper focus is rather the *global* state-relation structure described by the *universal* wave function, and perhaps this will be sufficiently rich to attribute to each quantum system a unique set of relations sufficient to distinguish it from all other particles.[16]

The problem with this initial reply is simply that symmetrization applies to all wave functions of composite systems, even the universal wave function. Unlike the space-time case, there *is* good reason to think the global state structure provided by QM is *perfectly* symmetric (or antisymmetric) because QM itself *requires* as much.

The correct reply to the problem of symmetric structures is to distinguish between the properties and relations that characterize a place in a structure and *that place itself*. Saunders's work on weak discernibility is of use here. If we return to the spin-singlet state for the moment, the fact that there is an *irreflexive* relation establishes that there are two *places* in the singlet state structure: place A and place B enter into the state-relation of *having opposite spin values along some axis*. The objects occupying places A and B may share all of their properties and relations, but they occupy distinct places in the relational structure nonetheless, and hence MSE prohibits swapping them by making the occupation of their places essential to them. In other words, distinct places in a structure may be characterized by the very same properties and relations.

The role of weak discernibility in establishing numerical diversity in this context is controversial (see Wüthrich 2009 for one important challenge). My point here is simply that weak discernibility allows us to distinguish *places* in structures from the properties and relations used to characterize them.

Admittedly, this notion of a *place* is in need of further development, but if it is intelligible, then the symmetry problem can be addressed for weakly discernible entities. Two electrons in the singlet state occupy distinct places in the structure of state-relations described by the quantum state of the system they compose. MSE holds that they occupy these places *essentially* (in the sense explicated above), and hence there is no distinct possible world in which the electrons swap places. This means that models related by such a swap—$\Psi = \frac{1}{\sqrt{2}} (|\uparrow\downarrow\rangle - |\downarrow\uparrow\rangle)$ and $\Psi' = \frac{1}{\sqrt{2}} (|\downarrow\uparrow\rangle - |\uparrow\downarrow\rangle)$—should be taken to represent *the same world*.

Of course, the symmetry problem remains for systems of absolutely indiscernible entities such as a system of bosons. It is unlikely, however, that the universal wave function will describe a global state structure that is perfectly symmetric (i.e., bosonic) given that we think fermions are included among the fundamental particles of our world.

11.5 Structural Individuality

There is a natural connection between (general) permutability and the issue of individuality in physics. E. J. Lowe (1998), for example, characterizes individuals ("individual objects" in his terms) as having determinate identity-conditions and determinate countability. GP, however, seems to undermine the determinate identity of space-time points and elementary particles, as Lowe suggests.

> The single electron shell of a neutral helium atom contains precisely two electrons: and yet, apparently, there is no determinate fact of the matter as to the identity of those electrons. This is because the two electrons in the atom's shell exist in a state of so-called "superposition," or "quantum entanglement." Our inability to say which electron is which is not merely

due to our ignorance, or inability to "keep track" of an electron in such circumstances: not even God could say which electron was which, because there is simply no fact of the matter about this. What this means is that an identity statement of the form "x is the same electron as y" may simply not be either determinately true or determinately false. (Lowe 1998, 194)

Thus, according to Lowe, electrons are not genuine individuals because there is no fact about "which is which," and hence they lack determinate identity. Presumably, the reason for thinking that there is no fact about the matter about which electron is which in a superpositional state has to do with the applicability of GP; the fact that we can permute the situation and nothing changes shows that there is no fact of the matter about which is which. But it is here that we must be very careful to distinguish permutations at the level of *models* from permutations of actual electrons. GP, I have claimed, should be understood as a claim about the *relation* between models and the world; permuting elements of a model results in a distinct model that represents the same world. This means that GP *does not* establish that there is no fact of the matter about which electron is which, and hence, electrons *may* be taken to be individuals after all.

MSE takes advantage of this possibility and argues that elementary particles (and space-time points) in fact *are* individuals. These objects have determinate countability and identity-conditions in virtue of their place in a relational structure in the manner illustrated in the previous section. For this reason, we may call them *structural individuals*. MSE thus offers a way to implement Stachel's aim of individuating these objects "entirely in terms of the relational structures in which they are embedded."

It is worth noting that structural individuals differ in important ways from traditional individuals. For example, Leibniz's Identity of Indiscernibles (PII) fails if formulated in the usual way: $\forall F[(Fx \leftrightarrow Fy) \rightarrow (x = y)]$, where F ranges over properties and x and y are objects. As we've seen, two electrons may share all of their properties while remaining distinct. However, if we understand F as ranging over *places* in structures rather than *properties*, the principle does hold for structural individuals.[17] Thus, we arrive at the surprising result that if MSE is the best explanation of GP in QM and GR, then the basic objects of these theories are individuals in Lowe's sense—albeit *structural* individuals for which the PII (as traditionally formulated) fails.

11.6 Conclusion

I have argued that MSE provides a metaphysics of objects capable of accounting for GP as it occurs in the two pillars of modern physics, GR and QM. I will conclude by briefly saying something about the sense in which this proposal is *structuralist*. Above it was remarked that structuralism takes the world to be "fundamentally structural." MSE takes objects to be essentially structural: to be a certain point or particle requires occupying a certain place in a relational structure. This bestows on structure an ineliminable role in determining what the world is like fundamentally. While MSE does not go so far as to replace objects with objectless structures, it does recognize irreducible structures as part of our fundamental ontology. To this extent, the view embraces the structuralist dictum that the world is "fundamentally structural."

Acknowledgments

I am grateful to Alexandre Guay, Jenann Ismael, Jonathan Schaffer, John Stachel, Oliver Pooley, Paul Teller, Richard Healey, Shamik Dasgupta, Shaughan Lavine, and audience members at Virginia Tech and the 2013 APA Pacific Division Meeting in San Francisco for helpful discussions on earlier versions of this chapter.

Notes

1. There is no uncontroversial conception of structure in this context. For our purposes, the informal idea of a structure as a *network of relations* will suffice, but the arguments that follow are intended to be compatible with other notions of structure as well.

2. In Stachel's (1989) reconstruction of Einstein's argument, the hole argument is formulated so that only g is changed. On this version of the argument the "hole" must be an empty region of space-time ($T_h = 0$). The presentation here follows the more general form of the hole argument developed in Earman and Norton 1987.

3. In this notation $d*$ is the drag-along that corresponds to the diffeomorphism d. For example, $d*g$ is a metric field that results from "dragging" the values of g at each point in the original model $\mathcal{M}$ to its image under d.

4. This notion of "collapse" is unrelated to the collapse of the wave function posited by some interpretations of QM.

5. This criticism is raised by Oliver Pooley (2006, 104) in response to Stachel and endorsed by Caulton and Butterfield (2012, 264–265).

6. Another potential route involves thinking of permutation as placing a constraint on the dynamics of systems (French and Krause 2006). From this perspective, GP might emerge as the principle that motivates the move from a full configuration space to a reduced one.

7. There is substantial disagreement on how to think about permutation invariance in QM, and many further issues would need to be addressed in a successful defense of GP. See French and Rickles 2003, Saunders 2006a for a sampling of the issues involved.

8. CB acknowledge this in a footnote (2012 n. 3), but argue that it is enough for their purposes to allow for the possibility of nonindividuals and objects with extrinsic individuating properties.

9. The term "bare modal claim" is due to Dasgupta (2011). Dasgupta criticizes solutions to the hole argument that proceed by simply denying that W and dW are distinct possible worlds. He calls this a "bare modal claim" and argues that views that deploy such claims have done nothing to alter the commitment to nonqualitative facts about the individuality space-time points that generate the hole argument. Of course, CB explicitly deny such nonqualitative facts play any role in "grounding the individuality" of physical objects, but they fail to offer any details of how individuality may be grounded qualitatively. Thus, their claim makes little progress from the bare modal claim $W = dW$.

10. Things get more interesting when we consider state-relations between more than two particles. For example, we may write the state of three indistinguishable bosons or fermions as follows (where m,n,s represent orthogonal states).

$$\Phi = \frac{1}{\sqrt{6}}\left[\left(|mns\rangle + |nsm\rangle + |smn\rangle\right) \pm \left(|nms\rangle + |snm\rangle + |msn\rangle\right)\right] \quad (11.2)$$

Φ already exhibits the state-relation of "occupying orthogonal states," but there may be additional structures present in addition. If we take the particles involved to be the electrons in a lithium atom in the ground state, for example, then we may add that two of them are in the lowest-energy state $1s$ (with opposite spins) and one is the next energy state $2s$. Such relations between states are difficult to describe in words, but the point remains that there is a certain relational structure in which such particles are embedded.

If we follow CB and allow for the possibility of paraparticles other than bosons and fermions, then Φ does not exhaust the possible configurations of three indistinguishable particles among the states m,n,s. This allows for further structures of state relations.

11. This view is inspired by the observation of Healey (1995) that defeating the hole argument doesn't require metric essentialism, but rather a weaker view he calls "minimal essence." According to minimal essence, if anything has the location properties space-time point p does in the actual world, then p does. In other words, there cannot be a point q that usurps p's location properties. As the next section will make clear, MSE is a generalization of this thesis.

12. I will focus on GR and the hole argument here for ease of presentation, but the points carry over equally well to permutations in QM.

13. This argument and its name are due to Jonathan Schaffer.

14. One may wonder whether the idea of replacing a manifold point with a "mole" really makes sense. After all, regardless of what we want to say about space-time points, manifold points are abstracta, and hence it is not at all obvious that one can regard them as having identity conditions robust enough to support this idea. If one were willing to claim that manifold points have haecceities, then this idea would make sense, but this does not seem to be a very natural view to adopt.

I will simply assume here that there is some framework in which replacement of this sort makes sense.

15. Here are two reasons one might think this is the case. First, highly symmetric space-time models actually used in physics—such as the Schwarzschild solution—are generally regarded as *idealizations* and, because they are, we should not assume that the world is as symmetric as they suggest. Second, if our space-time includes us, then it seems to follow that there will be unique geometrical relations between space-time points and us.

16. The reference to the "universal wave function" here is not meant to suggest that MSE is committed to the Everett interpretation of QM, where that term most often occurs. MSE takes QM to provide a structure of state-relations between particles. If the scope of QM is the entire world, then that structure will be determined by the quantum state of the entire world, whose representation is the universal wave function.

17. Alternatively, we may note that this version of the PII requires absolute discernibly for distinctness, and propose instead that it allow for weak discernibly as well: $\forall F \forall R([(Fx \leftrightarrow Fy) \land \neg(Rxy \lor Ryx)] \rightarrow (x = y))$, where R is an irreflexive relation. Either way, there is an important difference in the form of the principle that holds for structural individuals in comparison to the traditional PII.

References

Adams, Robert M. 1979. Primitive Thisness and Primitive Identity. *Journal of Philosophy* 76: 5–26.

Butterfield, Jeremy. 1989. The Hole Truth. *British Journal for the Philosophy of Science* 40: 1–28.

Caulton, Adam, and Jeremy Butterfield. 2012. Symmetries and Paraparticles as a Motivation for Structuralism. *British Journal for the Philosophy of Science* 63: 233–285.

Dasgupta, Shamik. 2011. The Bare Necessities. *Philosophical Perspectives* 25: 115–160.

Earman, John, and John Norton. 1987. What Price Spacetime Substantivalism. *British Journal for the Philosophy of Science* 38: 515–525.

Esfeld, Michael, and Vincent Lam. 2010. Holism and Structural Realism. In *Worldviews, Science and Us: Studies of Analytical Metaphysics. A Selection of Topics from a Methodological Perspective*, ed. Robrecht Vanderbeeken and Bart D' Hooghe, 10–31. Singapore: World Scientific Publishing.

French, Steven, and Dean Rickles. 2003. Understanding Permutation Symmetry. In *Symmetries in Physics: Philosophical Reflections*, ed. Katherine Brading and Elena Castellani, 212–238. Cambridge: Cambridge University Press.

Healey, Richard. 1995. Substance, Modality and Spacetime. *Erkenntnis* 42: 287–316.

Ladyman, James. 1998. What Is Structural Realism? *Studies in History and Philosophy of Science Part A* 29: 409–424.

Ladyman, James, and Don Ross. 2007. *Every Thing Must Go: Metaphysics Naturalized*. Oxford: Oxford University Press.

Lewis, David. 1986. *Philosophical Papers*. Vol. 2. Oxford: Oxford University Press.

Lowe, E. J. 1998. Entity, Identity and Unity. *Erkenntnis* 48: 191–208.

Lowe, E. J. 2007. Sortals and the Individuation of Objects. *Mind and Language* 22: 514–533.

Maudlin, Tim. 1988. The Essence of Space-Time. *PSA: Proceedings of the Biennial Meeting of the Philosophy of Science Association*, vol. 2, *Symposia and Invited Papers*, 82–91.

Maudlin, Tim. 1990. Substances and Space-Time: What Aristotle Would Have Said to Einstein. *Studies in History and Philosophy of Science Part A* 21: 531–561.

Pooley, Oliver. 2006. Points, Particles, and Structural Realism. In *The Structural Foundations of Quantum Gravity*, ed. Dean Rickles, Steven French, and Juha Saatsi, 83–120. Oxford: Oxford University Press.

Saunders, Simon. 2003. Physics and Leibniz's Principles. In *Symmetries in Physics: Philosophical Reflections*, ed. Katherine Brading and Elena Castellani, 289–307. Cambridge: Cambridge University Press.

Saunders, Simon. 2006a. Are Quantum Particles Objects? *Analysis* 66: 52–63.

Saunders, Simon. 2006b. On the Explanation for Quantum Statistics. *Studies in History and Philosophy of Modern Physics* 37: 192–211.

Stachel, John. 1989. Einstein's Search for General Covariance, 1912–1915. *Einstein and the History of General Relativity* 1: 63–100.

Stachel, John. 2002. The Relations between Things versus the Things between Relations: The Deeper Meaning of the Hole Argument. In *Reading Natural Philosophy: Essays in the History and Philosophy of Science and Mathematics*, ed. David Malament, 231–266. Chicago: Open Court.

Wüthrich, Christian. 2009. Challenging the Spacetime Structuralist. *Philosophy of Science* 76(5): 1039–1051.

CHAPTER 12 | Bohm's Approach and Individuality

PAAVO PYLKKÄNEN, BASIL J. HILEY,
AND ILKKA PÄTTINIEMI

12.1 Introduction

The usual interpretation of the quantum theory implies that we must renounce the possibility of describing an individual system in terms of a single, precisely defined conceptual model. We have, however, proposed an alternative interpretation ... which leads us to regard a quantum-mechanical system as a synthesis of a precisely definable particle and a precisely definable ψ-field. (Bohm 1952a, 188)

Perhaps the greatest challenge to the notion that objects are individuals with well-defined identity conditions comes from modern quantum and relativity physics. For ever since the early days of the quantum revolution, the identity and individuality of quantum systems have frequently been called into question (see, e.g., French 2011 and the references therein; French and Krause 2006; Ladyman and Ross 2007, chap. 3).

Many of the founding figures of quantum theory, and most notably Niels Bohr, held that it is not possible to describe individual quantum objects and their behavior in the same way as one can in classical physics, pointing out that the individual cannot be separated from the whole experimental context (for a recent penetrating discussion of Bohr's views, see Plotnitsky 2010). The idea that quantal objects might, in some sense, be "nonindividuals" was also considered early on by, for example, Born, Heisenberg, and Weyl (French 2011, 6).

One physicist who throughout his career emphasized the holistic features of quantum phenomena was David Bohm (1917–1992). For example, in his 1951 textbook *Quantum Theory*, which reflected the usual interpretation of quantum theory, he characterized individual quantum objects in strongly relational and contextual terms:

> Quantum theory requires us to give up the idea that the electron, or any other object has, by itself, any intrinsic properties at all. Instead, each object should be regarded as something containing only incompletely defined potentialities that are developed when the object interacts with an appropriate system. (1951, 139)

However, as is well known, soon after completing his 1951 textbook, Bohm discovered an alternative interpretation of quantum theory that gives individuals a much stronger status than the usual interpretation. His motivation stemmed from his dissatisfaction with the fact that the usual interpretation was not providing an ontology, a comprehensive view of quantum reality beyond the fragmentary experimental phenomena (Bohm 1987). Besides, discussions with Einstein in Princeton in the early 1950s strongly inspired him to start searching for a deterministic extension of quantum theory:

> In this connection, I soon thought of the classical Hamilton-Jacobi theory, which relates waves to particles in a fundamental way. Indeed, it had long been known that when one makes a certain approximation [WKB], Schrödinger's equation becomes equivalent to the classical Hamilton-Jacobi equation. At a certain point, I suddenly asked myself: What would happen, in the demonstration of this equivalence, if we did not make this approximation? (Bohm 1987, 35)

From the ontological point of view the puzzling thing about the WKB approximation is that we start from Schrödinger's equation, which according to the usual interpretation does not refer to an ontology, then remove something from Schrödinger's equation, and suddenly obtain the classical Hamilton-Jacobi equation that refers to a classical ontology. But how can *removing* something from a "nonontology" give us an ontology? Bohm's insight was to realize that if one does not approximate, one can see an unambiguous ontology that is hiding in Schrödinger's equation:

> I quickly saw that there would be a new potential, representing a new kind of force, that would be acting on the particle. I called this the quantum potential, which was designated by Q. This gave rise immediately to what I called a causal interpretation of the quantum theory. The basic assumption was that the electron *is* a particle, acted on not only by the classical potential, V, but also by the quantum potential, Q. This latter is determined by a new kind of wave that satisfies Schrödinger's equation. This wave was assumed, like the particle, to be an independent actuality that existed on its own, rather than being merely a function from which the statistical properties of phenomena could be derived.

Q is responsible for all quantum effects (such as the interference patterns of electrons and quantum nonlocality). However, whenever Q is negligibly small, quantum ontology gives rise to classical ontology. In 1952 Bohm published in *Physical Review* two papers that presented this interpretation (which independently rediscovered and made more coherent a theory that de Broglie had presented in the 1927 Solvay conference). To see the relevance of Bohm's interpretation to the question of individuality in quantum theory, let us consider how he contrasts his approach with that of Bohr:

> Bohr suggests that at the atomic level we must renounce our hitherto successful practice of conceiving of an individual system as a unified and precisely definable whole, all of whose aspects are, in a manner of speaking, simultaneously and unambiguously accessible to our conceptual gaze.... in Bohr's point of view, the wave function is in no sense a conceptual model of an individual system, since it is not in a precise (one-to-one) correspondence with the behavior of this system, but only in a statistical correspondence. (1952a, 167–168)

In contrast to this, Bohm's alternative interpretation regards

> the wave function of an individual electron as a mathematical representation of an objectively real field. (1952a, 170)

Thus for Bohm, an individual quantum-mechanical system has two aspects:

> It is a synthesis of a precisely definable particle and a precisely definable ψ-field which exerts a force on this particle. (1952b, 188)

Now, if the Bohm theory is a coherent option, it undermines the arguments of those who claim that nonrelativistic quantum theory somehow forces us to give up the notion that quantum objects are individuals with well-defined identity conditions. Ironically there is also a tension between Bohm's 1952 theory and much of his own other more anti-individualist (i.e., structuralist and process-oriented) work—both before and after 1952. Given this tension, it is not surprising that Bohm and Hiley developed Bohm's 1952 theory further in their later research (Bohm and Hiley 1987, 1993).

The question of whether quantum particles are individuals is also raised by the philosophers James Ladyman and Don Ross in their thought-provoking and important book *Every Thing Must Go* (2007, hereafter ETMG). They advocate the view that quantum particles are not individuals (or, at most, are weakly discernible individuals). They acknowledge that there seem to be individuals in the Bohm theory but go on to refer to research by Brown, Elby, and Weingard (1996) that they interpret as saying that in the Bohm theory, the properties normally associated with particles (mass, charge, etc.) are inherent only in the quantum field and not in the particles. It would then seem that there is nothing there in the trajectories unless one assumes the existence of some "raw stuff" of the particle. In other words it seems that *haecceities* are needed for the individuality of particles of the Bohm theory, and Ladyman and Ross dismiss this as idle metaphysics.

In what follows we will first give a brief account of Ladyman and Ross's views on individuality in quantum mechanics (section 12.2). We then introduce Bohm's 1952 theory, focusing on the way it seems to have room for quantum individuals (section 12.3). In section 12.4 we first present Ladyman and Ross's criticism of Bohmian individuality and then go on to challenge this criticism. We illustrate how puzzling quantum experiments such as the Aharonov-Bohm effect are explained in terms of the quantum potential approach. In section 12.5 we show how the symplectic symmetry forms a basis common to both classical and quantum motion and in section 12.6 bring out its relevance to the question of quantum individuality. In section 12.7 we show how some of the problems with the 1952 Bohm theory can be resolved by the radical proposal that the quantum potential functions as active information. We also note that while the Bohm theory allows us to retain the notion of individual particles, such particles have only a limited autonomy. In section 12.8 we briefly consider Bohm's other main line of research (the "implicate order"), which emphasizes the primacy of structure and process over individual objects. This research further underlines that while individuals have relative autonomy in Bohm's approach, they are not fundamental. Thus, in a broad sense, Bohm and Hiley's approach to quantum theory has interesting similarities to Ladyman and Ross's structural realism, even though the former gives quantum individuals a stronger status than the latter.

12.2 Ladyman and Ross on Individuality in Quantum Mechanics

In the third chapter of ETMG, Ladyman and Ross discuss identity and individuality in quantum mechanics. Following French and Redhead (1988), they first establish that indistinguishable elementary particles, that is, particles that have the same mass, charge, and so on, behave differently in quantum mechanics than they do in classical statistical mechanics. For quantum particles an "indistinguishability postulate" states that a permutation of indistinguishable particles is not observable, and thus those states that differ only by a permutation of such particles are treated as the same state with a different labeling. This might point to the view that quantum particles are not individuals.

Individuality is, however, an ontological property whereas (in)distinguishability is an epistemic one. So how are these two related? Ladyman and Ross identify three candidates in the philosophical tradition for individuality:

1. Transcendent individuality: the individuality of something is a feature of it over and above all its qualitative properties.
2. Spatiotemporal location or trajectory.
3. All or some restricted set of their properties (the bundle theory) (ETMG, 134).

Number 1 above is ruled out because it involves *haecceities*, and thus involves what Ladyman and Ross would consider idle metaphysical speculation. Granting this restriction for the sake of the argument, the interesting candidates are #2 and #3.

A connection between individuality and distinguishability is given by the Principle of the Identity of Indiscernibles (PII), which can be taken roughly to state that no two objects have exactly the same properties. It is easy to see that everyday objects satisfy both #2 and #3, while the point particles of classical mechanics satisfy #2. Then for both everyday objects and particles of classical mechanics PII is true and individuality and distinguishability can be taken to be the same thing. However, for certain quantum systems neither #2 nor #3 seems to hold. Ladyman and Ross take as an example of such a state the singlet state of two electrons orbiting a helium atom:

$$\psi = 1/\sqrt{2}\left[|\uparrow\rangle_1 |\downarrow\rangle_2 - |\downarrow\rangle_1 |\uparrow\rangle_2\right]. \tag{12.1}$$

Here any property that can be ascribed to particle 1 can also be ascribed to particle 2. So in this state the two electrons share all their extrinsic and intrinsic properties, thus falling foul of both #2 and #3. So it would seem that quantum particles are not individuals.

However, this conclusion might follow from a too strict a notion of discernibility. Following Saunders (2003a, 2003b, 2006), Ladyman and Ross give three notions of discernibility:

(1) Absolute discernibility
(2) Relative discernibility
(3) Weak discernibility

These can be defined as follows (ETMG, 137):

(1) "Two objects are absolutely discernible if there exists a formula in one variable which is true of one object and not the other." This holds for ordinary everyday objects.
(2) "Two objects are relatively discernible just in case there is a formula in two free variables which applies to them in one order only. ... An example of mathematical objects which are ... relatively

discernible include the points of a one-dimensional space with an ordering relation, since, for any such pair of points x and y, if they are not the same point then either $x > y$ or $x < y$ but not both."

(3) "Two objects are weakly discernible just in case there is two-place irreflexive relation that they satisfy." The fermions in a singlet state are discernible in this sense, as they satisfy the relation "is of opposite spin to."

Now since electrons in the singlet state are discernible, they can be viewed as individuals. But they are weakly discernible. This is a thoroughly structuralist view, "as individuals are nothing over and above the nexus of relations in which they stand" (ETMG, 138).

12.3 The Bohm Theory

Now let us turn to consider how the Bohm theory deals with these situations. Starting from the Schrödinger equation, we find the real part can be written as

$$\frac{\partial S}{\partial t} + \frac{1}{2m}(\nabla S)^2 + Q + V = 0 \tag{12.2}$$

under a polar decomposition of the wave function $\psi(r,t) = R(r,t)\exp[iS(r,t)/\hbar]$. This equation has a similar form to the classical Hamilton-Jacobi equation except for the appearance of a new term

$$Q = -\frac{\hbar^2}{2m}\frac{\nabla^2 R}{R}, \tag{12.3}$$

which is known as the quantum potential. Of course this means identifying the momentum of the particle by $p = \nabla S$, where S is the phase of the wave function. This relation, also known in some versions of this approach as the "guidance condition," enables the trajectory of the particle to be calculated. Note that the so-called Bohmian mechanics approach to Bohm's theory emphasizes that we get a deterministic particle mechanics directly from the first-order guidance equation involving the velocities of the particles (see Goldstein 2013). However, in this chapter we will be focusing on the way the Bohm theory arises from the above Hamilton-Jacobi type equation.

Figures 12.1 and 12.2 provide well-known visualizations.

So there is a version of quantum theory (the Bohm theory) according to which each particle has a definite and distinct trajectory at all times. This suggests that quantum particles are individuals, with position being

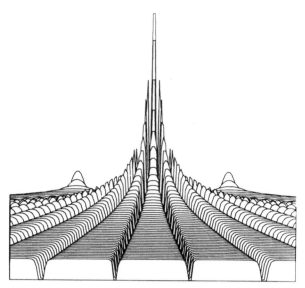

FIGURE 12.1 Quantum potential for two Gaussian slits

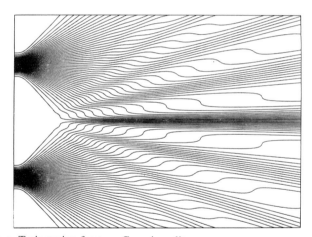

FIGURE 12.2 Trajectories for two Gaussian slits

the property in virtue of which particles are always different from one another.

The biggest problem for retaining the notion of individuality is the one posed by particles in entangled states described by equation (12.1). This is an entangled spin state that has been discussed in detail by Dewdney, Holland, Kyprianidis, and Vigier (1988), but for our purposes here it is sufficient to consider a general two-body wave function, $\psi(r_1, r_2, t)$, and use the two-body Schrödinger equation to find

$$\frac{\partial S}{\partial t} + \frac{(\nabla_1 S)^2}{2m_1} + \frac{(\nabla_2 S)^2}{2m_2} + Q(r_1, r_2, t) + V(r_1, r_2) = 0. \tag{12.4}$$

The second and third terms in this equation correspond to the kinetic energies of each particle, so once again the problem of individuality does not seem to arise in the Bohm theory. The entanglement is reflected in the nonlocal quantum potential energy term $Q(r_1, r_2, t)$. Furthermore, since the trajectories do not cross, we follow Brown, Sjöqvist, and Bacciagaluppi (1999, 233) and conclude that indistinguishable fermions will always have distinct trajectories (for further discussion see French 2011, 14–15; French and Krause 2006, 178). Thus individuality is preserved.

12.4 The Bohm Theory and Haecceities

So the Bohm theory seems to suggest, contra Ladyman and Ross, that quantum particles can be individuals in a stronger sense than they claim. They do acknowledge the existence of the Bohm theory in a footnote, but do not see it as a problem for their nonindividualistic view. They write:

> Of course, there is a version of quantum theory, namely Bohm theory, according to which QM is not complete and particles do have definite trajectories at all times. However, Harvey Brown et al. (1996) argue that the "particles" of Bohm theory are not those of classical mechanics. The dynamics of the theory are such that the properties, like mass, charge, and so on, normally associated with particles are in fact inherent in the quantum field and not in the particles. It seems that the particles only have position. We may be happy that trajectories are enough to individuate particles in Bohm theory, but what will distinguish an "empty" trajectory from an "occupied" one? Since none of the physical properties ascribed to the particle will actually inhere in points of the trajectory, giving content to the claim that there is actually a "particle" there would seem to require some notion of the raw stuff of the particle; in other words *haecceities* seem to be needed for the individuality of particles of Bohm theory too. (ETMG, 136n)

Actually, Ladyman and Ross are somewhat one-sided in their reporting of the views of Brown, Elby, and Weingard (1996). For in their paper Brown and coauthors are *not* arguing for the view that in the Bohm theory, properties like mass, charge, and so on, normally associated with particles are *only* inherent in the ψ-field and not in the particles. What they do argue for is that certain experiments (for example, certain types of interferometry experiments) rule out the possibility that these properties are associated with the Bohm particle *alone*.

They point out that there are two principles we can adopt here. First, there is the *principle of generosity*, according to which the properties can be attributed to both the ψ-field and the particle. Second, there is the *principle of parsimony*, according to which properties such as mass are attributes not of the particle but of the ψ-field alone. They do not take a definite stand on which principle we should adopt. However, they draw attention to reasons to adopt the principle of generosity, while at the same time indicating difficulties inherent in the principle of parsimony. It thus seems clear that, contra what Ladyman and Ross suggest, they are more in favor of the principle of generosity. To be fair, however, we should acknowledge that the issue is subtle and people's views on this vary—it seems that Harvey Brown himself was in favor of parsimony before opting for generosity. For Brown, Elby, and Weingard (1996) acknowledge that the principle of parsimony is implicit in Brown's earlier work; they also note that it is implied by some of Bell's suggestions (Bell 1990, 30).

Let us now examine in more detail the arguments that suggest that properties such as mass or charge are not inherent only in the particles. In order to bring the issue into focus Brown, Dewdney, and Horton (1995) introduce and define the localized particle properties thesis (LPP): particle properties (such as mass, charge, etc.) are attributes of the particle rather than the ψ-field. That is, the mass, say, of the particle is localized at the position of the particle at all times. They go on to point out that several experiments seem to violate the LPP.

In the neutron interferometry experiments of Colella, Overhauser, and Werner (1975), a neutron stream travels through a beam splitter along two routes, producing an interference pattern. Now, if the apparatus is tilted in such a way that one of the routes has a higher gravitational potential than the other, the interference pattern is shifted. According to the Bohm theory the particle travels one of the paths, while the ψ-field travels both paths. Brown, Dewdney, and Horton note that if we assume that all of the electron's gravitational mass is concentrated in the path where the particle is, it becomes difficult to understand intuitively why the interference pattern is shifted. For if the empty path ψ-field carries no gravitational mass, how could the difference in the gravitational potential integrated over the two paths be felt by the particle? So they argue that for gravitational mass, the LPP seems to be violated.

According to Bohm and Hiley the particle and the ψ-field are strictly speaking indivisible. However we can think of them as two different aspects of an underlying whole. Now, there seems to be no inherent reason in such ontology that, say, the mass should be entirely localized with the particles. Indeed the mathematics suggests that mass is implicated in both the particle and the field aspect. For mass appears both in the mathematical expression of the kinetic energy of the particle and in the mathematical expression of the quantum potential that reflects the ψ-field. This is in harmony with the principle of generosity; that is, it seems that mass resides in both the particle aspect and in the field aspect of the individual system.

In the Aharonov-Bohm (AB) effect a similar situation arises with charge. In the AB effect the normal two-slit experimental arrangement has in the geometrical shadow of the two slits a shielded region containing a magnetic field. The shield is such that the electrons cannot experience the magnetic field directly at all. Nevertheless one finds that the interference pattern is shifted by an amount that depends on the strength of the magnetic flux in the shielded region. Since the electrons do not experience this flux directly, it is difficult to understand why the interference pattern is changed. Indeed, Brown, Sjöqvist, and Bacciagaluppi write (1999, 234n):

> The expression for the phase shift due to the flux in the shielded solenoid depends on charge being present on spatial loops within the support of the wave function and enclosing the solenoid.

But again the trajectory of the Bohmian particle associated with the charged particle does not encircle the solenoid. So the LPP seems to be violated for charge.

The AB effect has been numerically analyzed in terms of the Bohm theory by Philippidis, Bohm, and Kaye (1982). This analysis shows that in the AB effect, the vector potential (from which the magnetic line of flux is derived) affects the phase of the wave function in such a way that the latter gives rise to an asymmetric quantum potential (see figure 12.3).

If we compare this QPE with that shown in figure 12.1, where no magnetic line of flux is present, we see that the pattern of the quantum

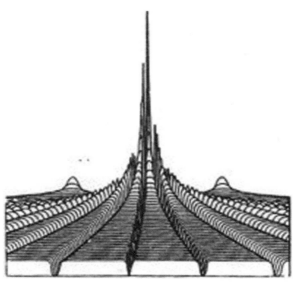

FIGURE 12.3 The quantum potential for the Aharonov-Bohm effect. Notice the asymmetrical shift

BOHM'S APPROACH AND INDIVIDUALITY | 235

potential has been shifted off the axis of symmetry by an amount that depends on the strength of the enclosed flux. This, in turn, produces a shift in the ensemble of trajectories as was shown in Philippidis, Bohm, and Kaye (1982). This results in an overall pattern shift that has actually been observed in experiment (see Bayh 1962). As with the gravitational example, it is reasonable to assume that the charge inheres in both the particle and the wave (see Bohm and Hiley 1993, 52–54 for more details).

12.5 Symplectic Symmetry as the Common Basis of Classical and Quantum Dynamics

There has been a common misconception that the use of the quantum potential implies a return to the classical paradigm and that quantum mechanics is a radical departure, so radical that any elements of the classical paradigm must be avoided at all costs. Indeed the way that the quantum Hamilton-Jacobi equation (12.2) is derived hides a much deeper relation between classical and quantum dynamics. Mathematically they have a common kinematic symmetry, namely, the symplectic group of transformations. As Melvin Brown (2006, v) has succinctly put it,

> This very general group of transformations maintains the fundamental relationship between position and momentum in mechanics, and its covering group (the metaplectic group) correspondingly transforms the wave function in quantum mechanics.

Assuming that the reader may not be familiar with the mathematics of covering groups, we will here approach the subject in a nonformal way (for a more technical presentation, see, e.g., de Gosson 2001, Brown 2006).

To bring out this deeper connection, let us return to examine the motivations that led Schrödinger to his equation. When the wave properties of electrons and atoms had been established, Schrödinger recalled Hamilton's discussions on the relation between ray optics and wave optics. Hamilton had shown that paraxial optical rays could be described by the same pair of dynamical equations that he had proposed for classical particles and was attempting to generalize these to capture the wave properties of light. (See Guillemin and Sternberg 1984 for more details.) What the Hamilton-Jacobi theory had shown was that the rays (classical particle trajectories) were perpendicular to a set of surfaces of constant action S. Although the action is well defined mathematically, namely, $S = \boldsymbol{p} \cdot \boldsymbol{x} - Et$, its physical meaning was not clear. Schrödinger noticed that if we could regard these surfaces as surfaces of constant phase, then perhaps we could find an equation that would form the basis of what he called "Hamiltonian undulatory mechanics."

Schrödinger's arguments to derive his equation were, at best, heuristic. Even Schrödinger (1926) himself writes, "I realise that this formulation is not quite unambiguous." But this should not be surprising as the necessary mathematics of the geometry that underlies both classical and quantum dynamics did not exist in the 1920s.

However the equation quickly gave results that agreed with experiment, so that the equation was taken as an a posteriori given, independent of its origins. It became *the* defining equation of quantum phenomena, and as a consequence of trying to understand this equation, the "wave function" became the center of attention (this attention continues to the present day; see, e.g., Ney and Albert 2013). With this position came the paradoxes of quantum theory that remain unresolved. Relatively few physicists or philosophers have attempted a sustained exploration of the deeper mathematical background from which the equation appears. Indeed the Schrödinger equation is taken as a given, arising as if by magic. Even Feynman (1965) acknowledged that the equation was not derived from anything known in physics or mathematics. As he remarked: "It came out of the mind of Schrödinger."

The connection between classical and quantum dynamics begins to emerge as we examine the common symmetry, the symplectic symmetry, underlying both Hamiltonian dynamics and the Schrödinger dynamics (de Gosson 2001 and 2010). These deeper connections have emerged relatively recently in the mathematics literature and are just beginning to become appreciated by the physics community. What one learns is that classical mechanics and ray optics emerge at the level of the symplectic group of transformations itself, but wave theory and quantum mechanics begin to emerge at a higher level, namely in the double cover of the group, that is, the metaplectic group and its generalization. The "double" here means, roughly, that there are always two elements in the metaplectic group representing one element in the symplectic group.

The double cover is a vital part of quantum mechanics, since it enables one to capture global properties of the type that occur in our discussion of the AB effect. These global properties play a significant role in Bohm's notion of "unbroken wholeness," which we will discuss in a later section.

We are already very familiar with this idea of a double cover in the case of rotational symmetry. Here the double cover is the spin group. This gives us the spinor with which we describe fermions, and these spinors form the mathematical basis of the entangled singlet state given by equation (12.1) above. Furthermore it is the relation between the group and its double cover that gives us a mathematical description of the experimentally confirmed difference between a 2π and a 4π rotation that shows up in experiments with fermions (see Werner et al. 1975).

This is not the place to go into mathematical details, and we will simply point out some interesting results to bring out the unexpected connection between classical and quantum mechanics. First, de Gosson and Hiley (2011) have shown that if we formally "lift" a Hamiltonian flow onto the double cover space, we find a unique flow in this covering space and this flow satisfies a Schrödinger-like equation, confirming the results in Guillemin and Sternberg (1984). It is "Schrödinger-like" because at this stage it contains an arbitrary parameter with dimensions of action to enable us to put position and momentum on the same footing. The mathematics alone does not enable us to identify this parameter with Planck's constant. Its value is determined by experiment as it is in the standard theory.

Second, de Gosson (2010) draws attention to a deep topological theorem in symplectic geometry known as the "Gromov no-squeezing" theorem. This states that even in classical mechanics, it is not possible to reduce a canonical volume such as $\Delta x \Delta p$ by means of a Hamiltonian flow alone. When this region is lifted into the covering space, it provides the source of the uncertainty principle. In this way one can say that the symplectic features of the classical world contain the footprint of the uncertainty principle even at the classical level. This structure provides a rigorous mathematical background for Schrödinger's "undulatory mechanics."

12.6 Bohmian Quantum Individuals in Light of the Metaplectic Group of Transformations

We cannot go into the mathematical details here, nevertheless we feel it is necessary to explain some aspects that emerge from the details. Since we want to focus on the relationship between the action surfaces and phase surfaces, it is convenient to describe the dynamical evolution in terms of a notion of a flow. For classical particles, the Hamiltonian flow is in phase space

$$z(t') = f_{t',t} z(t),$$

where $z(t) = (x(t), p(t))$ are the coordinates of the particles in the phase space. The Schrödinger flow is

$$\psi(t') = F_{t',t} \psi(t),$$

where $\psi(x, t) = (R(x, t), S(x, t))$. In each case the flow is determined by the Hamiltonian. By focusing on the flow, we have a different way of understanding the relationship between the two types of behavior.

We can illustrate how the flow determines the behavior of the individual by re-examining the AB effect. Here the flow must reflect the difference between those paths that encircle the enclosed flux and those that do not. What this means is that the Schrödinger flow itself is a global flow, not a local flow. In order to access the relevant global properties of the environment, the flow must carry the information about the global properties of the environment, which is encoded in the covering group. When we examine the effect of this covering group in the underlying group, we find that the energy associated with the particle splits into two parts:

$$E_{particle} = \frac{(\nabla S)^2}{2m} - \frac{\hbar^2}{2m}\frac{\nabla^2 R}{R} = KE + QPE.$$

In this sense the QPE is not the source of a force acting on the particle. Rather it is a potentiality for the behavior of the particle-like process that finds itself at a particular region in space (this may be somewhat similar to the ideas of Esfeld et al. [2013], who consider the prospects of grounding the motions of particles in dispositions in the Bohm theory).

The splitting of the energy of the individual is not mere speculation, as experiments using weak measurements are in progress to measure these parts (Flack and Hiley 2014). In view of this radical possibility let us look at the role of the individual in a different way. Recall that Hamilton considered the optic ray as the locus of some invariant independent and indivisible feature of the energy flow, namely the wave-packet that we now call the photon. When we apply the same idea to the Schrödinger particle, we can regard the "trajectory" as the locus of some indivisible but *not necessarily localized* energy that we call the "particle." In the case of the photon it is not possible to slow it down or give it a precise position, so the notion of a "photon trajectory" is questionable. On the other hand for the quantum particle, it is possible to slow it down and examine it in the classical limit, and show it becomes a point-like object. Such a situation arises whenever the quantum potential energy (QPE) is small compared with the classical energy.

Hiley and Mufti (1995) have given a simple model that nicely illustrates this feature. Suppose we have a situation in which the quantum potential is time dependent and becomes smaller as time progresses. One can then show that the ensemble of quantum trajectories merges smoothly into an ensemble of classical trajectories. Thus a particle following a trajectory in the quantum domain will become a particle obeying the rules expected of a classical particle. In our view these results present strong evidence that it is a coherent possibility that a particle keeps its identity and individuality in a quantum context, even though some of its energy is now involved in exploring its environmental neighborhood via the ψ-field producing the associated QPE.

The radically different nature of individuality in the quantum domain appears when two or more particles become "entangled" as illustrated in equation (12.1). In the Bohm approach these particles are coupled by a common quantum potential energy. This potential is *nonlocal* in the sense that the behavior of one particle is "locked" into the behavior of the other. This gives a time evolution that involves the two spatially separated particles behaving as a single entity. An example of such an object is the Cooper pair, the electron pair responsible for superconductivity. It is tempting to see such an entity as a new type of individual, where we find a "twoness" in an underlying individual whole. This is in contrast to two particles described by a product of two wave functions, which can be seen as separate individuals.

If we return to examine the details of the pair of particles described by equation (12.1), we find an ensemble of correlated trajectories that have been calculated by Dewdney and coauthors (Dewdney, Holland, and Kyprianidis 1986, 1987, Dewdney et al. 1988). There we see that if one of the particles enters the field of a Stern-Gerlach magnet, it is then deflected either "up" or "down" depending on the positions of each particle at the time just before the particle enters the magnetic field. The particle in the field has its trajectory changed, while the other particle continues in a straight line. At the same time both spin components become well defined. This is a surprising result, but shows quite clearly that the individual parts cannot be thought of as isolated "little spinning spheres," a point that was emphasized by Weyl (1931).

In the AB situation, an ensemble of incident particles would then give rise to a shifted interference pattern, the shift being determined by the enclosed magnetic flux. Nowhere do we lose the identity of the individual particle even though it is responding to a global situation. This notion of individuality is strengthened when it is realized that if the particle were not charged, then its Schrödinger evolution would be very different and the interference pattern would be unaffected by the presence of any enclosed magnetic line of flux. In this sense the quantum potential energy is a "private" energy; it is "individual" in the sense of belonging to the individual particle. This is why on detection the particle reveals its total energy.

Note that this is an example where quantum theory seems to involve stronger and more peculiar individuality than classical physics! This point is often left unnoticed in discussions of individuality in the quantum theory, where one typically emphasizes the nonindividualistic aspects of the quantum domain. Note in particular that if two particles are conventionally described by a product of two wave functions and the particles do not interact through a classical potential, they do not experience each other's quantum potential even though they may both be in a region of space where their wave functions have significant spatial overlap (see Brown, Elby, and Weingard 1996, 313–314).

12.7 Problems with the Bohm Theory and Their Solution via the Hypothesis of Active Information

Bohm was not entirely happy with his 1952 theory, and in his 1957 book *Causality and Chance in Modern Physics*, as well as in a 1962 article (republished in Bohm 1980), he summarized his criticisms. First, he admitted that the form of the quantum potential is "strange and arbitrary" and that unlike, say, the electromagnetic field, it has no visible source. He added that while the theory is logically consistent, the quantum potential should be seen as "at best a schematic representation of some more plausible physical idea to which we hope to advance later, as we develop the theory further" (Bohm 1980, 102). He also pointed out that for the many-body system the wave function lives in a $3N$-dimensional configuration space (where N is the number of particles). It is difficult to understand what such a multidimensional ψ-field means from a physical point of view (this is a common traditional criticism of the de Broglie–Bohm approach, see, e.g., Putnam 1965).

One way of responding to these problems is an approach to Bohm theory that has become known as "Bohmian mechanics," which we briefly mentioned above. Here one can say that the positions of particles are the "primitive ontology." What, then, is the wave function? Goldstein and Zanghi (2013) propose that we consider the wave function as nomological, something more in the nature of a law than a concrete physical reality. So if the wave function does not describe a physical field, the question of the $3N$-dimensional field for the many-body system does not arise. However, we believe that the quantum potential approach enables us to explore the ontological meaning of quantum theory in a deeper way, so let us consider whether it is possible to make this approach physically more viable.

Remember the two worries: the form of the quantum potential is strange and arbitrary, and it is difficult to give a physical interpretation to a multidimensional ψ-field. When Bohm re-examined his 1952 theory in the late 1970s, he realized that the quantum potential might be telling us something radically new about the nature of reality. For he noticed that the quantum potential depends only upon the form, or the second spatial derivative of the amplitude R, of the ψ-field. This form, in turn, reflects the form of the environment (such as the presence of slits, but also features such as magnetic flux lines and differences in gravitational potential).

Could it be that the ψ-field is literally "in-forming" or putting form into the activity of the particle, rather than pushing and pulling the latter mechanically? Bohm called such information "active information," because this is an instance where information acts to bring about changes in the behavior of the particle. Note also that this idea of the ψ-field as a field that encodes information provides a new way of understanding the multidimensionality of the ψ-field for the many-body system. Indeed, Bohm suggested that the wave function describes not a multidimensional field, but rather an information structure that can quite naturally be considered to be multidimensional, that is, organized into as many sets of dimensions as may be needed. Thus the quantum wave field is not regarded as a simple source of a mechanical force. He speculated that the information encoded in the quantum potential is carried in some much more subtle level of matter and energy that has not yet manifested in physical research (see Bohm and Hiley 1987, 336). Note that the two key anomalies of the 1952 theory (i.e., the arbitrary form of Q and the multidimensionality of the many-body ψ-field) thus became the cornerstones of a new interpretation of the ψ-field as a field of information (Pylkkänen 1993).

Let us next consider the individuality of Bohmian particles in the light of this active information approach. First of all, Bohm and Hiley were led to propose that, say, an electron has an internal structure that enables it to respond to the information in the ψ-field. So, according to this view Bohmian quantum particles are not the point particles of classical physics, but much more subtle entities. Note also that because of the holistic features of the quantum potential, these particles have only a relative autonomy. In the case of a single particle, because the quantum potential only depends upon the form of the ψ-field, it does not necessarily fall off with distance even if the intensity of the ψ-field becomes weak as the field spreads out. This means that even very distant features of the environment (e.g., slits) can have a strong effect upon the particle, thus underlining its lack of autonomy. Strictly speaking the entire experiment has to be treated as an undivided whole, which is reminiscent of Bohr's view. However, while Bohr suggested that this whole is unanalyzable, in the Bohm theory one can now analyze it in thought in terms of the movement of the particle acted on by the quantum potential.

In the two-body system the autonomy of the individual becomes weaker still, for the quantum potential depends on the position of both particles in a way that does not necessarily fall off with the distance. This means that there is the possibility of a nonlocal interaction between the two particles. We can generalize this to the N-body system, where the behavior of each particle may depend nonlocally on all the others, no matter how far away they may be. Nonlocality is an important new feature of the quantum theory, but Bohm was keen to emphasize that there is yet another feature that is even more radical. For in the Bohm theory there can be a nonlocal connection between particles that depends on the quantum state of the whole, in a way that cannot be expressed in terms of the relationships of the particles alone. This quantum state of the whole, described by the many-body wave function, evolves in time according to the Schrödinger equation, which led Bohm and Hiley to write:

> Something with this sort of independent dynamical significance that refers to the whole system and that is not reducible to a property of the parts and their inter-relationships is thus playing a key role in the theory. ... *this is the most fundamental new ontological feature implied by quantum theory.* (Bohm and Hiley 1987, 332)

The above quote reveals the holistic character of Bohm's interpretation of quantum theory. Even if his 1952 theory in a sense rediscovered the lost individuality of quantum objects, his quantum ontology was not a return to the individuals of classical physics. He thought that quantum theory was primarily about dynamical wholeness that is not reducible to the interactions between individuals. As Max Jammer has pointed out, this means that the individuals are not "constitutive" to the whole but rather depend on the state of the whole (1988, 696). Related to this, Tim Maudlin has commented:

> David Bohm has long contended that what is radically new about the quantum theory is the "undivided wholeness" that it posits, and if Bohm is right, philosophical commentaries on the quantum theory have long been preoccupied with the wrong features of the theory. (1998, 49)

What is also relevant here for the present volume is that the wholeness of Bohmian quantum systems seems analogous to the organic unity of biological systems, suggesting interesting links between physics and biology:

> The quantum potential arising under certain conditions has the novel quality of being able to organize the activity of an entire set of particles in a way that depends directly on the state of the whole. Evidently, such an organization can be carried to higher and higher levels and eventually may become relevant to living beings. (Bohm and Hiley 1987, 332)

The quantum potential approach thus provides potentially useful tools for a holistic approach in biology. There seems to be at least an analogy in the way the whole and the part are related in some quantum phenomena and some biological phenomena.

Our discussion above suggests a richer view of Bohmian particles than is presupposed by Ladyman and Ross, or Brown, Elby, and Weingard. Indeed, Bohm proposed that it is plausible that the behavior and structure of matter does not always become simpler as we go to lower dimensions. Radically, he suggested that a particle such as an electron may have a structure (somewhere between 10^{-16} cm and 10^{-33} cm). This structure is assumed to be complex and subtle enough to respond to the information described by the wave function. So according to this hypothesis, there is definitely more to the individuality of Bohmian particles than mere *haecceities*.

12.8 Bohm's Scientific Structuralism

We have seen above that while the 1952 Bohm theory in a sense rediscovered the quantum individuals and particle trajectories that got lost in the usual interpretation, the theory should by no means be seen as a return to a mechanistic ontology where individuals are fundamental and constitutive to the whole. Bohm's other main line of research, known as the "implicate order," likewise sees quantum theory as a guide toward a conception of a new holistic and dynamic order in physics (Jammer 1988, 696). This work, which begins to develop in the early 1960s, aims to develop a deeper underlying theory from which quantum theory and relativity can be derived as approximations, and their relation thus understood. This framework suggests a strongly structuralist, process-oriented way of understanding individual quantum systems that is in some ways similar to Ladyman and Ross's structural realism. At an early phase of this work Bohm wrote:

> In this theory ... the notion of a separately existing entity simply does not arise. Each entity is conceptually abstracted from a totality of process ... with the electron, what actually exists is a structure of underlying elementary processes or linkages supporting a pattern corresponding to an electron. (1965a, 291)

In the later implicate order view, an electron is not a little billiard ball that persists and moves, but should more fundamentally be understood as

> a recurrent stable order of unfoldment in which a certain form undergoing regular changes manifests again and again, but so rapidly, that it appears to be in continuous existence. (Bohm 1980, 194)

Finally, in the final chapter of their 1993 book *The Undivided Universe* Bohm and Hiley, when discussing quantum field theory and emphasizing the ontological primacy of movement required by relativity, summarize this nonindividualistic line of thought as follows:

> The essential qualities of fields exist only in their movement. ... The notion of a permanently extant entity with a given identity, whether this be a particle or anything else, is ... at best an approximation holding only in suitable limiting cases. (1993, 357)

Thus much of Bohm's work supports the idea that individuals are not metaphysically fundamental in the light of contemporary physics. His emphasis on notions such as "structural process" (1965b), "order," and "movement" (1980) as fundamental in physics suggests that the philosophical home of Bohm's (and Hiley's) more general approach to physics might well be found in some form of scientific structuralism that takes movement as fundamental, rather than in a metaphysics that takes individuals as basic (cf. Ladyman and Ross 2007). Indeed, Hiley's recent work on symplectic geometry can be seen as bringing Bohm's 1952 approach closer to scientific structuralism. For ultimately Hiley's work leads to the algebraic approach that was initiated by Bohm and Hiley (1993, chap. 15).

12.9 Concluding Remarks

We have above shown that the prospects of individuality in the Bohm theory are stronger than Ladyman and Ross imply. This suggests that there is an underdetermination of metaphysics by physics in nonrelativistic quantum theory when it comes to the question of individuality (as indeed has been emphasized by French and Krause 2006, 189–197). However, it is important to realize that the notion of an individual in the Bohm theory—especially in Bohm and Hiley's (1987, 1993) developed account of it—is very different from what we would expect from the classical perspective. For although the Bohmian quantum individual has a well-defined energy, that energy is not a local energy. This is consistent with Niels Bohr's views in two ways. First, as emphasized in Bohm and Hiley (1993), the particle is never separated from the quantum field. It is an invariant feature of the total underlying process. This is consistent with Bohr's notion of the "impossibility of subdividing quantum phenomena" in the sense that the whole experimental arrangement must be taken into account (Bohr 1958, 50–51).

Second, we have suggested that the Bohmian individual is not a localized point-like object. As Bohr remarks (1958, 73) the quantum process is a "closed indivisible phenomenon." The energy is not localized at a point. In fact complementarity can be taken to imply that energy transcends space-time. Nevertheless there is a center of energy, a generalization of the center of mass that can be given a position in space-time. It is this particle-like center that moves with the Bohm momentum.

These ideas are not consistent with a classical notion of a particle and, we feel, can only be given a more comprehensive meaning in terms of something like Bohm's (1965b) notion of "structural process." Thus the overall Bohmian approach to physics does not, from the metaphysical point of view, mean a return to the individuals of classical physics, but has instead strongly structuralist features. In particular, we noted that Bohm and Hiley have since the 1960s been developing a broader scheme they call "the implicate order," which goes beyond the 1952 Bohm theory (Bohm 1980; Bohm and Hiley 1993, chap. 15; Hiley 2011; Pylkkänen 2007; for Bohm's own attempt to reconcile "hidden variables" and the implicate order, see his 1987). We acknowledge that this scheme seems to have some relevant similarities to Ladyman and Ross's ontic structural realism, while there also may be some significant differences. The discussion of these similarities and differences will, however, be a subject of another study (some preliminary attempts have already been made by Pättiniemi 2011 and Pylkkänen 2012).

Acknowledgments

We received extensive and insightful comments on this chapter from Guido Bacciagaluppi and David Glick for which we are grateful. Bacciagaluppi's comments encouraged us to strengthen our characterization of the particle pair in an entangled state as an individual, while Glick's comments prompted us to emphasize the differences between Bohmian and classical particles. We also received valuable feedback from Alexander Guay, Tuomas Tahko, and an anonymous referee. An early version of this chapter was presented 14 May 2012 at the Philosophy of Science seminar of the University of Helsinki and 19 May 2012 at the "Individuals across Sciences" symposium in Paris. We thank the participants in these events for the many thoughtful comments we received.

References

Bayh, Werner. 1962. Messung der kontinuierlichen Phasenschiebung von Elektronenwellen im kraftfeldfreien Raum durch das magnetische Vektorpotential einer Wolfram-Wendel. *Zeitschrift für Physik* 169: 492–510.

Bell, John. 1990. Against Measurement. In *Sixty-Two Years of Uncertainty*, ed. Arthur I. Miller, 17–31. New York: Plenum Press. Also appeared in *Physics World* 3(8) (1990): 33–40.

Bohm, David. 1951. *Quantum Theory*. New York: Dover.

Bohm, David. 1952a. A Suggested Interpretation of the Quantum Theory in Terms of "Hidden Variables" I. *Physical Review* 85(2): 166–179.

Bohm, David. 1952b. A Suggested Interpretation of the Quantum Theory in Terms of "Hidden Variables" II. *Physical Review* 85(2): 180–193.

Bohm, David. 1957. *Causality and Chance in Modern Physics*. London: Routledge & Kegan Paul.

Bohm, David. 1965a. Problems in the Basic Concepts of Physics. In *Satyendranath Bose 70th Birthday Commemoration Volume*, vol. 2. Calcutta: Prof. S. N. Bose 70th Birthday Celebration Committee. An inaugural lecture delivered at Birkbeck College, University of London, 13 February 1963, with two appendices.

Bohm, David. 1965b. Space, Time, and the Quantum Theory Understood in Terms of Discrete Structural Process. In *Proceedings of the International Conference on Elementary Particles, 24th–30 September* 1965, 252–287. Kyoto: Publication Office, Progress of Theoretical Physics, Yukawa Hall, Kyoto University.

Bohm, David. 1980. *Wholeness and the Implicate Order*. London: Routledge.

Bohm, David. 1987. Hidden Variables and the Implicate Order. In *Quantum Implications: Essays in Honour of David Bohm*, ed. Basil J. Hiley and F. David Peat, 33–45. London: Routledge.

Bohm, David. 1988. A Realist View of Quantum Theory. In *Microphysical Reality and Quantum Formalism*, vol. 2, ed. Alwyn van der Merwe, F. Selleri, and G. Tarozzi, 3–18. Dordrecht: Kluwer.

Bohm, David, and Basil J. Hiley. 1987. An Ontological Basis for Quantum Theory: I. Non-relativistic Particle Systems. *Physics Reports* 144(6): 323–348.

Bohm, David, and Basil J. Hiley. 1993. *The Undivided Universe: An Ontological Interpretation of Quantum Theory*. London: Routledge.

Bohr, Niels. 1958. *Atomic Physics and Human Knowledge*. New York: Wiley.

Brown, Harvey, Chris Dewdney, and George Horton. 1995. Bohm Particles and Their Detection in the Light of Neutron Interferometry. *Foundations of Physics* 25: 329–347.

Brown, Harvey, Andrew Elby, and Robert Weingard. 1996. Cause and Effect in the Pilot-Wave Interpretation of Quantum Mechanics. In *Bohmian Mechanics and Quantum Theory: An Appraisal*, ed. James T. Cushing, Arthur Fine, and Sheldon Goldstein, 309–319. Dordrecht: Kluwer.

Brown, Harvey, Erik Sjöqvist, and Guido Bacciagaluppi. 1999. Remarks on Identical Particles in de Broglie-Bohm Theory. *Physics Letters A* 251: 229–235.

Brown, Melvin. 2006. *The Symplectic and Metaplectic Groups in Quantum Mechanics and the Bohm Interpretation*. Google Base.

Colella, Roberto, Albert W. Overhauser, and Samuel A. Werner. 1975. Observation of Gravitationally Induced Quantum Interference. *Physical Review Letters* 34: 1472–1474.

de Gosson, Maurice. 2001. *The Principles of Newtonian and Quantum Mechanics: The Need for Planck's Constant*. London: Imperial College Press.

de Gosson, Maurice. 2010. *Symplectic Methods in Harmonic analysis and in Mathematical Physics*. Basel: Birkhäuser Verlag.

de Gosson, Maurice, and Basil J. Hiley. 2011. Imprints of the Quantum World in Classical Mechanics. *Foundations of Physics* 41: 1415–1436. DOI:10.1007/s 10701-011-9544-5. http://arXiv:1001.4632v2.

Dewdney, Chris, Peter R. Holland, and Anastasios Kyprianidis. 1986. What Happens in a Spin Measurement? *Physics Letters A* 119: 259–267.

Dewdney, Chris, Peter R. Holland, and Anastasios Kyprianidis. 1987. A Causal Account of Non-local Einstein-Podolsky-Rosen Spin Correlations. *Journal of Physics A: General Physics* 20: 4717–4732.

Dewdney, Chris, Peter R. Holland, Anastasios Kyprianidis, and Jean-Pierre Vigier. 1988. Spin and Non-locality in Quantum Mechanics. *Nature* 336: 536–544.

Esfeld, Michael, Dustin Lazarovici, Mario Hubert, and Detlef Dürr. 2013. The Ontology of Bohmian Mechanics. *British Journal for the Philosophy of Science* 0 (2913), 1–24. doi: 10.1093/bjps/axt019.

Feynman, Richard P., Robert B. Leighton, and Matthew Sands. 1965. *The Feynman Lectures on Physics.* Vol. 3. Reading, MA: Addison-Wesley.

Flack, Robert, and Basil J. Hiley. 2014. Weak Measurement and Its Experimental Realisation. *Journal of Physics: Conference Series* 504: 012016.

French, Steven. 2011. Identity and Individuality in Quantum Theory. In *Stanford Encyclopedia of Philosophy*, Summer 2011 ed., ed. Edward N. Zalta. http://plato.stanford.edu/archives/sum2011/entries/qt-idind/.

French, Steven, and Décio Krause. 2006. *Identity in Physics: A Historical, Philosophical, and Formal Analysis.* Oxford: Clarendon Press.

French, Steven, and Michael Redhead. 1988. Quantum Physics and the Identity of Indiscernibles. *British Journal of the Philosophy of Science* 39: 233–246.

Goldstein, Sheldon. 2013. Bohmian Mechanics. In *Stanford Encyclopedia of Philosophy*, Spring 2013 ed., ed. Edward N. Zalta. http://plato.stanford.edu/archives/spr2013/entries/qm-bohm/.

Goldstein, Sheldon, and Nino Zanghi. 2013. Reality and the Role of the Wave Function in Quantum Theory. In *The Wave Function: Essays on the Metaphysics of Quantum Mechanics*, ed. Alyssa Ney and David Z. Albert, 91–109. New York: Oxford University Press.

Guillemin, V. W., and S. Sternberg. 1984. *Symplectic Techniques in Physics.* Cambridge: Cambridge University Press.

Hiley, Basil J. 2011. Process, Distinction, Groupoids and Clifford Algebras: An Alternative View of the Quantum Formalism. In *New Structures for Physics*, ed. Bob Coecke, 705–750. Berlin: Springer.

Hiley, Basil J., and Ali H. Aziz Mufti. 1995. The Ontological Interpretation of Quantum Field Theory Applied in a Cosmological Context. In *Fundamental Problems in Quantum Physics*, ed. Miguel Ferrero and Alwyn van der Merwe, 141–156. Dordrecht: Kluwer.

Jammer, Max. 1998. David Bohm and His Work: On the Occasion of His Seventieth Birthday. *Foundations of Physics* 18: 691–699.

Ladyman, James, and Don Ross. 2007. *Every Thing Must Go: Metaphysics Naturalized.* Oxford: Oxford University Press.

Maudlin, Tim. 1998. Part and Whole in Quantum Mechanics. In *Interpreting Bodies: Classical and Quantum Objects in Modern Physics*, ed. Elena Castellani, 46–60. Princeton, NJ: Princeton University Press.

Ney, Alyssa and David Z. Albert, eds. 2013. *The Wave Function: Essays on the Metaphysics of Quantum Mechanics.* New York: Oxford University Press.

Pättiniemi, Ilkka. 2011. Structural Realism and the Implicate Order. A paper presented in the Swedish Herbal Institute Research and Development symposium Aspects of David Bohm's Work in Science and Philosophy, Åskloster, Sweden, 7–9 July 2011.

Philippidis, Chris, David Bohm, and Robert D. Kaye. 1982. The Aharonov-Bohm Effect and the Quantum Potential. *Nuovo Cimento* 71B: 75–88.
Plotnitsky, Arkady. 2010. *Epistemology and Probability: Bohr, Heisenberg, Schrödinger and the Nature of Quantum-Theoretical Thinking*. New York: Springer.
Putnam, Hilary. 1965. A Philosopher Looks at Quantum Mechanics. In *Beyond the Edge of Certainty: Essays in Contemporary Science and Philosophy*, ed. Robert G. Colodny, 17–31. Englewood Cliffs, NJ: Prentice Hall.
Pylkkänen, Paavo. 1993. On the Philosophical Implications of Bohm and Hiley's Interpretation of Quantum Theory. A poster presentation at the Symposium on the Foundations of Modern Physics 1993, University of Cologne, Germany, 3 June 1993.
Pylkkänen, Paavo. 2007. *Mind, Matter and the Implicate Order*. New York: Springer Frontiers Collection.
Pylkkänen, Paavo. 2012. Cognition, the Implicate Order and Rainforest Realism. *Futura* [journal of the Finnish Society for Future Studies] 31: 74–83.
Saunders, Simon. 2003a. Structural Realism Again. *Synthese* 136: 127–133.
Saunders, Simon. 2003b. Critical Notice: "The Conceptual Development of 20th Century Field Theories," by Tian Yu Cao. *Synthese* 136: 79–105.
Saunders, Simon. 2006. Are Quantum Particles Objects? *Analysis* 66: 52–63.
Schrödinger, Erwin. 1926. Quantisierung als Eigenwertproblem I and II. *Annalen der Physik* 79: 361–376 and 489–527.
Werner, Samuel A., Roberto Colella, Albert W. Overhauser, and C. F. Eagen. 1975. Observation of the Phase Shift of a Neutron Due to Precession in a Magnetic Field. *Physical Review Letters* 35: 1053–1055.
Weyl, Hermann. 1931. *The Theory of Groups and Quantum Mechanics*. London: Dover.

CHAPTER 13 | Branch-Relative Identity

CHRISTINA CONROY

13.1 Introduction

Let us be optimistic and assume that we have established that there is a meaningful way to consider existent individual entities. Once we have done so, we will need a way to talk about not only their synchronic identity, but their diachronic identity. In this chapter, I will focus my discussion on the criteria of reidentifying persons over time. What I am most interested in considering are the criteria that are required when the question "With whom will I be identical in the future?" is asked in the context of a metaphysical picture that has some sort of branching structure. That such a structure is a reasonable context in which to ask such a question is justified by taking seriously the metaphysical implications of Everettian quantum mechanics (EQM).

As will be seen, EQM suggests that there is some branching structure to the world. In this context, familiar problems of diachronic identity arise. The problem with which I am concerned here is a Ship of Theseus–type problem, and I will argue that an answer to the question "With whom will I be identical postbranching?" can be found in analogy with a solution proposed by Derek Parfit in (1971). In keeping with Parfit's own solution, I also propose that we change the relation that we suppose stands between a person and her successor(s). Parfit suggests that one use either the relation of psychological continuity or psychological connectedness in place of the standard equivalence relation of identity. I, however, propose that we use a notion of what I am calling "branch-relative identity" to reidentify persons over time in a branching universe. Before turning to a discussion of this relation, let us first look at the basics of EQM.

13.2 Everettian Quantum Mechanics

Quantum mechanics is taken to be the best physical description of the fundamental level of our world. But depending on the interpretation of quantum mechanics one chooses, the metaphysical picture of the fundamental world that is implied by the mathematics differs wildly. To set the stage for the metaphysical picture I am proposing, let us consider very briefly how the Everettian interpretation of quantum mechanics differs from a standard collapse formulation of quantum mechanics.

The standard interpretation of quantum mechanics says that there are two ways of describing the evolution of a quantum system over time: (1) when the value of some property of a system is measured, or otherwise interacts with something outside the system, that property instantaneously and randomly takes one of its possible determinate values; (2) when no measurement is made of a system, the property remains indeterminately valued. But there is a problem: if we take measurement devices to be physical systems like any other, then the standard interpretation says that the quantum system that makes up the measuring device will evolve deterministically, but the second law says that it will take on a definite value. This is not logically possible. This, in short, is the quantum measurement problem.[1]

In 1957, Hugh Everett III wanted to apply quantum mechanical principles to the universe taken as a closed system. The problem that he saw with this is that in such a model, there is nothing external to the system that can cause the collapse; so the state of the universe, taken as a whole, will always and continually evolve in a deterministic way (Everett 1957, 1973). To solve this problem, Everett proposed that one drop the stochastic law. Once one does so, the measurement problem also vanishes. What results is a mathematical description of the world in which everything is always in an ever increasingly complex superposition with everything with which it interacts. Thus, every time a measurement is made, every time another particle becomes entangled with a system, there is some sort of branching event. In other words, branching is happening all the time, throughout the universe, with very short time intervals between events.

To try to put a more colloquial spin on the idea of branching, consider the following example. Say that we have a particle that can take any of several different values for some property. (To make the discussion as accessible as possible, let's call this property "color.") If the particle is in a superposition of having many different colors, then we can represent its state this way, where $|red>_p$ means the particle is red:

$$\psi_p = n_1 |red>_p + n_2 |orange>_p + n_3 |yellow>_p + n_4 |green>_p + n_5 |blue>_p + n_6 |indigo>_p + n_7 |violet>_p.$$

Now if an observer, let us call him "$Hugh_0$," interacts with the particle, then according to EQM, Hugh becomes entangled with it and enters into a superposition of seeing all the colors. Thus the state of the system of particle + Hugh (p + H), can be represented in the following way, where $|\text{"red"}>_H$ means that Hugh observed the particle as being red. Please note that in the equation that follows (and all others) I am suppressing the terms in which Hugh gets an incorrect measurement record, that is, one in which he gets "violet" even though the particle is red. This idealization does not affect the points made in the current argument.

$$\psi_{p+H} = n_1 |\text{red}>_p |\text{"red"}>_H + n_2 |\text{orange}>_p |\text{"orange"}>_H + \ldots \\ + n_7 |\text{violet}>_p |\text{"violet"}>_H$$

For ease of exposition, let us refer to the branch on which one of Hugh's potential successors sees some color by the successor's name. For example, let us call the branch $n_n |\text{color}>_p |\text{"color"}>_H$ by the name "$Hugh_n$." When we consider a simplistic picture of what the evolution of the system p+H might look like over time (figure 13.1), it resembles a tree with a single trunk, where we find $Hugh_0$ before the measurement, from which emanate multiple branches,[2] on each of which we find one $Hugh_n$, each a (potential) Hugh after the measurement.

It seems reasonable to ask the question, "With whom will $Hugh_0$ be identical postbranching?"[3] But if we take seriously Everett's claim that each of these branches is equally "actual," none more "real" than the rest (1957, 146, note added in proof), it would seem that $Hugh_0$ has multiple diachronic successors. The problem with this is that if $Hugh_0$ is identical (in the standard sense) to $Hugh_1$ and if he is also identical to $Hugh_2$, then by the transitivity of identity $Hugh_1$ must be identical to $Hugh_2$, but that is not generally the case. Thus, we seem to have a problem if we want to

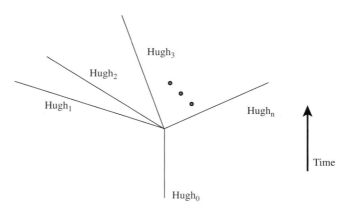

FIGURE 13.1 A simplistic picture of the evolution of the system p+H over time

use the standard equivalence relation of identity in the context of branching evolution.

That standard identity will not always work as a criterion of diachronic identity in a context in which there is branching (or "fission," as it is often referred to) is not a new problem. Those who have worked on the problem in recent decades have not generally been working with EQM in mind.[4] However, the problems with which they grapple should be familiar to those who are working in the context of EQM. And indeed, as will be seen below, there are several philosophers who have taken the lessons found in more garden-variety metaphysical pictures and applied them to questions about personal identity in the context of EQM. In this chapter I propose to do much the same. I will consider a promising analogy between what Parfit (1971) proposes about personal identity in the case of fission and the issues with personal identity in the case of branching as seen in some versions of EQM. Considering the problem of diachronic identity in analogy with Parfit is also not new (see, e.g., Saunders and Wallace 2008a, 2008b); however, as will be seen, what I hope to contribute to the conversation is another possible way one can follow the path down which the analogy leads.

13.3 Parfit's Argument

In his paper "Personal Identity," Derek Parfit argues that what matters in diachronic survival is not necessarily described by a one-one relation; hence, not necessarily identity (1971). He uses David Wiggins's example of a person whose brain is halved in such a way that a majority of normal functionality is maintained for each half (Wiggins 1967, 52–53). Each half is then housed in a different body, resulting in Brown I and Brown II, whose memory reports and intelligences are identical. Wiggins then goes on to consider the criteria of personal identity. He knows that if we say that each of Brown I and Brown II is identical to the original Brown we run into the familiar difficulties due to the transitivity of identity.

Parfit expands on this and proposes a metaphysical picture that should look remarkably familiar to anyone who has considered the metaphysical implications of EQM. After each branching event, Parfit takes there to be three possibilities for "my" survival:

(1) I do not survive.
(2) I survive as one of the two people.
(3) I survive as both.

Parfit argues that (1) is false since in basic brain-transplant thought experiments philosophers generally agree that a person can survive the transplantation of her entire brain, and medicine tells us that people can survive with just half a brain (Carabello et al. 2011; Cukiert et al. 2009); thus, for the purposes at least of a thought experiment, one seems able to survive the transplantation of half of one's brain (Parfit 1971, 5 and 9–10).[5] Parfit also considers possibility (2) to be highly implausible. Since each of the halves of my brain is exactly similar, neither of the resultants seems to have a better claim to being "me" after the experiment than does the other. So it is unclear how I could say that I survive as one, but not the other. All that remains as a plausible outcome is (3).

Parfit deals with (3) in two ways: he considers what happens if "survival" implies identity and if it does not. If "survival" implies identity, then to hold (3) one must claim either (*a*) that there is only one person with a mind divided between two bodies, or (*b*) that the two resultant people compose a third, in the same way that one pope's crown is composed of three crowns (Parfit 1971, 7–8; Wiggins 1967, 40). Parfit does not find the first absurd, merely unsatisfactory (1971, 5–7). The second requires that one retain the language of identity but change what one means by "person".

One might argue that this latter route is exactly that taken by most of those who have since considered the question of diachronic identity. Independent of EQM, David Lewis and Theodore Sider have suggested that we understand persons as "space-time worms" or as "person stages" such that people extend not only in space, but also through time (Lewis 1986; Sider 1996). In the context of EQM, these are also the two routes taken by those who have considered diachronic identity.[6]

Since various varieties of perdurantism have been used to try to solve the diachronic identity problem, let us instead consider whether we can retain a three-dimensional or endurantist notion of persons and still solve the problems with diachronic identity. This is the route taken by Parfit. He suggests that we should give up the language of identity (8). But if we do that, it seems to also suggest that "survival" does not imply identity. One might then raise the objection that possibility (3) is not apropos the problem of identity. Parfit rebuts this objection by arguing that what matters in questions such as "Do I survive over time?" is a relation between one's present self and one's future self that need not be transitive. He proposes that the relation we should consider in questions of survival is psychological connectedness.

Psychological connectedness can be defined as the relation that holds between two persons, separated in time, such that the later person is in the mental state that he is largely due to the fact that the earlier person is in the mental state that he is (Olson 2013). This relation comes in degrees: one can be more or less psychologically connected to a successor of oneself. For example, we might say that $Hugh_0$ is connected to a high degree to $Hugh_{100}$ but not as highly connected to $Hugh_{1,000}$ who would appear on a much later branch in the same tree. This might be illustrated with the following example: let us say that $Hugh_0$ had a bad experience with blueberry pie as a child. When we consider his successors, we find that $Hugh_{100}$, a successor of his five years later, does not like to eat blueberry pie, but $Hugh_{1,000}$, a successor of $Hugh_0$'s 20 years in the future, has largely outgrown that aversion to blueberry pie and now can enjoy it. Thus, we can say that $Hugh_{100}$ is psychologically connected to a large degree to $Hugh_0$ since his averse mental state is clearly directly related to $Hugh_0$'s mental state with regard to blueberry pie, but $Hugh_{1,000}$ is psychologically connected to only a small degree to $Hugh_0$ because $Hugh_{1,000}$'s mental state of relative enjoyment of blueberry pie is not as directly related to $Hugh_0$'s mental state of aversion.

Psychological connectedness is also not transitive. We can understand this in two ways. First, if we take the degree of psychological connectedness to potentially decrease the further one goes along one's descendent tree, then there will potentially come a point at which the degree of connectedness is zero. From the example above we can say that this happens at $Hugh_{10,000}$—the point at which $Hugh_0$'s aversion to blueberry pie has been completely forgotten. Then, we can say that $Hugh_0$ is psychologically connected to $Hugh_{1,000}$ and that $Hugh_{1,000}$ is psychologically connected to $Hugh_{10,000}$ but that $Hugh_0$ is not psychologically connected to $Hugh_{10,000}$.

A second way we can understand the relation to not be transitive is if we take the degree of connectedness to be an additional relata in the relation, turning psychological connectedness into a ternary rather than binary relation. And it is a familiar requirement of the property of transitivity that it apply only to binary relations.

Psychological continuity is another important relation between successors. According to Parfit, it is defined in terms of psychological connectedness: psychological continuity is a relation between two persons such that the mental state of one is related to that of the other just in case some of the mental states of one are psychologically connected, or related through a chain of psychological connections, to the mental states of the other (Olson 2013). Since, according to Parfit, judgments of identity supervene on psychological continuity, and since psychological continuity is defined via psychological connectedness, and since psychological connectedness comes in degrees and is not transitive, we can see that whatever relation it is that holds between a person and her multiple successors cannot be standard identity (1971).

Thus, Parfit argues, rather than asking with whom one will be identical, one can now ask about one's "descendants" (those who appear later in the tree) and "ancestors" (those who appeared earlier). Psychological continuity will be implied when one considers some individual anywhere on the tree as a descendant or ancestral self—this potentially gives us the kinds of transitivity claims that we want to be able to make. The nontransitive psychological connectedness is implicated when one considers "one of my future selves" or "one of my past selves."

Parfit's use of psychological connectedness and continuity seem to solve the problem that we are after. But it is my contention that while they may solve one problem, they do so at the expense of an aspect of diachronic identity that we want to retain. There are two things that one wants out of their theory of diachronic identity in the context we are considering: (1) One wants to be able to track one's self over time while (2) not running into problematic transitivity claims like those we are considering in this chapter. We have just rehearsed the way in which these relations will solve the problem that we are considering in this chapter—thereby satisfying (2). But if we take the claims about the two relations being intransitive seriously, then we lose the ability to satisfy (1).

To see this, consider again the discussion just a couple paragraphs back. It is easy to see that if we take psychological connectedness to be intransitive because it is a ternary relation, then we cannot claim that if $Hugh_0$ is psychologically connected to $Hugh_1$ to degree x and $Hugh_1$ is psychologically connected to $Hugh_2$ to degree y, then $Hugh_0$ is psychologically connected to $Hugh_2$ to any degree, even if they are on the same branch. In order to say that, a great deal more subtle work needs to be done to explain the degrees to which a person is psychologically connected to a successor and how that can be used to avoid the problematic transitivity claims while still retaining the desired ones.

If we consider that psychological connectedness is not transitive because there will come a point at which the degree to which one is related to a successor is eventually zero, then again, a great deal more work on the subtleties of how one's psychological connectedness drops off needs to be done. A measure for psychological connectedness needs to be described, and how it does not serve as a third argument place in the relation needs to be explained. Of course, these two concerns do not even begin to touch the rest of the objections that can be raised to Parfit's suggestion for using psychological connectedness and continuity to replace standard identity. But to go into all of these would take us well beyond the scope of this chapter. (For a good overview of the various arguments, see Olson 2013.)

It is my contention that a relation that I am calling "branch-relative identity" can be described that can satisfy both (1) and (2). In order to understand the justification for relativizing identity, it would be helpful to consider the interpretation of EQM that I take to be the most charitable, faithful, and conservative: the Relative Facts Interpretation (RFI) (Conroy 2010, 2012). However, regardless of the interpretation of EQM you favor, if that interpretation implies some sort of branching structure to the world (universe), then my proposal for relativizing identity will apply.

13.4 The Relative Facts Interpretation of EQM

The RFI differs from other interpretations in that it does not add any physics or mathematics to Everett's formulation, but merely asks, and attempts to answer, the question "If EQM is true, what does that mean for the metaphysical structure of the world?" It has a straightforward proposal: there is but one world and this world is populated by two types of facts: absolute and relative. Any reader familiar with the literature on the many-worlds interpretations (MWIs) will recognize similarities between what I am proposing and a proposal developed by Simon Saunders in the mid-1990s that I will call "E-relativism." These similarities are not accidental, but there are clear differences. One of these is the way the RFI and E-relativism treat the claims made by Everett in the note added in proof (1957). Though there is much to say about this note and the various interpretations to which it has given rise, suffice it here to say that the main difference between the RFI and E-relativism is the number of worlds it proposes.[7]

Saunders's E-relativism has been variously characterized as belonging to the many-worlds tradition (Laudisa and Rovelli 2008, Vaidman 2014) and the single-world tradition (Barrett 1999). Despite the fact that there are numerous places where Saunders explicitly differentiates E-relativism from a many-worlds interpretation,[8] Saunders has indicated that he considers E-relativism to be read in the context of many worlds.[9] I believe that the confusion regarding how to characterize Saunders's views comes from failing to note the distinction between what one might call a naive MWI, like that developed by Bryce DeWitt (1968), and a more modern MWI that considers decoherence phenomena, what one might call the "Oxford interpretation" (Wallace 2010, 2012). Saunders takes himself to be, and to have been, working in the latter tradition. This is the primary point at which E-relativism and the RFI part ways. While E-relativism proposes multiple real worlds, the RFI takes it that there is just one. To argue for the saliency of this view would take us far afield of the scope of this chapter. (That argument is being more fully fleshed out in Conroy 2015.) But to reiterate an earlier point, the proposal made in this chapter would suffice for any metaphysical context in which the world branches in some way. If one takes the proper interpretation of EQM to be some form of

MWI, then what follows trades heavily on the earlier work of Saunders while more fully developing the notion of branch-relative identity.

13.4.1 Absolute and Relative Facts

When a system is in an eigenstate of an observable, there is some absolute fact about the value of that observable for the system. Consider the example of a system that consists of two particles, p and q, in the singlet state—the state in which one finds the electrons in the lowest orbital of a helium atom:

$$|SSS\rangle = 1/\sqrt{2}\left(|\uparrow_x\rangle_p|\downarrow_x\rangle_q + |\downarrow_x\rangle_p|\uparrow_x\rangle_q\right).$$

In this equation, take "$|\uparrow_x\rangle_p$" to mean that particle p is x-spin up, "$|\downarrow_x\rangle_q$" to mean particle q is x-spin down and vice versa for the other terms. This equation describes two particles that are related in such a way that whenever one of them is x-spin up, the other is x-spin down.

If we consider the composite system consisting of p and q as isolated, then it is in an eigenstate of some observable. Let us call that observable *components with opposite x-spin*. If we maintain the eigenvalue-eigenstate link—and we do—then this gives us the fact that the system determinately has the absolute property *components with opposite x-spin*. So if we ask of the system whether it has the absolute property *components with opposite x-spin*, we will always get the answer yes.

However, if we do not want to consider the properties of the composite system taken as a whole, but instead want to consider the properties of the component parts, p and q (as we most often do), we run into difficulties. When a system is in a nonseparable entangled superposition like the singlet state, it is no longer possible to consider one element of the superposition without also considering the rest of the elements. We cannot then say that there are any absolute facts about the components of the system taken individually. Rather, there are only relative facts.

When p and q are in the singlet state, neither particle has any determinate absolute spin property independent of the spin property of the other. However, each does have determinate *relative* spin properties: p has the determinate *relative* property "x-spin up relative to q being x-spin down" and the determinate *relative* property "x-spin down relative to q being x-spin up," and vice versa for q. These relative facts do not supervene on any nonrelational properties of their relata, but instead supervene on some structural feature of the system such as a relation that obtains between the components of the system. We uniquely assign the relative state of one subsystem by choosing as a target of relativization the complement of it. We use the phrase *relative to* to indicate the chosen target.[10]

When we consider the ubiquity of entanglement for even the best-isolated systems, we will see that most properties that can be ascribed to systems are going to be relative facts. What I want to argue is that it is natural to extend this relativization to the property of diachronic identity. Once we have done this, we will see that we can avoid the problems associated with diachronic identity in a branching metaphysics.

13.5 Hugh and the Hugh$_n$s

Consider Hugh again: whenever he makes a measurement, the world branches. If we take there to be one observer, Hugh, before a measurement, then in analogy with Parfit's discussion we want to know which of the following are true after the branching:

(4) Hugh does not survive.
(5) Hugh survives as one of the many potential Hugh$_n$s.
(6) Hugh survives as all of the potential Hugh$_n$s.

Since Hugh is able to report to us the result of the measurement that he made, it seems that (4) is quite unlikely.

As Parfit does in (1989), one might want to argue that (4) is correct and that Hugh fails to survive. If this is the case then the problem with which we are currently concerned seems less worrisome. It would not be the case that Hugh is identical to any of his successors; so the problematic transitivity claim would not likely be made. Rather, we would make a claim about the psychological connectedness or about counterpart relations between Hugh and his "successors" (or perhaps some other relation between Hugh and his "successors"). Whether or not these claims can be made sense of or should be preferred is an adjudication that we will not make here. However, what I will say is that it seems at least reasonable to conclude that since someone whom I would pretheoretically take to be Hugh is giving me a report of his measurement result, Hugh has survived the experiment. Therefore, it seems reasonable to draw an analogy between Parfit's (1971) treatment of the question of diachronic identity and how one might try to do so in the context of EQM.

If (5) is taken to be a reasonable outcome, then postmeasurement one observer remains and the other potential observers do not actually exist. Then, that one observer "merely" has the difficulty of locating himself on one of the resultant branches. But to argue this seems to imply that the observer is not a part of the deterministically evolving system. This is not in keeping with the understanding of pure wave mechanics that I am here using—namely that an observer is a nonseparably entangled part of the system that he is measuring.[11] Thus, analogous to Parfit's discussion, (4) and (5) both are implausible outcomes.

Let us now consider (6). After a measurement, we can say that there are a number of resulting (potential) observers, $Hugh_1$, $Hugh_2$, ..., $Hugh_n$ who all bear some relation to $Hugh_0$. They will presumably all share the same memories,[12] have the same intelligence, and have the same physical attributes, at least for a short time. But we do not want to claim that the resulting observers hold to $Hugh_0$ the equivalence relation of identity since we will encounter familiar problems with transitivity. Even in the smallest time frame, postmeasurement, there will be some difference in the properties of $Hugh_m$ and $Hugh_n$; otherwise there would not be two of them. The difference could, of course, be in aspects of the world that $Hugh_m$ and $Hugh_n$ do not currently have epistemic access to, but then given that they are on two distinct branches, there will be, in principle at least, some way to differentiate them.

If we claim that each $Hugh_m$ is identical to $Hugh_0$, then we also have to claim that each $Hugh_m$ is identical to each $Hugh_n$ (where $m \neq n$). If we do not want to make the latter claim (and in general we do not), then we cannot make the former while still maintaining the transitivity of identity. So in the context of a metaphysical picture supported by pure wave mechanics, we need to be able to describe the relation between $Hugh_0$ and the $Hugh_n$s with some relation other than identity, but one that still captures the close connection between $Hugh_0$ and the various $Hugh_n$s that result after a measurement. What I propose is that rather than claiming that $Hugh_0$ is identical with any of the $Hugh_n$s, we instead say that he is identical relative to X to any of the $Hugh_n$s, where "X" picks out a suitable target of relativization.

13.6 Branch-Relative Identity

I propose to call this class of relations "branch-relative identity" relations in order to help separate them from the account of relative identity proposed by Geach (1962, 1967). That account suggests that two things are not identical simpliciter; rather they are only identical relative to some sortal. Thus with Geach-relativism we cannot say, "x is identical to y"; rather we have to assert, "x is an identical A to y," where A is some sortal term. I agree that the first phrase is incomplete, but in the context of EQM, I hold that the appropriate completion is "x is identical to y relative to A," where A is a target of relativization like the ones we use to define other relative facts; that is, it is the complement of the system that consists of y.

Consider again the system that consists of Hugh and a particle that can take on several different "color" properties:

$$\psi_{p+H} = n_1 |\text{red}\rangle_p |\text{"red"}\rangle_H + n_2 |\text{orange}\rangle_p |\text{"orange"}\rangle_H + \ldots + n_7 |\text{violet}\rangle_p |\text{"violet"}\rangle_H$$

Just as the RFI defines the relative properties of a subsystem relative to the state of its complement, so too we will define the branch-relative diachronic identity of $Hugh_0$ and his successors relative to the complement of the subsystem that consists of the successor of $Hugh_0$ that we are referencing. So we will say that $Hugh_0$ is identical to $Hugh_1$ relative to $|\text{red}\rangle_p$, and that $Hugh_0$ is identical to $Hugh_2$ relative to $|\text{orange}\rangle_p$, and that $Hugh_0$ is identical to $Hugh_7$ relative to $|\text{violet}\rangle_p$, and so on.

To work in place of standard identity, branch-relative identity needs to retain as many of the same properties of standard identity as possible, without falling prey to the difficulties with which we are concerned. And, as we will see, it does. Let us first look at reflexivity and symmetry before turning to the more subtle discussion of transitivity.

First, to see that branch-relative identity is reflexive consider that the claim "$Hugh_1$ is identical to $Hugh_1$ relative to X" will be true regardless of the target of relativization or the $Hugh_n$ we choose to consider. This claim is really just one of synchronic identity, and synchronic identity on a branch is no more problematic than in a nonbranching context.

To see that branch-relative identity is symmetric requires a bit more work. In order to recover this property, we must first fix the target of relativization. This target fixing is not only necessary, but also desired in this context because the symmetry claims in which we are interested are those along a particular branch. Once we have fixed our target of relativization, we can claim that $Hugh_m$ and $Hugh_n$ are identical relative to that target. By specifying, and fixing, the target of relativization, we carve up the state of the system such that only one branch is of interest to us. We then make our identity claim about $Hugh_m$ and $Hugh_n$, both of whom are on the branch in question. This also is no more problematic than similar claims in a nonbranching context.

To see that branch-relative identity is intransitive, we cannot fix our target of relativization. This is because we need the target to function as a third argument place in the relation in order to even generate true claims. While we say that $Hugh_0$ is identical to $Hugh_1$ relative to $|\text{red}\rangle_p$, we say that $Hugh_0$ is identical to $Hugh_2$ relative to $|\text{orange}\rangle_p$. In order to make the problematic identity claim with which we started this chapter, we would have to say that not only is $Hugh_0$ identical to $Hugh_1$ relative to $|\text{red}\rangle_p$ but also that $Hugh_0$ is identical to $Hugh_2$ relative to $|\text{red}\rangle_p$. Aside from the fact that the latter is not true and so does not generate the problematic transitivity claim, it is also the case that when we allow the target of relativization to vary and become one of the relata, branch-relative identity relations

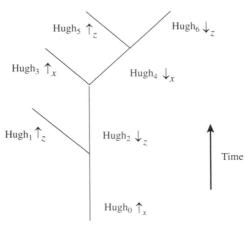

FIGURE 13.2 The evolution of the system when $Hugh_0$ measures the z-spin of a particle in an eigenstate of x-spin up

of this type are ternary rather than binary, and so no longer candidates for transitivity.

As we saw earlier in our critique of Parfit's use of psychological connectedness, there are other kinds of transitivity claims that we would like to make in the context of diachronic identity. Consider a picture like the one shown in figure 13.2.

This tree can be understood as representing what happens when $Hugh_0$ measures the z-spin of a particle in an eigenstate of x-spin up. This measurement results in $Hugh_1$ who gets "z-spin up" and $Hugh_2$ who gets "z-spin down." Then $Hugh_2$ makes an x-spin measurement of the same particle, resulting in $Hugh_3$, who gets "x-spin up," and $Hugh_4$, who gets "x-spin down." Finally $Hugh_4$ measures the z-spin, again resulting in $Hugh_5$, who gets "z-spin up," and $Hugh_6$, who gets "z-spin down."

We would like to claim that if $Hugh_2$ is branch-relative identical to $Hugh_4$ and if $Hugh_4$ is branch-relative identical $Hugh_6$ then $Hugh_2$ is branch-relative identical to $Hugh_6$. This kind of claim is necessary for a thoroughgoing theory of diachronic personal identity. It is also necessary to be able to successfully do empirical science and to have scientific knowledge. If $Hugh_6$ believes that his measurement of the z-spin of the particle resulted in the same value that he got for his first z-spin measurement, then we want him to be able to be as sure as possible that he is remembering correctly. If $Hugh_6$ cannot be sure that he was $Hugh_2$, rather than $Hugh_1$, who got "z-spin up," then he cannot be confident in his or anyone's reliability as an observer. We want it to be the case that when an observer reports a series of results, those results actually obtained for him.

One way of going about this is to make a type of transcendental argument that it must be the case that if $Hugh_2$ is branch-relative identical to $Hugh_4$ and if $Hugh_4$ is branch-relative identical $Hugh_6$, then necessarily $Hugh_6$ is branch-relative identical to $Hugh_2$; otherwise we will be incapable of doing empirical science. This can be done parallel to the route Barrett takes (1999, 198–201) when he is discussing how one might account for reliable memory records in EQM. Another way we can do this is to be careful about what our branch-relative identity claims actually entail in terms of the limitations one must set on their application. It is not true to claim that $Hugh_0$ is branch-relative identical to just any $Hugh_n$ we could imagine. For example, let's take the following to be the state of $Hugh_0$ and a particle, where $|Rx>_H$ represents the state of $Hugh_0$ being ready to make a measurement of the x-spin of a z-spin particle:

$$\psi_{p+H} = |Rx>_H \, 1/\sqrt{2}(|\uparrow_x>_p + |\downarrow_x>_p).$$

Assuming that our measuring device is accurate (as mentioned before, a harmless idealization for our present purposes), and assuming that $Hugh_0$ is a reliable and truthful observer (another harmless assumption), then there are certain limitations on which branch-relative identity claims we can make. By "reliable and truthful" I mean that when we ask an observer whether he has a measurement record, if he has one then he will answer that he does; if he does not have one, he will answer that he does not. He will also always truthfully report what he has as a measurement record—if he recorded x-spin down he will answer "x-spin down" when asked what he recorded.

If the following is the state of $Hugh_0$ and a particle after the measurement, where $|\text{``}\uparrow_x\text{''}>_H$ represents the state of $Hugh_n$ measuring "x-spin up" for the particle

$$1/\sqrt{2}(|\uparrow_x>_p|\text{``}\uparrow_x\text{''}>_{Hn} + |\downarrow_x>_p|\text{``}\downarrow_x\text{''}>_{Hm}),$$

then we would not be able to truly say of $Hugh_0$ that he was branch-relative identical to some successor who saw "y-spin up" when he read off the results of his x-spin measurement. The only truthful branch-relative claims that we can make are those related to the $Hugh_n$s who are represented in the expansion of the state of $Hugh_0$'s system postmeasurement.

A convention that will help here is to give names to the individual branches. When we represent the state of a system postmeasurement, we can give each summand a name. Thus in the above state, we can give the name "X_1" to the summand $|\uparrow_x>_p|\text{``}\uparrow_x\text{''}>_{Hn}$ and the name "$X_2$" to the summand $|\downarrow_x>_p|\text{``}\downarrow_x\text{''}>_{Hm}$. Once we do this, we can then rephrase the above limitation to indicate that we can only say that $Hugh_0$ is identical to $Hugh_n$

relative to X provided that X is the name of a branch that is a summand in the expansion of the postmeasurement state of the system of $Hugh_0$. So we can make the claims that $Hugh_0$ is identical to $Hugh_n$ relative to X_1 and that $Hugh_0$ is identical to $Hugh_m$ relative to X_2 because X_1 and X_2 are the names of branches in $Hugh_0$'s postmeasurement state.

Now, since quantum states can be written in any basis one chooses, we can represent the postmeasurement state of $Hugh_0$'s system in any basis we like. This will result in there being a different set and number of branches that go to make up the state of the system. If we took the above postmeasurement state and rewrote it in the z-spin basis, it would appear this way:

$$1/\sqrt{2}(1/\sqrt{2}(|\uparrow_z>_p + |\downarrow_z>_p)|``\uparrow_x"\!>_{Hn} + 1/\sqrt{2}(|\uparrow_z>_p - |\downarrow_z>_p)|``\downarrow_x"\!>_{Hm}).$$

With this representation, once we distribute all the factors, we have four summands to name. Let's call them "Z_1," "Z_2," "Z_3," and "Z_4." Note that in this case the branch we previously named "X_1" can be also named "$Z_1 \wedge Z_2$."

Now we have what we need to make sense of the truthfulness of the claim that if $Hugh_0$ is branch-relative identical to $Hugh_n$ and if $Hugh_n$ is branch-relative identical to $Hugh_k$ then $Hugh_0$ is branch-relative identical to $Hugh_k$. Consider again a picture in which we have added branch names (figure 13.3).

We can now say that $Hugh_2$ is identical to $Hugh_4$ relative to X_4 and that $Hugh_4$ is identical to $Hugh_6$ relative to X_6. But all we can say about $Hugh_2$ relative to $Hugh_6$ is that they are identical relative to something like $X_4 \wedge X_6$. But if we take the target of relativization to be an argument place,

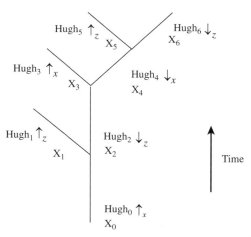

FIGURE 13.3 The evolution of the system when $Hugh_0$ measures the z-spin of a particle in an eigenstate of x-spin up, including branch names

then we seemingly lose transitivity; the loss of which you will recall was an earlier advantage of this view. However, we can reclaim the advantage, and the desired transitivity, by redefining transitivity in a branching context. We do so by making clear what was an implicit requirement in the application of transitivity, but which the following definition will make explicit:

Branch-Relative Transitivity
$\forall x, y, z, m, n[[(x =_{relm} y) \& (y =_{reln} z)] \rightarrow (x =_{relm+n} z)]$
provided that $m = \Sigma i + n$ where the is are also branch names.

What this is saying in (somewhat) plainer terms is this: take the two branches that make up the target of relativization in the consequent. Transitivity holds provided that one of these branches is one of the summands of the postmeasurement state of the other branch. This will allow us to claim truthfully that if $Hugh_2$ is identical to $Hugh_4$ relative to X_4 and $Hugh_4$ is identical to $Hugh_6$ relative to X_6 then $Hugh_2$ is identical to $Hugh_6$ relative to $X_4 \wedge X_6$ provided that X_6 is a name of one of the summands of the expansion of the postbranching state of a system on branch X_4 or vice versa. And when we look at the tree we see that this is exactly the case.

Branch-relative transitivity does not result in the problematic transitivity claims that we were concerned about at the beginning of this chapter. The following claim is false: If $Hugh_0$ is identical to $Hugh_1$ relative to X_1 and $Hugh_0$ is identical to $Hugh_2$ relative to X_2, then $Hugh_1$ is identical to $Hugh_2$ relative to $X_1 \wedge X_2$. If X_2 is not the name of one of the summands of the expansion of the postbranching state of a system on branch X_1, and X_1 is not the name of one of the summands of the expansion of the postbranching state of a system on branch X_2, then with the limitation on branch-relative transitivity, even with a true antecedent, we are not required to assert the truth of the consequent. The imposed limitation on transitivity claims is not unprecedented. Those working in more garden-variety metaphysics that do not include any branching structure often impose limitations on such relations. They impose restrictions on the truth of diachronic transitivity claims that are based on things like causal connections, psychological continuity, psychological connectedness, somatic continuity, and the like. The restriction proposed for branch-relative transitivity is by comparison relatively well-defined and easy to apply.

13.7 Conclusion

What we have done then is to provide a way in which philosophers working in the context of a branching metaphysics can discuss diachronic identity when there is a potential surfeit of successors. In analogy with

Parfit's attempt to retain a common-sense notion of person and replace the standard equivalence relation of identity with some relation that admits of degrees and is not transitive, I have here proposed that we use a relation that I am calling "branch-relative identity." This relation allows us to avoid the problematic identity claims like those seen in Ship of Theseus–type problems, but still retain the less problematic ones. By requiring that we make diachronic identity relative to a branch, we are allowing an observer to track his identity over time even when his universe evolves in a branching manner.

There is, for sure, still work to be done. Even if we can explain the metaphysics of diachronic identity in a branching context, we are left with epistemological problems. It is not immediately clear how one will explain the epistemological experience had by Hugh. He does not seem to have the experience of multiple successors. However, if we take seriously Everett's claim that "all elements of a superposition (all 'branches') are 'actual,' none any more 'real' than the rest" (1957, note added in proof), then we owe an explanation for why Hugh seems to have only one experience postmeasurement. While I have done some work to explain this (Conroy 2012), I merely provide a metaphysical picture that helps to make sense of it, not an epistemological one. Thus, such a picture is still needed.

A larger lacuna in what has been done in this chapter is the development of a clear way to make sense of probability claims in the context of an interpretation of EQM that includes branch-relative identity. David Wallace (2012) has recently employed decision theory in attempting to account for probability in the MWI that he champions. But whether or not that account will satisfactorily solve the problem in the current context is an adjudication that would require far more careful analysis than there is room for in this chapter.[13] Of course an acceptable theory of personal identity has to work in conjunction with one's interpretation of probability, and so one might argue that we cannot develop one independently of the other. However, one value to setting aside the question of probability in the current chapter is that it allows us to more carefully focus on a metaphysical picture that can be used independently of EQM, while the story one tells about probability is liable to depend heavily on the interpretation of quantum mechanics one chooses. Thus, while the theory of diachronic identity that is developed here is incomplete in that it does not clearly state how it will align with one's expectations regarding probability, it has the value of being applicable to any metaphysical picture that includes branching. One can then investigate whether it is consistent with the story of probability being told in that metaphysical context.

Additionally, work needs to be done to explain what role, if any, branch weights will play in branch-relative diachronic identity. They may turn out to be intimately linked to Hugh's rational decision-making, or they may serve an entirely different function in the overarching metaphysical and epistemological picture of the world as described by EQM—one not connected to personal identity. This question will likely be answered in the context of the account one gives of probability.

What has been done though, is to help make sense of the reidentification of an individual over time in the context of the picture of the world given by our best physical theory, interpreted as describing the world as having some type of branching structure. If we do in fact allow individuals into our ontology, this will give us a mechanism by which we can keep track of them over time.

Acknowledgments

The work on this chapter was funded by an AAUW Summer/Short-Term Research Publication Grant and a Summer Research Grant from Morehead State University. Many thanks go to Jeff Barrett, Kai Wehmeier, Paul Tappenden, Claudio Calosi, Mark Satta, Chip Hooper, Simon Saunders, and Cynthia Macdonald for comments on drafts of this chapter. Also thanks to those in attendance at the "Individuals across Sciences" conference in May 2012 and the Fifth Workshop on Philosophy of Science, Technology, Engineering and Mathematics in March 2013 for very helpful questions, comments, and conversation.

Notes

1. See Barrett 1999, 14–17 and Albert 1992, chap. 4 for excellent, accessible discussions of the measurement problem.

2. Sometimes in the literature philosophers use the term "branch" and other times they use the word "world." For a discussion of some of the philosophical differences between the two, see Butterfield 2001. Since nothing that I am here arguing depends upon those subtleties, I generally use the words interchangeably.

3. Many have argued quite persuasively that the problem of personal identity in EQM is worthy of serious consideration. For example, see Barrett 1999, Butterfield 2001, Saunders 1995, 1998; Saunders and Wallace 2008, 2008b, Vaidman 1998, 2014; Wallace 2006.

4. See, for example, Lewis 1983, Parfit 1971, 1989, Shoemaker 1984, Sider 1996.

5. It is interesting to note that Parfit comes to a different conclusion (1989). There he argues that in fact (1) is true. He still argues that personal identity is not what matters to survival (as he does in Parfit 1971), but in the later paper he is basing

his argument on a Humean conception of the "self," thus concluding that not only in the split-brain case, but at every moment, we are ceasing to exist. That (1) is true is also the route taken by others who attempt to solve the fission problem inherent in the psychological approach to personal identity. See Olson (2013) for a good overview.

6. Peter Lewis (2007) surveys the positions taken by and attributed to many-worlds theorists including himself, Hilary Greaves (2004), Simon Saunders (1998), and David Wallace (2006). The papers just listed are not meant to be exhaustive, merely representative.

7. For a more thorough and in-depth discussion of the note and the novel way the RFI interprets it (an interpretation informed by letters to and from Everett), see Conroy 2012.

8. See, for example, Saunders 1993, 1567; 1995, 253–254; 1996a, 127, 133; 1996b, 48; 1997, 45; 1998, 7–9.

9. From personal correspondence.

10. Floating in the back of some readers' minds may be the preferred basis question. I argue that the RFI does not suffer from a preferred basis problem in Conroy 2010, 2012.

11. For a discussion of the issue of self-location in the context of many-worlds interpretations see Lewis 2007, 2008, Saunders and Wallace 2008a, Vaidman 1998, Wallace 2005).

12. Or "q-memories," if we take the Hugh_ns to not be identical to Hugh_0. Parfit's definition of a q-memory is that "I am q-remembering an experience if (1) I have a belief about a past experience which seems in itself like a memory belief, (2) someone did have such an experience, and (3) my beliefs [sic] is dependent upon this in the same way (whatever that is) in which a memory of an experience is dependent upon it" (1971, 15). Parfit proposes this definition in order to avoid the objection that one can only have memories of one's own experiences, and so unless some Hugh_n is the same person as Hugh_0, he would not be able to remember the experiences of Hugh_0. We cannot assert that without begging the question.

13. Whether or not Wallace has been successful in his development of probability in the MWI is an open question. For criticism of his approach see Dizadji-Bahmani 2013 and Maudlin 2014. I want to thank an anonymous referee for this volume for calling my attention to these brand-new works of criticism.

References

Albert, David. 1992. *Quantum Mechanics and Experience*. Cambridge, MA: Harvard University Press.

Barrett, Jeff. 1999. *The Quantum Mechanics of Minds and Worlds*. New York: Oxford University Press.

Butterfield, Jeremy. 2001. Some Worlds of Quantum Theory. In *Quantum Mechanics, (Scientific Perspectives on Divine Action)*, vol. 5, ed. Robert John Russell et al., 111–140. Vatican City: Vatican Observatory Publications.

Carabello, R., M. Bartuluchi, R. Cersósimo, A. Soraru, and H. Pomata. 2011. Hemispherectomy in Pediatric Patients with Epilepsy: A Study of 45 Cases with Special Emphasis on Epileptic Syndromes. *Child's Nervous System* 27 (Dec.): 2131–2136.

Conroy, Christina. 2010. A Relative Facts Interpretation of Everettian Quantum Mechanics. Ph.D. dissertation, University of California, Irvine.

Conroy, Christina. 2012. The Relative Facts Interpretation and Everett's "note added in proof." *Studies in History and Philosophy of Modern Physics* 43: 112–120.

Conroy, Christina. 2015. Everettian Quantum Mechanics without Realism. Manuscript.

Cukiert, A., C. M. Cukiert, M. Argentoni, C. Baise-Zung, C. R. Forster, V. A. Mello, J. A. Burattini, and P. P. Mariani. 2009. Outcome after Hemispherectomy in Hemiplegic Adult Patients with Refractory Epilepsy Associated with Early Middle Cerebral Artery Infarcts. *Epilepsia* series 4, 50(6): 1381–1384.

DeWitt, Bryce S. 1968. Everett-Wheeler Interpretation of Quantum Mechanics. In *Battelle Rencontres*, ed. Cecile M. DeWitt and John A. Wheeler, 318–332. New York: Benjamin.

Dizadji-Bahmani, Foad. 2013. The Probability Problem in Everettian Quantum Mechanics Persists. *British Journal for the Philosophy of Science*. DOI 10.1093/bjps/axt035.

Everett, Hugh. [1957] 1973. "Relative State" Formulation of Quantum Mechanics. *Reviews of Modern Physics* 29: 454–462. Reprinted in *The Many Worlds Interpretation of Quantum Mechanics*, ed. Bryce S. DeWitt and Neill Graham (Princeton, NJ: Princeton University Press), 141–149.

Everett, Hugh. 1973. The Theory of the Universal Wave Function. Reprinted in *The Many Worlds Interpretation of Quantum Mechanics*, ed. Bryce S. DeWitt and Neill Graham, 3–140. Princeton, NJ: Princeton University Press.

Geach, Peter T. 1962. *Reference and Generality: An Examination of Some Medieval and Modern Theories*. Ithaca, NY: Cornell University Press.

Geach, Peter T. 1967. Identity. *Review of Metaphysics* 21(1): 3–12.

Greaves, Hilary. 2004. Understanding Deutsch's Probability in a Deterministic Multiverse. *Studies in History and Philosophy of Modern Physics* 35: 423–456.

Laudisa, Federico, and Carlo Rovelli. 2008. Relational Quantum Mechanics. In *Stanford Encyclopedia of Philosophy*, Summer 2013 ed., ed. Edward N. Zalta. http://plato.stanford.edu/entries/qm-relational/.

Lewis, David. 1983. Survival and Identity. In *Philosophical Papers*, vol. 1. New York: Oxford University Press.

Lewis, David. 1986. *On the Plurality of Worlds*. New York: Wiley-Blackwell.

Lewis, Peter. 2007. Uncertainty and Probability for Branching Selves. *Studies in the History and Philosophy of Modern Physics* 38: 1–14.

Lewis, Peter. 2008. Probability, Self-Location and Quantum Branching. Philosophy of Science Association 21st Biennial Meeting, Pittsburgh, PA. http://philsci-archive.pitt.edu/archive/00004309/.

Maudlin, Tim. 2014. *Critical Study* David Wallace: *The Emergent Multiverse: Quantum Theory according to the Everett Interpretation*. *Noûs* 48(4): 794–808.

Olson, Eric. 2010. Personal Identity. In *Stanford Encyclopedia of Philosophy*, Winter 2010 ed., ed. Edward N. Zalta. http://plato.stanford.edu/entries/identity-personal/.

Parfit, Derek. 1971. Personal identity. *Philosophical Review* 80: 3–27.

Parfit, Derek. 1989. Divided Minds and the Nature of Persons. In *Mindwaves: Thoughts on Intelligence, Identity and Consciousness*, ed. Colin Blakemore and Susan Greenfield, 351–356. Oxford: Blackwell.

Saunders, Simon. 1993. Decoherence, Relative States, and Evolutionary Adaptation. *Foundations of Physics* 23(12): 1553–1585.

Saunders, Simon. 1995. Time, Quantum Mechanics, and Decoherence. *Synthese* 102(2): 235–266.

Saunders, Simon. 1996a. Relativism. In *Perspectives on Quantum Reality*, ed. Rob Clifton, 125–142. Boston: Kluwer Academic Press.

Saunders, Simon. 1996b. Time, Quantum Mechanics and Tense. *Synthese* 107: 19–53.

Saunders, Simon. 1997. Naturalizing Metaphysics. *Monist* 80(1): 44–69.

Saunders, Simon. 1998. Time, Quantum Mechanics and Probability. *Synthese* 114: 373–404.

Saunders, Simon, and David Wallace. 2008a. Branching and Uncertainty. *British Journal for the Philosophy of Science* 59: 293–305.

Saunders, Simon, and David Wallace. 2008b. Reply. *British Journal for the Philosophy of Science* 59: 315–317.

Shoemaker, Sydney. 1984. Personal Identity: A Materialist's Account. In *Personal Identity*, ed. Sydney Shoemaker and Richard Swinburne, 67–132. Oxford: Blackwell.

Sider, Theodore. 1996. All the World's a Stage. *Australasian Journal of Philosophy* 79: 433–453.

Vaidman, Lev. 1998. On the Schizophrenic Experiences of the Neutron or Why We Should Believe in the Many-Worlds Interpretation of Quantum Theory. *International Studies in the Philosophy of Science* 12(3): 245–261.

Vaidman, Lev. 2014. Many-Worlds Interpretation of Quantum Mechanics. In *Stanford Encyclopedia of Philosophy*, Spring 2015 ed., ed. Edward N. Zalta. http://plato.stanford.edu/entries/qm-manyworlds/.

Wallace, David. 2005. Language Use in a Branching Universe. http://philsci-archive.pitt.edu/archive/00002554/.

Wallace, David. 2006. Epistemology Quantized: Circumstances in Which We Should Come to Believe in the Everett Interpretation. *British Journal for the Philosophy of Science* 57(4): 655–689.

Wallace, David. 2010. Decoherence and Ontology (or: How I Learned to Stop Worrying and Love FAPP). In *Many Worlds? Everett, Quantum Theory, and Reality*, ed. Simon Saunders, Jonathan Barrett, Adrian Kent, and David Wallace, 53–72. New York: Oxford University Press.

Wallace, David. 2012. *The Emergent Multiverse: Quantum Theory According to the Everett Interpretation*. New York: Oxford University Press.

Wiggins, David. 1967. *Identity and Spatio-temporal Continuity*. Oxford: Blackwell.

PART III | Beyond Disciplinary Borders

CHAPTER 14 | The Metaphysics of Individuality and the Sciences

MATTEO MORGANTI

14.1 Introduction

By and large, contemporary philosophers agree that metaphysics should pay close attention to the a posteriori input of science. Yet, naive naturalism, leading to the idea that metaphysics can be "read off" directly from our best theories, doesn't seem very promising: indeed, there seems to be simply nothing we can extract from scientific theories without a previously defined theoretical framework not itself coming from empirical science. The alternative represented by a radically empiricist eliminativism toward metaphysics is, clearly, even less appealing from the present perspective. Indeed, a unified methodology and shared conceptual background underpinning naturalized metaphysics and the metaphysics of science is still lacking. Thus, while it is correct to think that metaphysicians interested in the specific topic of identity and individuality should consider the status of identity and individuality in actual scientific domains and theories, it doesn't come as a surprise that little agreement has been reached on what the sciences can and do teach us with respect to the topic at hand.

Without entering into the thorny debate concerning the correct characterization of naturalism,[1] in this chapter I will try to remedy—at least partly—this situation. In particular, by making reference to the recent debate about identity and individuality in nonrelativistic quantum mechanics, I will examine the assumptions underlying recent defenses of "reductive" accounts of identity and individuality (both of the *Leibnizian* sort, according to which something is an individual if and only if it is qualitatively unique; and of the *contextual* sort, according to which the identity of something might be nonqualitative but is in any case determined by networks of relations—more on this later). I will suggest that the idea that these accounts are more congenial to a naturalistic approach to philosophical questions is not as compelling as it may seem at first. Even if we wish our metaphysics to be informed by our best science, I will argue, the non-reductive, "primitivist" account of identity and individuality as something fundamental and non-further-analyzable can be positively re-evaluated. The overall picture that will emerge, however, is that of a pluralistic approach, according to which different scientific domains and theories are likely to admit of, and in fact call for, different forms of individuality. While the focus of the chapter is on physics, and quantum theory in particular, this conclusion is believed to apply, and indeed to gain its real significance when applied, across the sciences—each science, or even theory, allowing for, if not requiring, a specific notion of individuality. Albeit indirectly, this seems to lend support to the more general view that, although it must be scientifically informed, metaphysics should preserve an important degree of autonomy: for only via a preliminary, essentially a priori, analysis of possibilities can one then go on to evaluating different alternatives as to the interpretation and consequences of science in general and of specific scientific disciplines and theories in particular.

The structure of the chapter is the following: Section 14.1 introduces the issue of identity and individuality, first in its general terms and then as it presents itself in quantum mechanics; in particular, the way in which recent results concerning so-called weak discernibility are supposed to buttress reductionism is briefly illustrated. Section 14.2 puts forward an alternative, more "conservative," primitivist proposal and motivates it on the basis of general methodological considerations. Naturalistic methodology turns out, as a matter of fact, to be largely neutral with respect to the issue. Section 14.3, however, taking its cue from considerations of physical theories going beyond nonrelativistic quantum mechanics, suggests that different conclusions should be drawn there. This leads one to a more cautious pluralism about identity and individuality: generalizing even more, to sciences other than physics, it is argued that a "level-relative" (yet objective) differentiation between forms of (non)individuality is made highly plausible by considerations having to do with both the practice of science and the nature of scientific theories and ontologies. A brief concluding summary follows.

14.2 Identity and Individuality in Quantum Mechanics

At least as a first approximation, individuality can be assumed to consist in the possession of determinate self-identity and numerical distinctness from other things[2] (synchronically and, possibly, diachronically—only the former will be our focus here). It has often been argued that, so understood, individuality can be reduced to something more fundamental. The only serious such "reductionist" option is, it would seem, the Leibnizian-Quinean view according to which every individual is qualitatively unique, that is, differs from other things with respect to at least one qualitative aspect. Leibniz took the qualitative uniqueness of objects as a metaphysical necessity, while Quine, following Hilbert and Bernays, argued that the identity sign is dispensable in the language (provided that some conditions are met). The two views are clearly complementary. Indeed, this position has become a very authoritative one in metaphysics—and, importantly for our present purposes, probably the dominant one among scientifically minded philosophers.

Why this is the case and whether this tendency is actually justified will be discussed later. For the time being, let us first assess what will from now on called "reductionism" in more detail. A consequence of reductionism is the truth of the *Principle of the Identity of Indiscernibles* (PII), which is the metaphysical principle according to which no two things can have all the same properties.[3] The question, then, is whether PII is true in a nontrivial form, one not putting numerical identity- and difference-involving properties on a par with all the others, doing which reductionism would become based on an unacceptable circularity. This restriction on admissible properties is, in particular, the reason why one speaks of "qualitative uniqueness." (What it means for a property to be "qualitative" is not obvious. Here we will assume that those properties that do not involve a noneliminable reference to some identity fact[s] are ipso facto qualitative, and only those).

To begin with, Leibniz subscribed to a strong version of PII, according to which any two things differ with respect to some intrinsic property. Nowadays, however, this restriction appears both unmotivated and such that it leads to a false principle: surely two electrons (say) do not differ in their intrinsic qualitative properties. What about a form of PII ranging over all monadic intrinsic and relational properties?

In classical mechanics (CM), a presupposition of impenetrability is generally made, from which it follows that classical entities always differ at least with respect to location (i.e., a relational property involving either material objects and substantial points/regions of space, or objects only, depending on whether space(-time) substantivalism or relationism is correct). As a consequence, PII seems vindicated in the classical domain. Impenetrability is not, however, assumed in an unquestionably more fundamental theory: namely, quantum mechanics (QM)—at least in its *standard, nonrelativistic* interpretation.[4] There it can be shown that whenever they are part of the same physical system, particles that share all their state-independent (i.e., roughly, essential) properties also share all their monadic and relational state-dependent (i.e., roughly, accidental) properties, as exactly the same probabilities for measurement outcomes involving their properties are attributed to them—including location (this was first explicitly shown by French and Redhead (1988)). Since it is widely agreed that quantum probabilities mirror objective facts about the world and not just our partial knowledge of things, one appears consequently entitled to draw metaphysical conclusions from the above evidence. In particular, one seems forced to conclude that PII is violated.

This is not the end of the story, though. Since Pauli's "exclusion principle" (EP) bans two indistinguishable fermions from having all the same quantum numbers, it seems to entail their discernibility. As first pointed out by Margenau, this doesn't determine a difference with respect to individual observables as intrinsic properties. Yet *if one assumes that relations exist that are not reducible to monadic properties*, and has PII quantify over these relations too, then a weaker form of the principle than those considered so far emerges that seems satisfied by quantum objects. In fact, considerations about fermions seem actually generalizable so as to apply to bosons too, thus making discernibility independent of EP. This line of argument was first taken up by Saunders (2003, 2006a), elaborated upon by Muller and Saunders (2008) and Muller and Seevinck (2009), and recently refined by Caulton (2013) and Huggett and Norton (2014). Differences notwithstanding, all these works assume that quantum particles can be discerned on the basis of fundamental *irreflexive* relations defined by the theory.

A first thing that must be noted for the purposes of the present discussion is that Muller, Saunders, and Seevinck distinguish *individuals* (which they take to be objects that are discernible on the basis of monadic properties) and *relationals* (weakly discernible entities). Consequently, they classify quantum particles as relationals, removing them from the class of individuals. We will get back to this point later, but this shouldn't be given too much weight here: for the time being, suffice it to say that Muller, Saunders, and Seevinck work with a different, more restrictive, concept of individuality. Indeed, if one sticks with the definition of individuality proposed at the beginning of this chapter, there is no need to talk about relationals, and the results concerning weak discernibility can be taken to support reductionism about individuality.

As our next step, let us introduce another metaphysical option by looking at an additional bit of significant empirical evidence, having to do with the statistical behavior of quantum systems. Indeed, especially if one is not persuaded by Leibnizian views on qualitative uniqueness, the most serious blow to the entrenched belief that the world is ultimately constituted by tiny individual objects (more or less) like those we experience on a daily basis comes from a consideration of the statistical behavior of quantum systems. Very briefly, if we consider N particles distributed over

M possible single-particle microstates, in classical mechanics (with distinguishable particles) the number of possible distributions W is $W = M^N$. Just consider two fair coins, for which $M = N = 2$, and thus $W = 4$; that is, the calculation gives the expected four possible outcomes for a simultaneous toss of the two coins. This is not true in the case of quantum particles, however, for which a smaller number of arrangements is available. For bosons, one has $W = (N + M - 1)! / N!(M - 1)!$, and in the case of fermions the number of possible states is even smaller (due to EP), and amounts to $W = M! / N!(M - N)!$. The key difference between the classical and the quantum case is that *permutations* of qualitatively identical particles lead to statistically distinct configurations in the former but not in the latter. In particular, classical systems can be in *nonsymmetric* states (that is, states in which individuals have definite but different values separately and for which, consequently, permutations do make a difference), while quantum systems cannot (but can instead be, as we have seen, in entangled states—which are also [anti]symmetric).

Why does this happen? Doesn't this contradict what we take to constitute the essential core of the very concept of an individual? The traditional explanation of the peculiarities of quantum statistics, which some call the "Received View" based on the fact that it was accepted by some of the founding fathers of the theory such as Schrödinger, is indeed that *quantum particles are not individual objects*, and this is why we shouldn't expect many-particle states to be sensitive to which object has which property. If an object is not an individual, the argument goes, it doesn't have a well-defined identity, distinct from that of other objects, and there consequently are no relevant permutations to be made in the first place. Nor can two objects have definite, distinct, and separate properties, as required for a system to be in a nonsymmetric state: it simply makes no sense to attribute a property in a two-particle system to *this* particle instead of *that* one. On this construal, then, quantum particles would be *nonindividuals* in the sense of being entities that are countable but not such that identity-involving formulas concerning them are well formed.

Other explanations of quantum statistics have been proposed that do not dispense with individuals, however. They are based on an assumption of classical indistinguishability (Saunders 2006b), on the rejection of the principle of equiprobability of states (Belousek 2000) or on the idea that there are fundamental state-accessibility restrictions in the quantum case (French and Redhead 1988, French 1989). Besides being controversial in themselves, these proposals need not be dealt with in detail here, as we are primarily interested in the issue of individuality and only tangentially in what the correct account of quantum statistics is. On the other hand, something more about this will have to be said later, when it will come to critically assessing primitivism about identity and individuality.

Getting back to our main discussion, the idea of nonindividuality might be taken to be what the abovementioned concept of a "relational" amounts to. Notice, however, that while relationals cannot be told apart from each other on the basis of their properties (they are *merely* weakly discernible) and yet possess well-defined identities, nonindividuals are particulars that lack identity conditions and may well be indiscernible—at least in the sense that one endorsing an ontology of nonindividuals might coherently reject the notion of weak discernibility, and PII more generally.

Getting back to our main discussion, that quantum statistics leads to nonindividuality is a very important conclusion, for it is tantamount to stating that, given nonrelativistic QM, there is no room for individuals even independently of whether PII, and thus "traditional" Leibniz-Quine reductionism, is true. More generally, one may go as far as to suggest that the above is a case of *experimental metaphysics* in the sense of Shimony (1981). Very roughly, the idea behind the notion of experimental metaphysics is that, in some cases, a metaphysical hypothesis together with a scientific theory entails some empirical predictions that can then be used to "test" the metaphysical hypothesis by a posteriori means. In the present case, metaphysical construals according to which individuals have to differ with respect to their monadic qualities and be "classical" in their group behavior are straightforwardly shown to be a nonstarter. Consequently, one seems forced to opt for ontologies without individuals.

Is this all there is to say? Can a more "canonical" metaphysical view of individuality, or at any rate, a view different from those considered so far, be defended after all? To provide an answer to these questions, it is useful to proceed in two stages, critically assessing the two options that we have found so far (weak discernibility and nonindividuality) as a first step, and then moving onto a more general evaluation of the available ontological views.

Let us begin with the first task. An objection that is usually leveled against the weak discernibility of quantum particles is that relations cannot discern because they ontologically depend on their relata; and that indeed the arguments under scrutiny presuppose what they have to demonstrate. I think this objection misses its target. To begin with, it is a more or less explicit assumption of the proposed view (as well as of a pretty popular more general view that is independent of—but closely related to—it, namely, ontic structuralism) that relations need not be ontologically derivative with respect to objects and monadic properties. Although debatable, this assumption should be accepted at least in the sense that it points to a logical and metaphysical possibility. There is no contradiction in taking relations as prior to their relata, especially in view of the fact that there are formal constructions in which relations are *not* essentially defined in terms of particulars. On a related, but more specific note, defenders of the view claim that the argument for weak discernibility does presuppose that there is a countable number of objects in the domain, but it looks as though such countability need only be assumed *at the formal level*. Since the relevant conclusions concern the concrete, physical domain and they by no means follow trivially from the fact that the relevant language includes certain individual constants, the circularity objection appears misplaced. Pending a (nevertheless much needed) characterization of the underlying metaphysics, as well as a more precise definition of the criteria for deeming relations and properties that are "constructed" at the formal level "physical," then, we can set the criticism based on circularity aside for the time being. That there are some bumps in the carpet that require flattening remains true, however, as can be seen based on the extant literature. Consider the following example. Muller and Seevinck explain that their proof leads one to take two identical bosons in a symmetric direct product state as discernible, while the state in question is one in which two completely similar, but noninteracting, entities are simply considered together. Muller and Seevinck simply accept this fact as "brute." Caulton (2013) and Huggett and Norton (2014), instead, seem to be in a position to avoid the above undesirable consequence, thanks to the fact that they propose different criteria for counting predicates as physically meaningful. But what is the correct criterion of physical meaningfulness? And, in case of conflict, how should we choose? More generally, what gets priority between such criteria and supposedly brute, and possibly counterintuitive, facts about identity?[5] A different line of criticism is also worth mentioning (Dieks and Versteegh 2008, Ladyman and Bigaj 2010): it is based on the thought that weak discernibility does not vindicate PII, because it does not correspond to the possibility of actually telling particles apart from each other through physical means, and is therefore completely extraneous to the spirit of the Leibnizian tradition and of the later empiricist/naturalistic endorsement of the Leibniz-Quine view. Whether or not one is impressed by this line of argument, it is a fact that there is a crucial question to be dealt with concerning the ontological status of the allegedly discerning relations.

This leads us to what I take to be the real problem with weak discernibility. Even assuming, as we did, that it is coherent to rely on already established facts of numerical distinctness, *one is left wondering whether countability truly is a merely formal matter.* That is, one wonders whether formal countability, crucial—as we have seen—for the proofs of the weak discernibility of quantum particles to get started, doesn't have a direct counterpart in the actual world. This, in the sense that they correspond to basic facts of identity and distinctness that in fact *ground*, rather than *depend on*, the qualitative facts that have been so emphasized by supporters of weak discernibility, and of reductionism about individuality more generally. In other words, couldn't QM be taken to suggest that countability is in fact metaphysically prior to discernibility, so that what one may call "primitivism" about identity and individuality (i.e., the view that, as mentioned in the introduction, takes at least some facts of identity and individuality as fundamental and not further analyzable) should in fact be preferred to Leibniz-Quine reductionism?

Similar remarks hold for the non-individuality option. For it is true that the peculiarity of quantum statistics *can* be taken to indicate that we should move to the label-less context of quantum field theory in Fock space, that is, to a context in which the basic ontological units are field "perturbations," to be understood as *properties* rather than *property-bearers* of some sort (so that, for instance, we should really say "two-bosonness is instantiated" rather than "there are two bosons"). And it is true that in this latter domain countability facts are insufficient for attributing intrinsic identities to particles, as there are no "labels" that putatively refer to specific objects. But recall that the question we are dealing with is squarely one concerning the metaphysical consequences for identity and individuality *of nonrelativistic quantum mechanics*, where in fact we *do* have labels that are naturally interpreted as referring to individual physical systems. Of course, that one has *both* labels and individuals, or at any rate objects, and not just empty names, might be disputed: for instance, by ontic structural realists, or by those who think the key entities in quantum theory are wave functions, which are normally spread out in a way that prevents one from establishing a one-to-one correspondence between labels and more or less well-defined "things." However, the point here is just that the same facts about the Hilbert space formalism of QM that are exploited to argue in favor of the weak discernibility of quantum particles could be exploited to support the more radical primitivist view. Thus, pending a precise individuality-friendly explanation of quantum statistics, the move to an ontology of nonindividuals doesn't appear particularly well-motivated in this case. One might object that it is the violation of PII under the assumption that distinct particle-like objects exist that leads one to drop the latter assumption and move to the field-theoretic setting. However, first, that this is the dialectic is not clear in the extant literature; second, the cogency of this argument cannot just be taken for granted, as one may equally infer the falsity of PII (or the need to endorse a "lighter" version of it, allowing for weak discernibility) from its failure to account for the identity conditions of the entities in a given domain. Thus, this latter counter-argument might be regarded as question-begging, and pointing to a conclusion (the priority of fields) that should be established independently (and certainly can be so established on physical grounds), with a view to consequently drawing metaphysical conclusions.

Of course, the question arises at this point as to which of the available ontologies should be preferred in the pre-quantum-field-theoretic case, especially in light of the peculiarities of quantum statistics. Both supporters of weak discernibility/structuralists and supporters of the nonindividuality option, not surprisingly, argue that their ontology provides the least committing and most plausible and straightforward metaphysical explanation of the evidence provided by nonrelativistic QM (see, for instance, Ladyman and Ross 2007, Arenhart and Krause 2013, 2014), but more is certainly needed: in particular, a well-defined set of criteria for critically assessing and comparing the various alternatives. In the next section, I will try to identify some such criteria and, based on them, I will put forward considerations in support of the idea that primitivism about identity and individuality should indeed be deemed preferable to the extant alternatives in the case of nonrelativistic QM *even from a broadly understood empiricist/ naturalistic point of view.*

14.3 A Conservative Proposal

Let us now move to the second task identified in the previous section, that of extending our previous discussion to other ontological accounts of non-relativistic QM, and providing a more general assessment. A crucial question to ask now is this: why should one feel compelled to be a reductionist about individuality in the first place, or at any rate to use PII as a principle of individuation?

A common answer is that science is, at root, *exclusively concerned with general, qualitative descriptions of reality*. That is, scientists are only interested in providing in principle testable descriptions of actual states of affairs and events, and consequently have no use for entities "playing the same role" in the world. Because of this, the reasoning continues, naturalistic metaphysicians should consider qualitative facts fundamental, for otherwise they would leave the existence of in principle dispensable entities as an open possibility. Advocates of nonindividuals argue along similar lines, although, of course, they do not need to explicitly defend a reductive account (or any particular account, for that matter) of individuality. Rather, they claim that nonindividuality is strongly suggested by the general features of the quantum domain—where, in particular, identity facts play no role whatsoever when detached from qualitative facts about particles and aggregates thereof.

However, even granting the plausibility of some naturalistic approach to metaphysics, it is in no sense obvious that "science is only about qualities." And even if it is, it is far from clear what this entails from the metaphysical viewpoint. To begin with, an essential fact to be pointed out in the present context is that indiscernible objects do, or at least can, in fact make a qualitative, empirical difference! Consider the very simple case of a world with two exactly similar material objects: clearly, such a world exhibits twice the mass of a world with only one such object. Generalizing from this type of thought-experimental scenarios, it seems right to claim that the naturalist, aiming to only formulate scientifically grounded metaphysical claims, need not, and should not, take indiscernibility as a criterion of individuation for *objects* but, rather, for *worlds*. An analogous claim can be found in Hawley (2009) and, in more indirect form, in Earman's claim that PII "lead[s] to a plausible methodological rule for scientific theorizing: in choosing two theories, where one theory employs a descriptive apparatus that implies the existence of distinct possible states between which we cannot distinguish by means of an observation, while the other theory has no such implication, then, all other things being equal, the latter theory is to be preferred to the former" (1979, 268). The core thought is that one ought to eliminate indiscernible "description of states," that is, actual or possible systems of objects (what I call "worlds"), not (necessarily) indiscernible individual things that are part of acceptable descriptions of those systems (i.e., that inhabit nomologically possible worlds). If this is correct, though, the requirement of "empirical relevance" rightly set by naturalists on metaphysical speculation can also be met by an ontological interpretation of QM that allows for numerically distinct indiscernible individuals. Indeed, recalling that, as was pointed out above, countability is a fundamental fact about the quantum formalism, and one that plays a crucial role in arguments in favor of the weak discernibility of quantum particles, nonrelativistic QM can be taken to naturally lend itself to a primitivist interpretation—primitive identity and individuality not being obviously, physically dispensable any longer—even in a science-oriented, naturalistic framework.

It might be objected that primitivism is in any event implausible from the naturalistic viewpoint because it posits the existence of primitive, intrinsic facts of identity that cannot but coincide with mysterious, purely metaphysical factors. Setting aside that it is not clear why the same should not be said with respect to the identity of properties (in particular, of universals, which it appears necessary to postulate in order to apply PII, at least in its traditional philosophical context), the response is available that science is compatible with, and in fact justifies (if not "requires"), the postulation of a "thin" form of primitive identity. That is, primitive identity facts can be intended not as constituting metaphysical additions, but rather as coinciding with fundamental facts about the existence of certain entities as *those* entities. This would fit an "Ockhamian," rather than "Scotian" view, according to which, to use Ockham's own words, "Whatever is singular is singular through nothing added to it, but by itself" (Ockham, *Ordinatio* I.2.6, 86).

This, I take to mean that, in a nominalistic fashion, individual objects must be said to have their unique identity primitively, but no commitment has to be made with respect to an actual "ontological component" of them that is responsible for such identity—that is, an ontologically "thick" *haecceitas* or primitive thisness.

Before drawing any conclusions, though, other considerations must be taken into account. According to some, identity facts may not be grounded on qualitative difference (as the primitivist says), but must in any case be *contextual*, that is, extrinsic and determined by the system in which the object possessing it is "inserted" (as the reductionist says). That this "structuralist" approach is in fact the only way to obtain, so to put it, the best of both worlds is argued on the basis of the following reasoning. If identity facts were not extrinsic, then the truth of haecceitism (the doctrine that there can be differences between what distinct worlds say *de re* about certain individuals that do not correspond to overall qualitative differences) would follow; but haecceitism is contradicted by contemporary science and appears in general in conflict with naturalism. In particular, this seems to be the case for general relativistic space-time models in view of the hole argument—which, as argued by Stachel (2004), forces one to identify representations of space-time that only differ by a permutation of points, thus suggesting the falsity of space-time substantivalism. And the same goes, as we have seen, for quantum statistics, which is manifestly permutation-symmetric (for this extension of Stachel's contextualist/structuralist approach to the quantum domain, see Ladyman 2007; for criticism, see Pooley 2006; for a reply to the criticism, Caulton and Butterfield 2012). At a first glance, this appears to be a compelling argument, striking a perfect balance between explanatory power and revisionary nature of the proposed ontological framework. Indeed, even if we insist on countability facts and their being directly mirrored by the Hilbert space formalism, the traditional primitivist appears unable to explain the sort of permutation invariance exhibited by quantum as well as general relativistic systems of entities. As a matter of fact, though, things are not so straightforward: for contextualism is by no means a direct consequence of the observed permutation invariance, and primitivists can in fact explain the relevant evidence.

To begin with, primitive identity need not entail haecceitistic differences. For what is true of distinct worlds is not univocally determined by the nature of the identity of the objects that inhabit them. A counterpart-theoretic treatment of possible worlds, for instance, may allow one to assume primitive intraworld identities (i.e., primitive identity as the ground for the individuality of specific objects in a particular world) together with anti-haecceitism. And the same goes for the sort of "Leibnizian superessentialism" according to which individuals have all their properties essentially. One might reply that primitive identities must necessarily also be transworld, that is, determine that distinct worlds may differ merely in terms of which entity is which, so in fact entailing haecceitism. But while, of course, a more detailed examination of the relevant logic and metaphysics is in order, to simply state this would definitely be question-begging in the present context. For the time being, the relatively uncontroversial claim that intra- and transworld identities could be related in several different ways, and permutation invariance is consequently not enough for drawing conclusions concerning the identity of things, is sufficient.

In addition to this, it is crucial to notice that, even if haecceitism holds, there might be (equally or more forceful) explanations of the evidence alternative to the contextualist view. That is, plausible ways of accounting for the fact that, even if haecceitistic permutations are genuinely possible, this doesn't have empirically relevant consequences in the case at hand.[6]

That this is indeed what happens in QM has been argued in Morganti (2009), which expands on Teller (1986) and suggests that all state-dependent properties of quantum particles in many-particle systems (entangled *and nonentangled*) are holistic properties, and are consequently independent of specific facts about specific particles and their monadic properties. From this it follows that particle exchanges do not and cannot give rise to relevant new states. *This, notice, fully accounts for the peculiarities of quantum statistics, including the impossibility of nonsymmetric states*: the latter require separate, well-defined (and distinct) properties for the separate components of the total system, which is exactly what the proposal rules out.

The price of this is, of course, an extension of Teller's holism to all state-dependent properties, and not just those of entangled systems. One might take this to be too high a price to pay, and/or to justify the move to a relational metaphysics where, in the end, objects play no role whatsoever. Note, however, that (*a*) in the case of nonentangled states the properties of the whole do supervene on those of its components, so the proposed form of holism is not as extreme as one may think; (*b*) radically modifying our entrenched beliefs about properties seems less costly that modifying our entrenched beliefs concerning the bearers of those properties—although perhaps Leibnizian reductionists, and bundle theories of objects more generally, aim to diminish it, there is a clear order of priority between objects and their properties; and (*c*) the move from an ontology where there are irreducible, fundamental relations and yet objects are noneliminable (e.g., because they are the bearers of the essential, state-independent properties that determine what kind of entity an entity is) to one where there are only relations certainly requires some additional argument.

The upshot of the foregoing discussion of identity and individuality is, thus, the following. There are clear indications coming from nonrelativistic quantum mechanics that the *classical* concept of an individual object cannot be preserved in that domain. However, there are no compelling reasons for the naturalist to endorse PII as a guide for the ascription of individuality to things or, alternatively, to opt for an ontology of nonindividuals or merely contextually identifiable things. In particular, it is possible to regard quantum countability (mirrored by the particle labels and names that are an integral part of the quantum formalism) as sufficient for endorsing primitivism; and to extend certain ideas that are normally applied to at least some many-particle quantum systems (i.e., entangled systems) so as to make the peculiar features of quantum statistics depend on the property-structure of the domain it describes rather than on the identity conditions of the objects inhabiting that domain and exhibiting that property-structure.

While a claim of possibility would already be a relevant result, let us conclude by suggesting that, as far as the nonrelativistic quantum mechanical context is concerned, the resulting primitivist framework is preferable in terms of simplicity, amount of revision required, and explanatory power. This judgment depends on whether or not one agrees that it is natural to think that the individuality of x does not depend on x's relations with the world, and on how one evaluates the virtues of simplicity, conservativeness, and explanatory power. As to the first point, conclusive arguments cannot be provided here, and I can only ask the reader to think carefully and say whether she or he would be ready to agree that an entity could stop or begin being an individual without anything changing at the level of its intrinsic, essential features, and only as a consequence of changes external to it. As for the second point, again there is no space to argue for this at length, but it is not absurd to think that conservativeness plays, or at any rate should play, a major role in theory choice. Ceteris paribus, that is, we can and should (and in fact do) always opt for the least amount possible of conceptual revision. Notice that the ceteris paribus clause neutralizes the often heard (but misguided) objection according to which science has taught us that maximizing revision is the best thing to do: even, say, special relativity was regarded by Einstein as the best way to preserve the greatest portion of our established beliefs while also accounting for the problematic evidence![7] As for simplicity and explanatory power, here we will just conform to the traditional view in philosophy according to which such virtues should indeed be sought: there is no need to take a stance with respect to issues such as whether these virtues are truth-conducive, whether they can be quantified with exactness, whether the world can be expected to be simple rather than complex, and the like. Taking stock, table 14.1 summarizes the main results and claims of the discussion so far.

TABLE 14.1 A Schematic Representation of the Costs Associated to Each Metaphysical Reading of Quantum Entities

	FIT WITH EVIDENCE	AMOUNT OF REVISION REQUIRED	DEGREE OF PLAUSIBILITY
Nonindividuality	Okay with respect to statistics, but not clear what stance is taken with respect to facts of (in)discernibility	Maximal at the level of language and ontology	Seems to unwarrantedly switch to a different theory and formal structure (quantum field-theoretic framework). Essentially ignores countability facts
Weak discernibility	√	Maximal if "*relationals*" lack the identity conditions of individuals, much less if the same as individuals. Irreducible relations are posited and identity is made extrinsic	Same as in the case of nonindividuality; or, at least not entirely clear whether and, if so, to what extent discerning relations are physically meaningful. Countability facts presupposed
Contextualism	√	Irreducible relations posited, (although possibly only to establish facts of numerical distinctness)	√
Primitivism	√	Minimal, only involving properties of *all* many-particle systems as holistic (yet supervenient on the properties of the component systems in the case of nonentangled states)	√

14.4 Quantum Fields and Beyond

It goes without saying that—as we mentioned already—nonrelativistic quantum mechanics is but one of the theories that the scientifically inclined metaphysician should look at. With a view to drawing some additional, and even more wide-ranging, conclusions and then closing the chapter, let us, therefore, now look at other parts of physics and of science more generally.

Let us start by looking in more detail at the quantum field-theoretic domain, which we mentioned only in passing in the previous sections. In the 19th century, the likes of Young, Fresnel, Stokes, and Thomson worked toward a thorough definition of the *ether* as a continuous medium constituted by dimensionless points through which light propagates. Faraday modified the notion of a physical field from that of an ad hoc device for representing the actions of bodies to that of a full-blown physical entity, ontologically prior to material objects. With Einstein, the ether was set aside. The notion of field, however, remained central. In the late 1920s, Heisenberg, Born, Jordan, and Dirac applied the methods of quantum mechanics to electromagnetism, replacing the classical variables taken to correspond to components of the field with quantum mechanical operators. The slow, but steady development of quantum electrodynamics and chromodynamics then led, from the late 1960s onward, to a physics in which fields certainly possess the status of fundamental entities. It is this theoretical framework that led to the Standard Model of elementary particles and to the remarkable achievements of experimental physics in the last few decades, culminating in the recent discovery of the Higgs boson.

But what are quantum fields? An intuitive answer would consist in taking a physical counterpart of fields as they are defined mathematically (very roughly, that is, as entities with an infinite number of degrees of freedom, provided with some physical content at all points) as ontologically basic; a more conservative one, and one in line with our earlier proposal, would suggest instead a particle-based ontology. Alas, particle-based ontologies and field-based ontologies seem to be equally in trouble when it comes to nonrelativistic and, in particular, relativistic quantum field theory (QFT). On the basis of a number of by now well-known technical results, on the one hand, Fraser (2008) forcefully argued against particles in QFT. Without for a moment aiming to provide a complete list and/or a satisfactory explanation of all the relevant facts, let us briefly look at some of the key novelties and results one is presented with when one looks at quantum fields. First of all, (1) "creation-" and "annihilation-events" play a crucial role in QFT, and this entails that (what would seem to be) continuous paths across space and time cannot be attributed to particles, and the number of putatively existing particles continuously changes.

Moreover, (2) it is possible, and indeed common, in QFT to encounter quantum states that are not eigenstates of the number operator. Such states cannot be handled unless they are described as quantum superpositions of states having different number of "things" in them. The situation becomes even worse when one brings relativity into play. For then one has to take into account that (3) local number operators are not well defined; (4) the objectivity of total number operators can be questioned (due to the Unruh effect, amounting to the fact that—roughly put—a change in the frame of reference causes a change in the number of particles); (5) such operators may even fail to exist (this is Haag's theorem, entailing that in an interacting relativistic QFT the customary representation of free particles cannot be used to describe interacting particles); and, last but not least, (6) expectation values for certain quantities do not vanish for the vacuum state, so that energy is not zero and there are physical happenings even when no "thing" is there. The foregoing seems to rule out a particle-based interpretation of QFT. However, as argued by Baker (2009), the relevant results seem to carry over to the field-ontological setting.[8]

If this is correct, there seem to be two options remaining: either a form of skepticism/agnosticism about the possibility of defining an ontology for (relativistic) quantum field theory; or the search for a revisionary ontological framework truly capable of making sense of the theory. Setting aside the former alternative as a live option that is, nonetheless, something like a last resort for the naturalistic metaphysician, here we can only state that the latter is indeed an intriguing possibility, but a lot of work is yet to be done. Ontologies of tropes, events, processes, states of affairs, structures, dispositions and factors have been proposed, but it is still unclear how these, if at all, deal with the problems besetting particle- and field-based ontologies. Perhaps this invites one to explore yet further options: say, an ontology based on two-dimensional extended simples modeled on strings; or, maybe, a monistic/stuff-based ontology vaguely along the lines suggested by Sidelle (1989) or Horgan and Potrč (2000).

Be this as it may, for the time being it cannot but be concluded that our earlier claims concerning nonrelativistic quantum mechanics do not carry over to quantum field theory. But, notice, the same goes for the other alternatives: for instance, how is one to account for the status of number operators in relativistic QFT from the viewpoint of weak discernibility, or contextual identity, or nonindividuality, all of which do not deny that "particles" come in countable aggregates? Generalizing on this, it can be argued that the expectation that a unified, all-encompassing account of identity and individuality—or any other ontological category, for that matter—be given is only reasonable if one assumes a strong form of reductionism, that is, a microreductionism à la Putnam-Oppenheim. But such an approach is rightly regarded nowadays as unpopular on independent grounds, and our results here cannot but further detract from it. Thus, any project of strong ontological reduction would seem to be unmotivated at the very start, and more flexible viewpoints should be preferred. This is demanded, we have seen, by a consideration of two theories within the same discipline (although arguably not at the same level of theoretical fundamentality)—that is, physics. And the same is likely to happen when one moves to other disciplines. Take, for instance, biology. There, not only is it the case that the concept of an "individual" is defined quite independently of what physics says. It is also the case that there is no agreement on *what* should be identified as the (or, "a") biological individual, and *why*. Is it the gene, or the species? Is it the organism as it is traditionally intended, or the organism together with the symbiotic bacteria inhabiting them? Does it have to be identified with whatever the fundamental unit of selection is? Or with something close to our common-sense intuitions about ourselves and living things in general?[9]

To be sure, more can and should be said, especially in a naturalistic context. In particular, even if one sets aside naive reductionism, important questions must be answered: how does the ontology of one discipline or theory relate to that of another? Is there a sense in which an ontology "emerges" out of another, perhaps in a way that we can try to reconstruct on the basis of science? Does metaphysics, like physics (or perhaps more), ultimately rest on the idea of the "one true description of the world," or does it not? For the time being, we must leave these questions open. However, based on our discussion, it looks as though a serious form of "pluralism," allowing for different conclusions at different levels of scientific inquiry, remains an option to be taken seriously given the indications coming from current science. And it certainly doesn't look like such pluralism is an obvious nonstarter when it comes to the "big picture." Independently of the abovementioned failure of traditional scientific reductionism, whether the naturalistic project would be more fruitfully grounded on the idea of a unique, all-encompassing ontology, albeit a noncanonical one (e.g., one of structures), is definitely open to discussion. To be sure, extant attempts to provide a general, unified ontology are in their early stages of development, and require a lot of additional work in the future.[10]

For the moment, then, the take-home message is that ontological questions such as that concerning individuality are best asked in a level- and context-dependent fashion, that is, always with specific and explicit reference to the entities belonging to the domain of inquiry of a specific science and a specific theory, and to the relevant formal apparatus and language. "Naturalistic pluralists" about individuality, in particular, should contend that the opposition between primitivism and reductionism can and should be overcome in favor of a more comprehensive and flexible view, allowing for different "forms of individuality" to be evaluated and assessed on the basis of their applicability and usefulness in the various specific theoretical contexts. A possible conceptual scheme resulting from all this might look something like the following:[11]

QFT: no individuals/stuff
⇓
QM: individuals with *primitivism*
⇓
CM: individuals with *reductionism* (based on extrinsic properties/relations)
⇓
Biology: individuals with *reductionism* (based on intrinsic properties)
⇓
. . .

14.5 Conclusions

In this chapter, it has been argued that nonrelativistic quantum mechanics can and should be regarded as a theory where the classical notion of individuality (with discernibility on the basis of monadic properties only) fails, but not that of individuality in general. Specific suggestions as to how to conceive of primitive identity and individuality in view of the relevant evidence have been offered, and primitivism has been defended as the most plausible ontological perspective on the theory, all things considered. On the other hand, it has also been argued that relativistic and nonrelativistic quantum field theories force one to undertake more radical conceptual changes, and ontological inquiry there might well lead to different conclusions. In light of the overall dynamics, the naturalistic metaphysician is well advised to be a pluralist and allow for the possibility that science gives us different metaphysical indications at different levels (or even in different domains at the same level and/or different representation of the same level/domain). Far from leading to relativism or constructivism, this was taken to be a natural consequence of the well-established fact that old-fashioned theoretical and ontological reductionism is unworkable, that is, of the fact that reality is complex and doesn't lend itself to a simple, unitary description. A sensible naturalistic methodology (i.e., one that, by taking science seriously, doesn't ipso facto endanger the autonomy of philosophical claims) is thus compatible not only with primitivism, but also with a degree of "ontological flexibility" that may open new philosophical problems but is certainly worth exploring in detail. More generally, the complexity and significance of the issue of identity and individuality shows that the prospects for research in naturalized metaphysics appear important but, in addition to a careful look at our best current science, they call for serious work on methodology. In particular, one should be always fully aware of the assumptions and intuitions that inevitably come into play when it comes to exploring the boundaries between philosophy and science and trying to put forward conjectures as to the essential structure of reality.

Acknowledgments

This chapter develops material presented at the "Individuals Across the Sciences" conference in Paris, 18–20 May 2012. I thank the audience there, as well as audiences in Amsterdam, Norwich, Urbino, and Valencia for their valuable feedback on related material, and Jonas Arenhart, Steven French and Décio Krause for their thoughtful comments on a previous draft. Lastly, I also gratefully acknowledge support from the Italian Ministry for University and Research, in the form of the three-year FIRB research grant F81J12000430001.

Notes

1. For various proposals on how to naturalize metaphysics, and how to understand naturalism more generally, see Ladyman and Ross 2007, Ritchie 2008, and Morganti 2013.

2. Here I will focus only on material objects broadly understood, in spite of the fact that, for instance, an event or a universal could be reasonably considered an individual.

3. It is worth pointing out that, while usually PII is taken to be a logical consequence of the so-called bundle theory of objects, according to which everything is composed of bundles of instances of universals, things are more complicated than this. First, one might have an ontology with properties only, but be a nominalist about universals. The result of this would be a "trope theoretic" view in the context of which, however, individuality would derive from the primitive identity of properties. More or less the same goes for those attempts to decouple PII and the bundle theory along the lines proposed by Rodriguez-Pereyra (2004): for, these attempts rely on a distinction between universals and their instances that, again, ultimately attributes a unique identity to the latter. Interestingly, the prospects for a qualities-only ontology without an explicit commitment to PII are currently being explored in connection to the ontology of quantum mechanics (see, e.g., Da Costa, Lombardi, and Lastiri 2013). Since, however, these approaches do not yet include an (explicit) theory of individuality alternative to those considered here, there is no need to deal with those in what follows.

4. In Bohmian mechanics, for instance, the uniqueness of trajectories is again warranted. However, in what follows, we will not deal with interpretations and theories

other than "standard" QM, if only to conform to the recent literature. A careful consideration of the alternatives is, however, certainly in order, especially in view of the pluralistic conclusions that will be defended later in this chapter.

5. It is interesting to note that the latest (to date) critical assessment of the extant methods for constructing weakly discerning relations concludes with the claim that all these attempts fail. This is, it is contended, because identity is presupposed not only at the formal level but in the very predicates that are used to refer to alleged physical properties. See Bigaj (2015), who also suggests a novel approach to the issue, suggesting that it might restore the 'traditional' discernibility of particles (that is, the 'absolute' sort of discernibility granted by the strong version of PII).

6. Recall our earlier reference to accounts of quantum statistics alternative to that provided by the "Received View" on quantum particles.

7. In other words, I am not claiming that being the least revisionary option is a decisive virtue in itself, nor that it is the most important. Rather, my claim is that, in science as in our everyday lives, we in fact proceed by changing as little as we can to make sense of the evidence, and this is fully compatible with changing a lot when the evidence (together with considerations of simplicity, unification, explanatory power, etc.) urges us to do so. I see no reason for not doing the same in philosophy.

8. For more details, and specific references, see Kuhlmann (2012).

9. As a matter of fact, the practice of biologists and philosophers of biology seems to suggest that (unlike, or, at any rate, even more than, in the case of physics) they don't start with a precise a priori definition of individuality with a view to seeing whether it works in the relevant domain, but rather from the assumption that whatever the central concept for theory-building will turn out to be, that will count as an individual.

10. Two additional questions for the 'anti-pluralist': if one is a naturalist, doesn't the failure of reductionism in the sciences lend straightforward support to the pluralist stance in metaphysics? And wouldn't it be better if the search for a unified ontology waited for the definition of something like a physical "theory of everything" (or at least a physics without internal inconsistencies)?

11. For more details, see Dorato and Morganti (2013). Generalizing, and to hint at another currently popular debate in metaphysics, ontological basicness/fundamentality might also be made discipline-, theory-, and/or problem-relative. And this potentially leads to an even larger project aiming to lend support to what one may call "Moderate Metaphysical Pluralism": that is, a view whereby objective structures of reality coexist with more flexible theory-relative structures. But this is just an indication for future work.

References

Arenhart, Jonas R. B., and Décio Krause. 2014. From Primitive Identity to the Non-individuality of Quantum Objects. *Studies in History and Philosophy of Modern Physics* 46B: 273–282.

Arenhart, Jonas R. B., and Décio Krause. 2013. Why Non-individuality? A Discussion on Individuality, Identity, and Cardinality in the Quantum Context. *Erkenntnis* 79: 1–18.

Baker, David J. 2009. Against Field Interpretations of Quantum Field Theory. *British Journal for the Philosophy of Science* 60: 585–609.

Belousek, Darrin W. 2000. Statistics, Symmetry, and the Conventionality of Indistinguishability in Quantum Mechanics. *Foundations of Physics* 30: 1–34.

Bigaj, Tomasz. 2015. Dissecting Weak Discernibility of Quanta. *Studies in History and Philosophy of Modern Physics* 50: 43–53.

Caulton, Adam. 2013. Discerning "Indistinguishable" Quantum Systems. *Philosophy of Science* 80: 49–72.

Caulton, Adam, and Jeremy Butterfield. 2012. Symmetries and Paraparticles as a Motivation for Structuralism. *British Journal for the Philosophy of Science* 63: 233–285.

Da Costa, Newton, Olimpia Lombardi, and Mariano Lastiri. 2013. A Modal Ontology of Properties for Quantum Mechanics. *Synthese* 190: 3671–3693.

Dieks, Dennis, and Marijn Versteegh. 2008. Identical Quantum Particles and Weak Discernibility. *Foundations of Physics* 38: 923–934.

Dorato, Mauro, and Matteo Morganti. 2013. Grades of Individuality. A Pluralistic View of Identity in Quantum Mechanics and in the Sciences. *Philosophical Studies* 163: 591–610.

Earman, John. 1979. Was Leibniz a Relationist? *Midwest Studies in Philosophy* 4: 263–276.

Fraser, Doreen. 2008. The Fate of "Particles" in Quantum Field Theories with Interactions. *Studies in History and Philosophy of Modern Physics* 39: 841–859.

French, Steven. 1989. Individuality, Supervenience and Bell's Theorem. *Philosophical Studies* 55: 1–22.

French, Steven, and Michael Redhead. 1988. Quantum Mechanics and the Identity of the Indiscernibles. *British Journal for the Philosophy of Science* 39: 233–246.

Hawley, Katherine. 2009. Identity and Indiscernibility. *Mind* 118: 101–119.

Horgan, Terry, and Matjaž Potrč. 2000. Blobjectivism and Indirect Correspondence. *Facta Philosophica* 2: 249–270.

Huggett, Nick, and James Norton. 2014. Weak Discernibility for Quanta, the Right Way. *British Journal for the Philosophy of Science* 65: 39–58.

Kuhlmann, Meinard. 2012. Quantum Field Theory. In *The Stanford Encyclopedia of Philosophy* (Winter 2012 ed., ed. Edward N. Zalta. http://plato.stanford.edu/archives/win2012/entries/quantum-field-theory/.

Ladyman, James, and Tomasz Bigaj. 2010. The Principle of the Identity of Indiscernibles and Quantum Mechanics. *Philosophy of Science* 77: 117–136.

Ladyman, James. 2007. On the Identity and Diversity of Objects in a Structure. *Proceedings of the Aristotelian Society* Supplementary Volume 81: 23–43.

Ladyman, James, and Don Ross. 2007. *Every Thing Must Go: Metaphysics Naturalized*. Oxford: Oxford University Press.

Morganti, Matteo. 2009. Inherent Properties and Statistics with Individual Particles in Quantum Mechanics. *Studies in the History and Philosophy of Modern Physics* 40: 223–231.

Morganti, Matteo. 2013. *Combining Science and Metaphysics: Contemporary Physics, Conceptual Revision and Common Sense*. Houndmills, Basingstoke: Palgrave Macmillan.

Muller, Fred A., and Simon Saunders. 2008. Discerning Fermions. *British Journal for the Philosophy of Science* 59: 499–548.

Muller, Fred A., and Michiel P. Seevinck. 2009. Discerning Elementary Particles. *Philosophy of Science* 76: 179–200.

Pooley, Oliver. 2006. Points, Particles and Structural Realism. In *The Structural Foundations of Quantum Gravity*, ed. Dean Rickles, Steven French, and Juha Saatsi, 83–120. Oxford: Oxford University Press.

Ritchie, Jack. 2008. *Understanding Naturalism*. Durham: Acumen.

Rodriguez-Pereyra, Gonzalo. 2004. The Bundle Theory Is Compatible with Distinct but Indiscernible Particulars. *Analysis* 64: 72–81.

Saunders, Simon. 2003. Physics and Leibniz's Principles. In *Symmetries in Physics: Philosophical Reflections*, ed. Katherine Brading and Elena Castellani, 289–308. Cambridge: Cambridge University Press.

Saunders, Simon. 2006a. Are Quantum Particles Objects? *Analysis* 66: 52–63.

Saunders, Simon. 2006b. On the Explanation for Quantum Statistics. *Studies in History and Philosophy of Modern Physics* 37: 192–211.

Shimony, Abner. 1981. Critique of the Papers of Fine and Suppes. *Philosophy of Science Proceedings 1980* 2: 572–580.

Sidelle, Alan. 1989. *Necessity, Essence and Individuation: A Defense of Conventionalism*. Ithaca, NY: Cornell University Press.

Stachel, John. 2004. Structural Realism and Contextual Individuality. In *Hilary Putnam*, ed. Yemima Ben-Menahem, 203–219. Cambridge: Cambridge University Press.

Teller, Paul. 1986. Relational Holism and Quantum Mechanics. *British Journal for the Philosophy of Science* 37: 71–81.

CHAPTER 15 | The Biological and the Mereological

Metaphysical Implications of the Individuality Thesis

MATTHEW H. HABER

15.1 Introduction

In philosophy, the term 'individual' is often treated as synonymous with 'mereological sum'. This reflects Henry Leonard and Nelson Goodman's construction of their "calculus of individuals" (1940) as formally indistinguishable from the logician Stanisław Leśniewski's axiomatic treatment of part-whole relations in his system of mereology (1927–1931), and their influence in actively encouraging the interchangeability of these systems. For example, they endorse the application of Leśniewski's mereology in J. H. Woodger's *The Axiomatic Method in Biology* (1937) as exemplifying their "convenience of the calculus of individuals" (46). In the 20th century these systems were adopted as alternatives to set theory, with concrete individuals (mereological sums) standing in place of abstract sets. Metaphysicians later adopted (and adapted) mereology as a theory of parts, wholes, and individuals, further encouraging this synonymy.

Biologists also use the term 'individual'. Though sometimes used synonymously with `organism`, it is generally regarded more broadly, and has been applied, for example, to social insect colonies, kin groups, homologies, holobionts, units of selection, species, and lineage-generating entities. Famously, the utility of this more expansive treatment was articulated in a series of articles by Michael Ghiselin and David Hull, asserting that species are neither natural kinds, nor sets, nor classes, but, much like organisms or cells, biological individuals (Ghiselin 1974, Hull 1976, 1978). This is the *individuality thesis*, and has since been applied widely in biology—though both the criteria and breadth of application of 'individual' in biology are still matters of debate.

It is easy to see how the individuality thesis (*that species are individuals*) has been interpreted as a mereological claim (*that species are mereological sums*). I will argue that this is a mistake, and that conflating these is both a conceptual and strategic error. It confuses the central issues at stake in both claims, and makes the job of evaluating (or defending) either unnecessarily burdensome. Advocates and critics alike should welcome a careful distinction of these projects.

Reconciling these applications of the term 'individual' requires drawing on multiple intellectual traditions that have not always been in close contact. Though this volume goes some way toward addressing this, readers may be more or less familiar with parts of that literature and it will be useful to start with a common jumping off point. Section 15.2 does this with a brief description of the individuality thesis and the axioms of mereology. Since both centrally concern parthood relations, it will also be useful to introduce a convention. I will use "part/whole" when referring to biological parthood concerns, and "part-whole" for mereological ones. Whether these are, in some sense, equivalent is just what is at stake.

My aim is also to establish that though the individuality thesis is not about mereology, it does carry metaphysical implications. These are more narrowly focused and sensitive to empirical theories than the general, sweeping metaphysics familiar to many of us trained in a philosophical tradition. Nonetheless, we ought to recognize these metaphysical commitments for what they are. These will be flagged as the details of the individuality thesis are unpacked, and represent an exemplar for doing *scientific* or *naturalistic* metaphysics.

15.2 Individuality: Biological and Mereological

The literatures on the individuality thesis and on adapting mereology for application to metaphysical problems of individuality are well developed and specialized, though they represent different intellectual traditions and research interests. So it will be useful to provide some entry points into each. I first offer my interpretation of the individuality thesis, presenting it as composed of three primary commitments. Following this is a standard presentation of the core axioms of mereology, along with enough discussion of the surrounding debates to motivate what follows. These will serve as common jumping-off points for the central argument below, though neither is intended as a full survey.

15.2.1 The Individuality Thesis

A brief statement of the individuality thesis will help get things started. Initially offered in terms of species, that is, that species are individuals (and not natural kinds or classes) (Ghiselin 1974, Hull 1976, 1978), the individuality thesis has been understood as a claim of ontological parity, as a rejection of essentialism, and as a characterization of species as spatiotemporally located entities with parts, not members. The focus here will be on the last of these, though it is worth taking a quick look at examples of each.

First, the individuality thesis is a commitment to ontological parity across the biological hierarchy. Hull captures this commitment in terms of how the names of biological individuals should be interpreted (1978, 338; see also Ghiselin 1966, 1974):

> Regardless of whether one thinks that "Moses" is a proper name, a cluster concept or a rigid designator, "*Homo sapiens*" must be treated in the same way.

On its own, this parity claim is neutral with respect to how we fill in the metaphysical details. It merely expresses the parity evident in Willi Hennig (1966, 1975) and G. C. D. Griffiths's (1974) treatment of biological hierarchy and taxonomy as a matter of diversification of systems ("systematics") as opposed to a logical ordering of classes ("classification"). Though the systems become recursively nested as a product of biological diversification, and though diversification itself may become highly specialized at different levels, the resulting hierarchy is itself a system of the same (ontological) category as its component systems (Okasha 2011 defends a similar view, casting it as a rank-free ecological hierarchy). In this regard, the individuality thesis is an expression and defense of the phylogenetic project (Hamilton 2012).

The individuality thesis is also often understood as a rejection of essentialism. One important way this has been interpreted—especially by systematists—is to understand identity in terms of history rather than intrinsic properties, traits, or characters. For example, in a phylogenetic system identity is associated with lineages, rather than characters expressed by parts of that lineage. Those characters may transform, after all, though the lineage persists, so conceptually divorcing these avoids a serious conflation (of defining a taxon for characters that may help diagnose it at some time). A stark example of this is *Crotalus catalinensis*—the Santa Catalina Island Rattleless Rattlesnake—which, as the name suggests, is a rattlesnake without a rattle (Beaman and Wong 2001).

Finally, Ghiselin and Hull both hold that biological individuals are spatiotemporally located concrete objects, and, importantly, have parts rather than members (in contrast to classes or sets). Though these sorts of claims are often framed in the context of a rejection of essentialism, they should also be understood as a consequence of the biology of individuals, rather than a logical stipulation (Ghiselin 1997, 40):

> Unfortunately the efforts of logicians along such lines have not produced the sort of logic (mereology) that would seem useful for our purposes (Leonard & Goodman 1940; Clarke 1981). We need something that treats individuals as more than just the sums of their parts.

And though Hull makes no mention of mereology (or Leonard and Goodman's "Calculus of Individuals"), he does describe Woodger's *The Axiomatic Method in Biology* as an application of classes in a Linnaean system, and as inadequate for an evolutionary one (1989, 131). In contrast, he consistently articulates the view that species have parts, not members, as a biological claim (typically in terms of evolution, e.g., 1989, chap. 6). Determining how the parts of a biological individual are related (as parts) is presented as an empirical research problem. It is something we may discover about, say, the products of evolution. In section 15.3.2 I will unpack this further, demonstrating how it subsequently generates the commitment that species are spatiotemporally located concrete objects.

On my interpretation, the individuality thesis may be described as a theoretical perspective with core commitments corresponding to the three claims above:

1. *The Parity Thesis*
   - Species, like organisms or cells, are biological individuals.
2. *The History Thesis*
   - Traits and characters play a diagnostic, not definitional, role in biological individuals.
3. *The Part/Whole Thesis*
   - The relevant part/whole relations are biological, not logical.

We can measure the depth of commitment to this theoretical perspective by evaluating the strength of commitment to each component (see Griesemer 2000a for more on theoretical perspectives; for the remainder of the chapter I will simply use "individuality thesis" as the name of this perspective). So on this formulation, biologists (or philosophers) may be more or less committed to the individuality thesis.

An advantage of this interpretation is that it incorporates an important advancement in recognizing that the original formulation in terms of species is an exemplar of a more general claim about biological individuality as an evolved character (see Wilson and Sober 1989, Ereshefsky 1991, and Baum 1998 for examples of broader applications of the thesis; De Monte and Rainey 2014 for treatment of this as a research problem in its own right). It also permits a broad interpretation of biological individuality, one that allows that it may be the result not merely of evolution, but of other biological processes (e.g., development; see Haber 2013). Though we might push deeper into each commitment, or ask whether the three commitments together are a rearticulation, revision, or expansion of Ghiselin and Hull's thesis, here that will be left aside and simply noted as the sort of research problem that could be (and has been) pursued.

This formulation of the individuality thesis leaves a lot unspecified, but that's by design. Part of its utility is in requiring theoretically motivated and empirically supported specification of details to be of use, and that it is largely neutral to *how* those details get filled in. This makes it useful as a general theoretical framework within which biological disputes can take place (e.g., over which relations between entities constitute them as biological parts of a more inclusive individual). This, of course, is a compelling reason to be clear about just what is included in the individuality thesis, so we can carefully identify what is part of the background assumptions for these sorts of debates and what is genuinely at stake.

More specifically, the focus here is on whether the individuality thesis entails (or, perhaps, even compels) a mereological interpretation of biological individuals. Let's briefly sketch out the contours of mereology before proceeding.

15.2.2 The Axioms of Mereology

Mereology is a formal theory of parts and wholes; it is an axiomatic system of parthood introduced by the Polish logician Stanisław Leśniewski (1927–1931). Leonard and Goodman (1940) developed (and helped popularize) the system as a calculus of concrete individuals, in direct competition with (yet analogous to) abstract set theory. It was adopted (and adapted) by logical positivists (who appreciated its nominalism [Goodman 1972b]) and analytic metaphysicians (who saw an avenue for a precise characterization of parts, wholes, and individuals). For many philosophers, the synonymy of 'individual' and 'mereological sum' has become entrenched; to speak of parts is just to speak of mereology (e.g., Simons 1987).

Offering even a brief survey of mereology is well beyond the scope of this chapter, and better left to those with a more focused expertise (for good entry points, see Varzi 2011 and Cotnoir 2013). My intent here is to provide a common jumping-off point that will motivate the arguments that follow. The aim is threefold: (1) state the axioms of mereology, (2) identify some points of agreement and (3) identify an ongoing dispute about those axioms. At stake is the scope of application of those axioms, and the intimacy of the relation between parthood and identity.

First, a standard formulation of the core axioms of mereology (adapted from Varzi 2011), taking '*Pxy*' as a primitive parthood relation, *x is part of y*:

(1) **Reflexivity**
Everything is part of itself;
Pxx.

(2) **Transitivity**
Any part of any part of a thing is itself part of that thing;
(Pxy & Pyz) → *Pxz*.

(3) **Antisymmetry**
Two distinct things cannot be part of each other;
(Pxy & Pyx) → *x = y*.

This gives us what is commonly referred to as *classical extensional mereology* (though see Hovda [2009], who identifies classical mereology in terms of its theorems rather than its axioms, providing five alternative axiomatizations).

Let's start with points of agreement. Including *Reflexivity* and *Transitivity* as core axioms or principles is generally uncontroversial; Gilmore (forthcoming) calls these the "non-negotiable core" of mereology (though see Wilson 2009). We might also include a background commitment to *Categorical Composition*: that parts belong either wholly or not at all to a mereological sum, or that parthood is logically bivalent. This rejects the possibility of mereological sums having parts that belong "more or less." Smith (2005) identifies a similar background commitment—the *Orthodox View*—framing it as a denial of vague composition. (I take these to be overlapping though distinct commitments, for reasons I will not go into.)

Most disputes about mereology may be set aside here. Some reflect matters of convention (e.g., over what should be treated as an axiom rather than a derived theorem); others are less germane to the topic at hand. More relevant are disagreements over the inclusion of *Antisymmetry*, what form it might take, and what other definitions, predicates, or theorems might accompany or be derived from it (Gilmore [forthcoming] provides a useful summary of the contours of this debate and what it entails). Among these is an important resource for the arguments below: the dispute over whether we ought to interpret mereology as an extensional or non-extensional system.

At stake is the strength of connection between parthood and identity. Extensional principles take the latter to be defined in terms of the former, reflecting an analogous principle found in set theory (that sets are defined by their members, such that two sets are identical if and only if they have exactly the same members). Given Leonard and Goodman's intention of developing a system that mirrored set theory, the appeal is straightforward. It permits a mereology that offers a nominalist alternative to all that set theory provides (see Varzi 2000, 2008 for a more contemporary defense).

What motivates the move to a non-extensional interpretation? Cotnoir (2010) provides a representative answer, cast as resolving a worrisome conflict. If two objects share all the same parts, then they are identical on an extensional account. If two objects are identical, then what is true of one must be true of the other (and vice versa). Yet there are many cases where this does not appear to be the case, for example, a statue and its clay share all the same parts, yet though the clay may persist despite being flattened, the same may not be said of the statue. One way to resolve these and other classic counterexamples is to abandon extensionality. This is not to say that mereological essentialists lack the resources to resolve such tensions (again, see Varzi 2000, 2008), but there are serious trade-offs on either option. (For those more familiar with the jargon, Cotnoir casts this as a conflict between the *extensionality of parthood principle* and the law of *indiscernibility of identicals*. In contrast, Jubien 2001 embraces mereological essentialism, identifying the tendency towards "object fixation" as the culprit. See Rea 1997 for a nice collection with discussion of the statue and clay and other classic counterexamples.)

Rather than getting bogged down in details, let's try to draw out the dispute over extensionality in broad contours by considering the proposition 'x is a mereological sum' (where x stands for some familiar object). Failing to specify whether 'mereological sum' is to be interpreted as extensional or not generates an ambiguity over how to read 'is' here. On an extensional interpretation, 'is' is an *is of identity*. The familiar object is identical to *a* mereological sum, and satisfies the axioms of mereology. x could not have had different parts, for then it would have been a different mereological sum and not identical to itself. The intimacy of parthood and identity is deep.

In contrast, on a non-extensional interpretation, the 'is' is an *is of predication*. x is not identical to any particular sum, though some such sum must coincide with it. x could have had different parts without any threat to x's identity, though it would coincide with a different mereological sum. On non-extensional accounts, identity and parthood are disentangled (to various degrees, depending on the particulars of the system, e.g., Sider 2007 advocates for a "moderate composition of identity").

For many philosophers, to speak of parts at all is just to speak of mereology (e.g., Simons 1987, though see Varzi 2006 and Fine 2010 for pluralist accounts of 'parthood'). So it is understandable why the individuality thesis—which holds that species have parts, not members—has been interpreted as being about mereology. I contend that this is a mistake, and that we should treat the biological project (*that species are individuals*) as distinct from the mereological one (*that species are mereological sums*). To understand how these projects are (and are not) connected, let's take a look at the metaphysical implications of the individuality thesis.

15.3 Metaphysical Implications of the Individuality Thesis

The individuality thesis is not metaphysically inert. It contains metaphysical implications, either directly in its core commitments or in terms of the research projects generated by those commitments. The central question here is whether a mereological interpretation is among those implications.

First, let's establish that the individuality thesis *has* been read as a claim of mereology. This despite good evidence that neither Ghiselin nor Hull intended this reading, and even resist it. At the very least this ought to give us pause before too quickly adopting the mereological interpretation. Rather, we ought to read the part/whole thesis as a biological claim.

This pause permits space to push a bit deeper into what it means to say that the part/whole thesis is a biological claim, keeping track of the metaphysical implications of this commitment along the way. If the part/whole thesis is equivalent to (or best interpreted as) a mereological claim, then the biological relations that confer individuality should also satisfy the axioms of mereology. As will be shown, they do not (both transitivity and categorical compositionality are violated), and so should not be evaluated as mereological commitments.

15.3.1 Mereological Interpretations of the Individuality Thesis

Interpreting Ghiselin and Hull's view as a biological claim remains controversial, and philosophers have routinely misread it as a mereological one. For example, Philip Kitcher (1984a, 1984b, 1987) argues that the individuality thesis is simply a preference for mereological treatments over set-theoretic ones, though, as he rightly notes, they are effectively equivalent (1984b, 620):

> I want to begin by distinguishing two versions of the species-as-individuals doctrine. The first version is a bare ontological proposal, rooted in a predilection for mereology rather than standard set theory. I don't doubt that the same biology and philosophy of biology can be done using either of these ontological alternatives. My critique of Hull's thesis was directed at a more exciting doctrine which connected the mereological approach to claims about the existence of laws and the impossibility of historically disconnected species.

Or consider Berit Brogaard's description of the individuality thesis (in a defense of it and species pluralism on mereological grounds):

> According to [the individuality thesis], which was first suggested by Ghiselin and later developed by Hull and others, the organisms of species taxa are joined together by the part-whole relation rather than by that of class membership. Species taxa are not classes but mereological sums that are cohesive the way organisms are. (2004, 223)

And:

> The species-as-individuals thesis makes a claim about the ontological status of species taxa. The claim is the part-whole relation rather than that of class membership joins together the organisms of species taxa. Thus, species taxa are mereological sums rather than classes of organisms. (2004, 240)

These and other mereological readings of the individuality thesis are too quick. As mentioned above, though Hull never explicitly rejected a mereological interpretation, there is good evidence that this is not what he had in mind. (Ghiselin, in contrast, disavows a mereological interpretation [1997].) For example, he dismisses Woodger's axiomatization project as "extremely deficient" and, with characteristic flourish, expresses a general skepticism toward logically formal treatments of biology (1974, 660):

> Scientists are not rushing to flesh out the formal analyses of theory reduction presented by philosophers. I do not see them even strolling in that general direction.

Furthermore, Hull is no nominalist, and is content to retain classes as a perfectly suitable category, even permitting that there is a species category (whose members are not classes, but individuals) (1976, 174–175):

> On the view being urged in this paper, both particular species and the species category itself must be moved down one category level. Organisms remain individuals, but they are no longer members of their species. Instead an organism is part of a more inclusive individual, its species, and the names of both particular organisms (like Gargantua) and particular species (like *Gorilla gorilla*) become proper names. The species category itself is no longer a class of classes but merely a class. As a class, it can continue to be defined in the usual way.

From Hull's dissatisfaction with Woodger's project, and of formal analyses of biology more generally, it seems reasonable to at least pause before too quickly reading mereological claims off of the individuality thesis.

Still, this pause is not enough on its own to resist a mereological interpretation of the individuality thesis. We might discover that the relations that confer biological individuality also satisfy the axioms of mereology. This sets the task ahead plainly, treating as an empirical problem whether or not the strategies employed by biologists for individuating entities are mereological.

15.3.2 Biological Not Mereological

On a mereological interpretation biological wholes are also mereological ones by definition; biological parthood must be a reflexive, transitive, and bivalent relation because it is mereological. (On an extensional interpretation, additional criteria must be satisfied.) Brogaard is explicit about this (2004, 226):

> The parthood relation is transitive. So if species taxa are mereological sums, then organisms are not the only parts species have; any part of, or combination of parts of, the organisms would also be parts of the species taxon. I do not believe this is a problem. It might seem odd to treat your elbow as belonging to the species of human beings. But I believe the oddness of this view stems from our habit of treating species as classes with members. Since your elbow is part of you, and you are part of the species of human beings, why not say that your elbow too is part of the human species?

Well, why not say this? The problem is that the question as posed by Brogaard is ill formed and imprecise. This is a result of conflating the biological and mereological projects, and confusing what is centrally at stake in either. It encourages a derivation of biological commitments from a logical one. To address the question, we must take a closer look at what it means to be a biological part, and whether this also entails (or is even consistent with) being a mereological part.

Though all three of the individuality thesis's core components carry metaphysical implications, a good place to focus is the part/whole thesis. Biologists have typically interpreted this as an empirical claim (e.g., Rowe 1987, 209):

> Hull (1976) pointed out that populations, species, and higher taxa need not be made up of similar organisms, but that in a genealogical system they must be made up of related organisms. He also pointed out that taxonomists traditionally have not imposed this requirement upon taxa; rather it has followed from the nature of the evolutionary process itself.

Two things to notice. First, the criteria of identity here is relational, not intrinsic. An organism is part of a taxon by virtue of how it relates to other organisms in that taxon, not due to any intrinsic property or character it has. This is as much a biological as metaphysical claim. Second, discovering what it means for organisms to be related as parts is an empirical research program that will be informed by theory, concepts, and empirical facts. Both of these points generalize (as expressed by the parity thesis). It will be as true for a cell of a multicellular organism as it is for a species of a clade. What it means to be a biological part of an individual is not imposed from a priori metaphysical commitments, but derived from empirically informed biological theory. Let's unpack this a bit, starting with one of Hull's original formulations (1976, 185–186):

> There is no fertilization at a distance. Either the organisms or their propagules must come into contact. But ancestor-descendant contact is required in all species. Regardless of the mode of reproduction, offspring come into existence in reasonably close proximity to their parents.

This leaves as an open empirical question how to characterize (or even recognize) reproduction, just what things will generate lineages, and how (or whether) this might happen recursively. There are entangled biological and metaphysical commitments here, both implicit and explicit, that constrain how to answer these questions in the context of the individuality thesis. For example, reproduction requires material contact or overlap, and is not measured in terms of similarity but development (see Griesemer 2000b for a worked out account of this). This is a rather radical claim, and still quite controversial (see Godfrey-Smith 2009). Regardless, these metaphysical commitments are a consequence of the biology, rather than derived from an a priori metaphysics or axiomatic system. That species are spatiotemporally located, concrete entities, with a beginning and an end, and have parts, not members, is a function of how lineages are generated, and of the facts of reproduction.

If our theories about these facts are wrong, then we may, in turn, need to adjust our metaphysical commitments. So be it. Furthermore, the individuality thesis must be sensitive to these facts, and either accommodate them or risk losing its utility or being replaced. It is worth pausing to note that this is exactly what we should expect in a scientific metaphysics; our metaphysical commitments must be able to accommodate the very same empirical facts that our best theories do, and our attitudes toward them should be similar.

Given the close ties between the individuality thesis and the development of phylogenetics (Hamilton 2012), it is hardly surprising that an account of biological parthood may be provided in terms of parent-offspring relations (e.g., as recursive, nested lineages [Haber 2012b]). Yet my interpretation of this thesis leaves open the possibility that other sorts of biological relations might confer individuality. Accounts of these typically include a specification of participating entities, and detailed diagnoses of how they may relate (often in terms of embodied processes or mechanisms). These may be general accounts or specific to particular levels or scales, reflecting a particular evolutionary path individuality may have taken.

For example, biologists might employ metabolic scaling theory to individuate an entity. West, Brown, and Enquist (1997) describe a model of allometric scaling that provides a general account of body size: "Our model provides a theoretical, mechanistic basis for understanding the central role of body size in all aspects of biology" (125). The model is a tool for identifying (individuating, in this case) the structurally and functionally interconnected parts of a metabolic distribution system. Hou and coauthors (2010) apply this to social insect colonies (3636):

> Our results also imply that colonies are groups of individuals that are functionally organized to capture and use energy in ways that are remarkably similar to those of unitary organisms. Indeed, the similarity in the scaling relationships for both colonies and unitary organisms suggests that the physiology and life history of colonies and unitary organisms follow the same "rules" with respect to size.

This study of how the parts of a social insect colony are functioning with regard to each other, satisfying West and coauthors' criteria of allometric scaling, is taken as strong evidence for treating social insect colonies as a unitary metabolic distribution system. That is, an individual whose parts are functionally interconnected though scattered.

Let's pause for a moment to appreciate the metaphysical significance here, especially with respect to the mereological interpretation. Recall that a nonnegotiable core of mereology is that parthood is transitive. That biologists routinely employ a theoretically motivated individuating strategy that relies on a functional account of parthood threatens to undermine a mereological interpretation (Varzi 2006, 142, emphasis added):

> It is obvious that if the interpretation of 'part' is narrowed by additional conditions (e.g., by *requiring that parts make a functional, direct, or otherwise distinguished contribution to the whole*), then *transitivity may fail*. In general, if x is a φ-part of y and y is a φ-part of z, x need not be a φ-part of z: the predicate modifier 'φ' may not distribute over parthood. But that shows the non-transitivity of 'φ-part', not of [mereological]-'part'.

This appears to be just the case here. Biological parts are φ-parts, not mereological ones, requiring, at least, some additional conditions to count as proper biological parts. Assessing these as parts in terms of mereology is to make a category error.

Furthermore, this problem will not be constrained to metabolic individuals. *All* biological accounts of individuality (or biological parthood) will reflect some biologically relevant condition. That's simply what they are designed to do. This is not to say that transitivity will fail in all such cases, only that it is empirically determined rather than logically derived. (Slater [2013] offers a similar observation.)

For example, Okasha (2006) provides another exemplar, specifying biological parthood in terms of the Price equation (a mathematical model capturing shared evolutionary fate, more or less). Same for Godfrey-Smith (2009, 2014), whose criteria are expressed in terms of Darwinian Populations (e.g., marked by reproductive bottlenecks). Other options are available, some offering variations on the theme of shared evolutionary fate, others identifying metabolic, developmental, ecological, or immunological relations (or some combination thereof) as ways that objects may stand to each other as parts of a biological individual (in some cases, these individuals, or some intersection of them, will also be organisms).

This failure of transitivity will be exhibited at higher levels as well. A powerful example of this is at the level of clades. Genealogical discordance describes cases where the phylogenetic pattern at one level of lineage (say, population or species lineages) is discordant with the phylogenetic patterns of parts of that lineage (say, gene or organismal level lineages). The upshot is that just because a species is part of a monophyletic group, this does not entail that all of its parts (e.g., organisms, genes, or their lineages) will be too. Indeed, under certain conditions a species tree is guaranteed to be discordant with its most frequent gene tree (Degnan and Rosenberg 2006). Furthermore, this failure of transitivity is predicted by theory, and many patterns of discordance are systematically diagnosable. This feature of biological systems may be powerfully exploited to extract other facts; for example, the amount of discordance between a lineage and its component parts may be used to calculate divergence time or other phylogenetic signals (Kingman 1982, Avise and Robinson 2008).

Genealogical discordance may be a bit esoteric to effectively make this point to a broad audience, but we can see the failure of transitivity at a more familiar level as well, in species. Brogaard recognizes this, noting that species concepts relying on interbreeding relations will violate transitivity (the ability to interbreed is not transitive). Her response is to propose modifications to these views, solely in order to salvage a mereological interpretation (2004, 233–234):

> What we need is a relation akin to those of reproduction and gene flow but transitive and symmetric. ... The revised biological species concept may be formulated as follows: two populations a and b are parts of the same species taxon if and only if a and b contribute to the same gene pool.

This strikes me as backward and unnecessarily constraining of biological theory (and it is not obvious that this formulation will be consistent with our best theories about biological species concepts, e.g., Coyne and Orr 2004). Rather, our metaphysics should follow the biology here, and simply acknowledge the failure of transitivity as a fact and consequence of how interbreeding groups (or other notions of species) emerge, persist, and diverge. If our current metaphysical tools have difficulty adequately accounting for this, then the proper response is to develop new metaphysical tools, not constrain the science. The latter leads to irrelevance. Better to consider the former as a research problem for the scientifically informed metaphysician, much as working out the details of speciation or ring species is a research problem for population biologists, not an inconvenient fact to be defined away. This is an embodiment of *naturalized* or *scientific* metaphysics (Ladyman 2007, Sarkar 2007, Ladyman and Ross 2007, Ross, Ladyman, and Kincaid 2013).

It is not just the violation of transitivity that undermines the mereological interpretation. Categorical composition will be violated as well. Biological parthood will, more often than not, be gradient, not bivalent. Furthermore, this attenuation of parthood will be biologically relevant, making it all the more difficult to define away. As parts become more or less strongly cohesive with the other parts of an individual, they will play different functional, structural or other roles *as parts* of that individual.

For example, as human skin cells detach from the basal layer of the epidermis they begin a program of terminal differentiation, moving toward the outermost surface of the skin. As the cell becomes less integrated with other skin cells (physiologically, metabolically, etc.) it will take on different functional roles in that organ. This persists even after cell death, when it will become enucleated and flattened but still be part of an important regulatory barrier. As this connection becomes further attenuated, the cell will eventually slough off (Fuchs and Green 1980, Fuchs and Raghavan 2002).

The symbiotic relationship between the Hawaiian bobtail squid *Euprymna scolopes* and the bioluminescent bacteria *Vibrio fischeri* provide several dimensions along which parts belong to individuals along a gradient. *E. scolopes* are nocturnal foragers that rely on counterillumination to camouflage themselves from predators against the moonlight (Jones and Nishiguchi 2004). Illumination is achieved through a symbiotic relationship with the bioluminescent bacteria *V. fischeri*, which are housed in the squid's light organ. Each dawn *E. scolopes* expels ~95% of the *V. fischeri*, both to aid daytime hiding (buried under the sand) and to avoid the symbiont overgrowing the host. This flushing out also helps seed the environment with *V. fischeri*, which may then colonize newborn squid (transmission between generations is not transovarian, but environmental; newborn squid actively acquire *V. fischeri* from the environment after birth) (Wei and Young 1989, McFall-Ngai and Ruby 1998, Nyholm and McFall-Ngai 2004). The remaining *V. fischeri* recolonize the squid's light organ, so that by dusk it may emerge and safely forage again. The relationship between these two lineages is stable, persisting not merely over individual life histories, but multiple generations, and evolutionary time. And though each species may develop in absence of the other, when *V. fischeri* is present it has specific influences on the development of *E. scolopes* (McFall-Ngai 2002, Koropatnick et al. 2004). This has become a model symbiosis, particularly for the study of how symbionts become actively integrated yet compartmentalized, how microbes influence the developmental programs of their hosts, and for how stable environmentally transmitted associations may persist over various timescales.

V. fischeri's influence on *E. scolopes*' development is not the only example of how gradation of parthood may occur along a developmental dimension (see West-Eberhard 2003, or the numerous studies of how the microbe *Wolbachia* influences arthropod reproduction and sex-selection as transovarian symbionts [O'Neill et al. 1997]). Fagan (this volume) provides another example, arguing that from a developmental perspective we ought to "consider stem cells as 'transformative entities,' mediating between two different levels of biological individuality: cell and organism." Fagan argues this is complicated by the fact that stem cells may only be identified retrospectively, and relative to the lineage they generate. In this case, the gradation is over levels of hierarchy as the cells become parts of a multicellular organism, in addition to temporal and spatial boundaries.

Fagan's example is reminiscent of the way populations of organisms gradually diverge from their parent lineages to form a distinct lineage in their own right. A nascent species might only be identified retrospectively as a stem lineage, after its descendant populations have progressed through enough stages of speciation. Species boundaries, in general, strain the criterion of categorical composition to its breaking point (Coyne and Orr 2004, 30):

> We feel that it is less important to worry about species status than to recognize that the process of speciation involves acquiring reproductive barriers, and that this process yields intermediate stages when species status is more or less irresolvable.

Most recent approaches to navigating the species problem embrace this to some degree, with some even arguing that the various species concepts are better recognized as species criteria that capture important stages along the individuation process at the species or population-lineage level (de Queiroz 1998, 1999 and Harrison 1998 provide examples of this in a phylogenetic context, in contrast to the interbreeding approach favored by Coyne and Orr). Furthermore, gradation may also span spatial (geographic) taxonomic boundaries, though, for a variety of reasons, these may have been lost to time (Darwin 1859). The result is a gradient model of speciation whose stages are themselves marked by gradient boundaries. The latter undercuts at least one mereological response: that species boundaries are simply overlapping mereological sums corresponding to the stages of speciation.

Metaphysicians may be tempted to see this as a problem of vagueness. Perhaps this is right, though, for reasons similar to my hesitation in too quickly reading a mereological interpretation of the individuality thesis, I think a pause is warranted here as well. The philosophical problem of vagueness is often cast as a problem of predication or irresolvability. In contrast, many biological predicates of individuation demand a gradient—rather than categorical—interpretation to reflect the theories of individuation and parthood on which they are based. This can be a bit confusing, as many of the same terms are used colloquially without appreciation for the scientific details. As a quick example, asserting that an organism o belongs to a species s is often misinterpreted as asserting a property of that organism, formally representing 'o is an s' as 'So'. This mistakes 'belonging to species s' as a one-place, rather than two-place predicate. That is, we ought to recognize that asserting 'o is an s' is equivalent to 'o belongs to s', and represent it as instantiating a biological *belongs to* relation, 'Bos' which may be satisfied more or less strongly (see Haber 2012a, for a more worked out account).

Let's summarize the metaphysical implications of the individuality thesis described above. There were quite a number, though a commitment to the mereological interpretation was not among them. We began with two, that biological parthood is relational, not intrinsic, and that genealogical relations require material contact or overlap, even if brief or indirect. The former has been built into metaphysical accounts drawn from our best biological theories; for example, Paul Griffiths (1999) and Samir Okasha (2002) both develop relational accounts of natural kinds and essentialism in response to this commitment. Other candidate biologically informed metaphysical commitments include that objects will be marked as much by how their parts vary as how they display similarity; that being a biological part or whole will often be expressed along a gradient, rather than categorically; and that parthood is a product of evolution and other biological processes, and may be selected for (or against). Though these are not the grand, sweeping metaphysical generalizations of old, they are, in many regards, more useful and impactful. If we adopt a scientifically informed stance toward metaphysics, it should come as no surprise that many of the conclusions we draw will be narrower but deeper in scope. Nor should it be surprising that these metaphysical implications will be controversial in a slightly different regard, that is, more sensitive to empirical testing and hinging on the theories from which they were derived (Ladyman 2007, Sarkar 2007).

Some might resist treating these as metaphysical commitments. Well, why shouldn't we? They are assertions about what things there are in the world, and why and how they are these things. The scope may be narrower than we (as philosophers) are used to, and some scientists may not be very comfortable embracing the metaphysical aspect of their sciences, but if we are to be doing scientific or naturalistic metaphysics, then we ought to follow the lead where it takes us.

15.3.3 Whither Mereology?

The mereological interpretation fails. Both transitivity and categorical composition are violated. (Were we to consider an extensional account, modal problems would further undermine it.) This amounts to an empirical claim, that our best biological strategies, theories, and models of individuation are in conflict with the axioms of mereology.

Yet not all hope is lost for the mereologist. Let's not lose sight that mereology is an axiomatic system. The question is not whether we are able to model biological systems with it, but whether it models biological systems *well*. Answering that requires that we know to what end we are applying these models; what research problems are these models advancing? If the goal is to capture a realistic treatment of biological individuals, or to cohere with our other biological models or concepts (drawn from our best biological theories), then it will fall short for the same reasons Woodger's axioms were left behind—a failure to relevantly bear on the research interests of biologists.

It may be useful to come clean on my pragmatic attitude toward metaphysical commitments. What concerns me are questions like these: Is the proposed system of metaphysics *consistent* with the individuality thesis? Does it generate useful research questions? Is it internally consistent or coherent? Are its resources and commitments useful in a biological framework? Some metaphysicians and realists might also wish to add, "Is it true?" Though this would certainly be a virtue, for reasons I will not go into here I do not view this as a necessary consideration. (Briefly, if we are considering how well a mereological system models a biological one, then we know that certain trade-offs will be necessary in the construction of that model, e.g., between precision, generality, and realism [Levins 1966, Cartwright 1983, Odenbaugh 2006, Wimsatt 2007, Weisberg 2013].) This, then, is the task for the mereologist: Demonstrate the *utility* of the mereological interpretation; establish that the models it generates are on par with or better for studying some biological feature than those from other hypotheses of biological part/whole relations.

15.4 Conclusion

Is the biological individuality thesis about mereology? Above I offered reasons to think not. Biological parthood may fail to be transitive. In some cases this discordance is systematically diagnosable and predicted by theory. Biological parthood is also typically not bivalent. Categorical biological boundaries are the exception, not the rule; gradients are the norm in biology, and these gradients are often biologically relevant. As parts become more or less part of an individual, they may play different functional or structural roles.

This is not a denial of the thesis that we ought to consider biological individuals mereological sums, only that this metaphysical commitment is not entailed or compelled by the individuality thesis. Conflating these positions (*a*) encourages a focus on the wrong issues; (*b*) distracts from the central project of articulating biological individuality; (*c*) may lead to erroneous rejection of either thesis; and (*d*) draws attention away from the genuinely metaphysical implications of the individuality thesis. All these may be avoided by carefully distinguishing the mereological from the biological project.

Acknowledgments

I thank Thomas Pradeu and Alexandre Guay for their great patience and guidance through this chapter, and challenging me to take on a deeply interesting task. They should be thanked also for the wonderful organization of the stimulating conference in which it was originally presented, along with the other contributors, presenters, and audience, who provided insightful feedback. Deep thanks are also due to the very thorough and helpful anonymous referees, whose comments and advice vastly improved this chapter.

References

Avise, John C., and Terence J. Robinson. 2008. Hemiplasy: A New Term in the Lexicon of Phylogenetics. *Systematic Biology* 57(3): 503–507.

Baum, David A. 1998. Individuality and the Existence of Species through Time. *Systematic Biology* 47(4): 641–653.

Beaman, Kent R., and Nelson Wong. 2001. *Crotalus catalinensis* Cliff Santa Catalina Island Rattleless Rattlesnake. *Catalogue of American Amphibians and Reptiles* 733: 1–4.

Brogaard, Berit. 2004. Species as Individuals. *Biology and Philosophy* 19(2): 223–242.
Cartwright, Nancy. 1983. *How the Laws of Physics Lie*. Oxford: Oxford University Press.
Clarke, Bowman L. 1981. A Calculus of Individuals Based on "Connection." *Notre Dame Journal of Formal Logic* 22(3): 204–218.
Cotnoir, Aaron J. 2010. Anti-symmetry and Non-extensional Mereology. *Philosophical Quarterly* 60(239): 396–405.
Cotnoir, Aaron J. 2013. Strange Parts: The Metaphysics of Non-classical Mereologies. *Philosophy Compass* 8(9): 834–845.
Coyne, Jerry A., and H. Allen Orr. 2004. *Speciation*. Sunderland, MA: Sinauer Associates.
Darwin, Charles. [1859] 1964. *On the Origin of Species*. Facsimile ed. Cambridge, MA: Harvard University Press.
De Monte, Silvia, and Paul B. Rainey. 2014. Nascent Multicellular Life and the Emergence of Individuality. *Journal of Biosciences* 39(2): 237–248.
de Queiroz, Kevin. 1998. The General Lineage Concept of Species, Species Criteria, and the Process of Speciation. In *Endless Forms: Species and Speciation*, ed. Daniel J. Howard and Stewart H. Berlocher, 57–75. Oxford: Oxford University Press.
de Queiroz, Kevin. 1999. The General Lineage Concept of Species and the Defining Properties of the Species Category. In *Species: New Interdisciplinary Essays*, ed. Robert A. Wilson, 49–89. Cambridge, MA: MIT Press.
Degnan, James H., and Noah A. Rosenberg. 2006. Discordance of Species Trees with Their Most Likely Gene Trees. *PLoS Genet* 2(5): e68.
Ereshefsky, Marc. 1991. Species, Higher Taxa, and the Units of Evolution. *Philosophy of Science* 58(1): 84–101.
Fine, Kit. 2010. Towards a Theory of Part. *Journal of Philosophy* 107(11): 559–589.
Fuchs, Elaine, and Howard Green. 1980. Changes in Keratin Gene Expression during Terminal Differentiation of the Keratinocyte. *Cell* 19(4): 1033–1042.
Fuchs, Elaine, and Srikala Raghavan. 2002. Getting under the Skin of Epidermal Morphogenesis. *Nature Reviews: Genetics* 3(3): 199–209.
Ghiselin, Michael T. 1966. On Psychologism in the Logic of Taxonomic Controversies. *Systematic Zoology* 15(3): 207–215.
Ghiselin, Michael T. 1974. A Radical Solution to the Species Problem. *Systematic Zoology* 23(4): 536–544.
Ghiselin, Michael T. 1997. *Metaphysics and the Origin of Species*. Albany: SUNY Press.
Gilmore, C. Forthcoming. Quasi-supplementation, Plenitudinous Coincidentalism, and Gunk. In *Substance: New Essays*, ed. Robert Garcia. Munich: Philosophia Verlag.
Godfrey-Smith, Peter. 2009. *Darwinian Populations and Natural Selection*. New York: Oxford University Press.
Goodman, Nelson. 1972. A World of Individuals. In *Problems and Projects*, 155–172. Indianapolis, IN: Bobbs-Merrill.
Griesemer, James. 2000a. Development, Culture, and the Units of Inheritance. *Philosophy of Science* 67(S1): S348.

Griesemer, James. 2000b. The Units of Evolutionary Transition. *Selection* 1(1–3): 67–80.

Griffiths, Graham C. D. 1974. On the Foundations of Biological Systematics. *Acta Biotheoretica* 23(3): 85–131.

Griffiths, Paul E. 1999. Squaring the Circle: Natural Kinds with Historical Essences. In *Species: New Interdisciplinary Essays*, ed. Robert A. Wilson, 209–228. Cambridge, MA: MIT Press.

Haber, Matthew H. 2012a. How to Misidentify a Type Specimen. *Biology and Philosophy* 27(6): 767–784.

Haber, Matthew H. 2012b. Multilevel Lineages and Multidimensional Trees: The Levels of Lineage and Phylogeny Reconstruction. *Philosophy of Science* 79(5): 609–623.

Haber, Matthew H. 2013. Colonies Are Individuals: Revisiting the Superorganism Revival. In *From Groups to Individuals: Evolution and Emerging Individuality*, ed. Frédéric Bouchard and Philippe Huneman, 195–217. Cambridge, MA: MIT Press.

Hamilton, Andrew. 2012. From Types to Individuals: Hennig's Ontology and the Development of Phylogenetic Systematics. *Cladistics* 28(2): 130–140.

Harrison, Richard G. 1998. Linking Evolutionary Pattern and Process: The Relevance of Species Concepts for the Study of Speciation. In *Endless Forms: Species and Speciation*, ed. Daniel J. Howard and Stewart H. Berlocher, 19–31. Oxford: Oxford University Press.

Hennig, Willi. 1966. *Phylogenetic Systematics*. Urbana: University of Illinois Press.

Hennig, Willi. 1975. "Cladistic Analysis or Cladistic Classification?": A Reply to Ernst Mayr. *Systematic Zoology* 24(2): 244–256.

Hou, Chen, Michael Kaspari, Hannah B. Vander Zanden, and James F. Gillooly. 2010. Energetic Basis of Colonial Living in Social Insects. *Proceedings of the National Academy of Sciences* 107(8): 3634–3638.

Hovda, Paul. 2009. What Is Classical Mereology? *Journal of Philosophical Logic* 38(1): 55–82.

Hull, David L. 1974. Informal Aspects of Theory Reduction. *PSA: Proceedings of the Biennial Meeting of the Philosophy of Science Association* 1974: 653–670.

Hull, David L. 1976. Are Species Really Individuals? *Systematic Zoology* 25(2): 174–191.

Hull, David L. 1978. A Matter of Individuality. *Philosophy of Science* 45(3): 335–360.

Hull, David L. 1989. *The Metaphysics of Evolution*. Albany: SUNY Press.

Jones, Bryan, and Michele Nishiguchi. 2004. Counterillumination in the Hawaiian Bobtail Squid, *Euprymna scolopes* Berry (Mollusca: Cephalopoda). *Marine Biology* 144(6): 1151–1155.

Jubien, Michael. 2001. Thinking about Things. *Philosophical Perspectives* 15: 1–15.

Kingman, John F. C. 1982. On the Genealogy of Large Populations. *Journal of Applied Probability* 19: 27–43.

Kitcher, Philip. 1984a. Species. *Philosophy of Science* 51(2): 308–333.

Kitcher, Philip. 1984b. Against the Monism of the Moment: A Reply to Elliott Sober. *Philosophy of Science* 51(4): 616–630.

Kitcher, Philip. 1987. Ghostly Whispers: Mayr, Ghiselin, and the "Philosophers" on the Ontological Status of Species. *Biology and Philosophy* 2: 184–192.

Koropatnick, Tanya A., Jacquelyn T. Engle, Michael A. Apicella, Eric V. Stabb, William E. Goldman, and Margaret J. McFall-Ngai. 2004. Microbial Factor-Mediated Development in a Host-Bacterial Mutualism. *Science* 306(5699): 1186–1188.

Ladyman, James. 2007. Does Physics Answer Metaphysical Questions? *Royal Institute of Philosophy Supplements* 61: 179–201.

Ladyman, James, and Don Ross. 2007. *Every Thing Must Go: Metaphysics Naturalized.* Oxford: Oxford University Press.

Leonard, Henry S., and Nelson Goodman. 1940. The Calculus of Individuals and Its Uses. *Journal of Symbolic Logic* 5(2): 45–55.

Leśniewski, Stanisław. 1927–1931. O Podstawach Matematyki [On the foundations of mathematics], I–V. *Przegląd Filozoficzny*, vols. 30–34.

Levins, Richard. 1966. The Strategy of Model Building in Population Biology. *American Scientist* 54(4): 421–431.

McFall-Ngai, Margaret J. 2002. Unseen Forces: The Influence of Bacteria on Animal Development. *Developmental Biology* 242(1): 1–14.

McFall-Ngai, Margaret J., and Edward G. Ruby. 1998. Sepiolids and *Vibrios*: When First They Meet. *BioScience* 48(4): 257.

Nyholm, Spencer V., and Margaret J. McFall-Ngai. 2004. The Winnowing: Establishing the Squid-*Vibrio* Symbiosis. *Nature Reviews Microbiology* 2(8): 632–642.

Odenbaugh, Jay. 2006. The Strategy of "The Strategy of Model Building in Population Biology." *Biology and Philosophy* 21(5): 607–621.

Okasha, Samir. 2002. Darwinian Metaphysics: Species and the Question of Essentialism. *Synthese* 131: 191–213.

Okasha, Samir. 2006. *Evolution and the Levels of Selection.* Oxford: Oxford University Press.

Okasha, Samir. 2011. Biological Ontology and Hierarchical Organization: A Defense of Rank Freedom. In *The Major Transitions in Evolution Revisited*, ed. Brett Calcott and Kim Sterelny, 53–64. Cambridge, MA: MIT Press.

O'Neill, Scott L., Ary A. Hoffmann, and John H. Werren, eds. 1997. *Influential Passengers: Inherited Microorganisms and Arthropod Reproduction.* Oxford: Oxford University Press.

Rea, Michael C., ed. 1997. *Material Constitution: A Reader.* Lanham, MD: Rowman & Littlefield.

Ross, Don, James Ladyman, and Kincaid Harold, eds. 2013. *Scientific Metaphysics.* Oxford: Oxford University Press.

Rowe, Timothy. 1987. Definition and Diagnosis in the Phylogenetic System. *Systematic Zoology* 36(2): 208–211.

Sarkar, Sahotra. 2007. *Doubting Darwin? Creationist Designs on Evolution.* Malden, MA: Blackwell.

Sider, Theodore. 2007. Parthood. *Philosophical Review* 116(1): 51–91.

Simons, Peter. 1987. *Parts: A Study in Ontology.* Oxford: Clarendon Press.

Slater, Matthew. H. 2013. *Are Species Real? An Essay on the Metaphysics of Species.* New York: Palgrave Macmillan.

Smith, Nicholas J. J. 2005. A Plea for Things That Are Not Quite All There: Or, Is There a Problem about Vague Composition and Vague Existence? *Journal of Philosophy* 102(8): 381–421.

Varzi, Achille C. 2000. Mereological Commitments. *Dialectica* 54(4): 283–305.

Varzi, Achille C. 2006. A Note on the Transitivity of Parthood. *Applied Ontology* 1(2): 141—146.

Varzi, Achille C. 2008. The Extensionality of Parthood and Composition. *Philosophical Quarterly* 58(230): 108–133.

Varzi, Achille C. 2011. Mereology. In *Stanford Encyclopedia of Philosophy*, Spring 2011 edn., ed. Edward N. Zalta. http://plato.stanford.edu/archives/spr2011/entries/mereology/.

Wei, S., and Richard Young. 1989. Development of Symbiotic Bacterial Bioluminescence in a Nearshore Cephalopod, *Euprymna scolopes*. *Marine Biology* 103(4): 541–546.

Weisberg, Michael. 2013. *Simulation and Similarity: Using Models to Understand the World*. New York: Oxford University Press.

West, Geoffrey B., James H. Brown, and Brian J. Enquist. 1997. A General Model for the Origin of Allometric Scaling Laws in Biology. *Science* 276(5309): 122–126.

West-Eberhard, Mary Jane. 2003. *Developmental Plasticity and Evolution*. New York: Oxford University Press.

Wilson, David S., and Elliott Sober. 1989. Reviving the Superorganism. *Journal of Theoretical Biology* 136(3): 337–356.

Wilson, Robert A. 2009. The Transitivity of Material Constitution. *Noûs* 43(2): 363–377.

Wimsatt, William C. 2007. *Re-engineering Philosophy for Limited Beings: Piecewise Approximations to Reality*. Cambridge, MA: Harvard University Press.

Woodger, Joseph. H. 1937. *The Axiomatic Method in Biology*. Cambridge: Cambridge University Press.

CHAPTER 16 | **To Be Continued**
The Genidentity of Physical and Biological Processes

ALEXANDRE GUAY AND THOMAS PRADEU

16.1 Introduction

In 1922, Kurt Lewin (a leading German-American psychologist, 1890–1947) proposed the concept of "genidentity" to better understand identity through time. Lewin's aim was to offer a conception of identity that would be relevant across different sciences, especially physics and biology (Lewin 1922). Later, philosopher Hans Reichenbach (1891–1953) distinguished different conceptions of genidentity, and applied them to physical cases (Reichenbach [1956] 1971). However, many philosophers and scientists have considered that Reichenbach's view of genidentity was imprecise and, consequently, failed to shed light on the identity of physical objects. In particular, Reichenbach's account does not seem to apply to particles in nonrelativistic quantum mechanics, a domain in which trajectories and causation are problematic (e.g., French and Krause 2006, 48–49). Recently, a few philosophers have resorted to the concept of genidentity to reflect on mereology (e.g., Smith and Mulligan 1982), or to understand biological entities (e.g., Boniolo and Carrara 2004), but their views have not aroused much discussion in the philosophical community. In this chapter, examining several specific examples taken from current classical physics and biology, we defend the genidentity view and show that it would be fruitful to adopt it in at least some areas of those sciences.

What does the concept of genidentity say? In a nutshell, it says that the identity through time of an entity X is given by the continuous connection of states through which X goes. For example, a "chair" is to be understood in a purely historical way, as a connection of spatiotemporal states from its making to its destruction. In this view, the individual X is never presupposed or given initially, because the starting point is the decision to follow a specific and appropriate *process* P, and the individual X supervenes on this process. For example (as detailed below), one can decide to follow the conservation of an internal physiological organization through time, and the effectuation of this process gives us our individual entity (in this case, an individual organism). In other words, for the genidentity view, what we single out as an "individual" is always the byproduct of the *activity* that is being followed, not its prior foundation (not a presumed "thing" that would give its unity to this activity). (Of course not every sequence of events is associated with an individual, as explained below.) A bacterium must be seen as a connection of spatiotemporal states: these states are "genidentical." Now, if a bacterium divides into two daughter bacteria, most biologists will say that this division marks a new start, which means that states of the mother bacterium and states of the daughter bacteria cannot be considered as "genidentical." The statement that "a new start" occurs, which is the key statement of the genidentity view, naturally depends on a *criterion of continuity* adopted by the observer. We show below how experimental sciences can help us define these criteria in specific contexts.

The genidentity view can be better understood by its opposition to other conceptions of temporal identity. First, the genidentity view is *antisubstantialist* since it says that the identity of X through time does not presuppose that "something" of X remains. (It is also—perhaps even more explicitly—*antiessentialist*, at least if essentialism is understood as the claim that X is the same if there exists a permanent "core" or "substrate" of X through time.) Indeed, in the genidentity view, the question "What *is* X, fundamentally?" is replaced by the question "How should I *follow* X through time?" Second, the genidentity view is also in opposition to the idea of *identity-resemblance*, according to which X remains through time if it sufficiently resembles itself. In other words, three general conceptions of identity through time are schematically distinguished here:

1. *Substantialism*: the identity of X is rooted in the idea that something of X remains through time (and *essentialism* is therefore one important form of substantialism, though not the only one: as will be explained below, Leibniz and Wiggins, for instance, are substantialists but not essentialists).
2. *Identity-resemblance*: the identity of X is rooted in the idea that X looks sufficiently like itself through time.
3. *Genidentity*: the identity of X is rooted in the idea that X is characterized by sufficiently continuous states through time.

As has often been recognized (e.g., Boniolo and Carrara 2004, 456 n. 1), the genidentity view draws on the conception of identity defended by

Locke ([1694] 1975). Exploring the problem of identity through time of a given entity (be it a tree, an animal, or a human), Locke considers that the most satisfying criterion for diachronic identity (the "principle of individuation") is continuity of states (1975, II, 27, §3, 330). Applied to the problem of the identity through time of a given man, this conception leads Locke to assert:

> This also shows wherein the Identity of the same *Man* consists; viz. in nothing but a participation of the same continued Life, by constantly fleeting Particles of Matter, in succession vitally united to the same organized Body. (Locke 1975, §6, 331)

Our intention is to build further on Locke's suggestion, in order to offer a precise definition of the notion of genidentity, applicable to physics and biology, and to lay the foundations of an ontology centered on processes and change rather than substances, invariance, and laws. Famously, Leibniz (in particular in his *New Essays*, written as a systematic response to Locke), disagrees with Locke on identity viewed as continuity:

> By itself continuity no more constitutes substance than does multitude or number. ... *Something* is necessary to be numbered, repeated and continued. (Leibniz [1765] 1916, Gerhardt II, 169)

In contemporary metaphysical debates about identity, David Wiggins (2001, 57) explicitly endorses Leibniz's view, and rejects the idea that identity could be defined as bare continuity (see also Wiggins 1968). The dispute between Locke and Leibniz, therefore, is far from being extinguished. Here we defend a view close to Locke's, in some specific cases taken from physics and biology (a similar defense, applied specifically to immunology, can be found in Pradeu and Carosella 2006).

Our focus in this chapter is on diachronic, not synchronic, identity: we are not asking what makes the identity of a physical or biological object *at time t*—for instance how it can be distinguished from other things, what its physical boundaries are, and so on. Instead, we are asking what makes X the "same" at two different moments in time. This will naturally involve questions about distinguishability, boundaries, or individuality, but always in a temporal, historical context. In other words, to use the language of relativity, we will not discuss why a particular event can be associated to a particular individual, but how different events can be related and seen as characterizing the same individual.[1]

Though the intention of its instigators was to make a useful contribution to experimental sciences, applications of the concept of genidentity to real science have been scarce and, usually, unsatisfying. In this chapter, we try to take up Lewin's objective by demonstrating that a better-defined concept of genidentity sheds an important light on identity through time both in physics and in biology.

We are aware that a direct comparison between physics and biology regarding the way they conceptualize identity may seem surprising. Physics and biology both raise questions about individuality and identity through time, but they do so in a markedly different way. These differences include the four following aspects:

1. In biology, parts-whole questions seem crucial, probably because most, if not all, biological entities appear as constituted of smaller biological entities (as, for example, when one asks to what extent the cells constituting a multicellular organism are themselves "individuals"), and most biological entities appear to be constituents of larger biological entities, while part-whole questions play a less important role in physics.[2]
2. In physics, distinguishing *one* particle among many identical particles is a key issue, while in biology even individuals that are said to be "identical" express, most of the time, some significant differences and, at the very least, can usually be spatially distinguished.
3. In physics, the principle of indiscernibles is critical in discussions about synchronic identity, but not in biology. Biologists often say that two living things are "identical" even when they do not share all their properties (in particular their position in space), as with clonal organisms.
4. In physics, discussions over structuralism are extremely important, as one crucial aim is to determine what remains invariant under transformations; in biology, structuralism plays a lesser role, if any (for exceptions to this trend, see French 2011, French and Ladyman, this volume).

Despite these important differences, we intend to show in this chapter the utility of a comparison between physics and biology in their understanding of identity through time, and the fruitfulness of the concept of genidentity in both domains. From our point of view, the gain is more significant in biology than in physics, but some important cases taken from physics can nonetheless benefit from the adoption of a genidentity view. In other words, we suggest that the genidentity view is useful both in biology and in physics, and, perhaps even more importantly, that it can be pivotal to fostering a dialogue between these two major scientific fields. The outline of the chapter is as follows. In section 16.2, we examine different concepts of genidentity and the difficulties they raise. In sections 16.3 and 16.4, we explore several examples, respectively in physics and in biology, that demonstrate the utility of adopting the genidentity view. In section 16.5, we lay the foundations of an ontology of *processes* and *change*, in contrast to an ontology of *substances, invariance*, and *laws*.

16.2 Approaches to Genidentity

16.2.1 Lewin's Conception of Genidentity

The term "genidentity" is typically regarded as coined by Lewin in 1920 (and later analyzed in Lewin 1922), though he was not the first to use it in a published work. For Lewin, the concept of genidentity is expressed in mereological terms. A temporally extended entity has to be understood as a multiplicity of entities, all parts of the genuine individual, and the relation between these entities is genidentity.[3] Since the kind of mereology involved in each scientific discipline is not exactly the same, the appropriate genidentity relation is therefore context dependent.

Lewin defines *simple* and *complete* genidentity. Two collections of temporal parts (sets of entities at two distinct times) are *simply genidentical* if they are genidentical but one of the collections (let us say the earlier one) could also be said to be simply genidentical to another collection at the second time. The two collections are said to be *completely genidentical* if that is not the case. This distinction will not play any role in this chapter but is important in Lewin's conception because simple and complete genidentity definitions are related. As shown by Smith and Mulligan (1982), in the context of physics, Lewin's definition of *simple* genidentity on the one hand and *complete* genidentity on the other can be formalized as shown in figure 16.1 (if we represent complete genidentity by $^P\equiv$, simple genidentity by $^P=$, is a proper or improper part of by $\leq$ and, is discrete from as /).

We may illustrate this definition with an example. Let us imagine that we have initially a piece of iron and a bucket of strong acid. The piece of iron falls in the bucket and dissolves, emitting heat. According to the above definition (iron + acid) $^P\equiv$ (acid solution + heat) and (iron + acid) $^P=$ (heat). The mereological definitions (simple and complete genidentity) imply that after the action of the acid on the iron piece, the iron is still in the solution in a certain form. And before this action the heat was in a certain form present. All this is of course true: the iron is still present as atoms in the solution and the heat was present in the initial state as chemical binding energy. But obviously what is meant by "to be present" is very different in each case. This is an important disadvantage of any mereological definition of genidentity. In this approach, if two states

$$a^P \equiv b := \neg\exists x\,(x/a\,\&\,x^P = b)\,\&\,\neg\exists x\,(x/b\,\&\,x^P = a)$$
$$a^P \equiv b \rightarrow \forall x\,(x/b \rightarrow a^P \neq x)\,\&\,\forall x\,(x/a \rightarrow b^P \neq x)$$
$$a^P = b \rightarrow \exists a'\exists b'\,(a' \leq a\,\&\,b' \leq b\,\&\,a'^P \equiv b')$$

FIGURE 16.1 Formalization of simple and complete genidentity according to Smith and Mulligan (1982)

TO BE CONTINUED | 321

are genidentical, they must be conceived as different parts of the same temporally extended entity. This assertion does not shed any light on the relation that makes these two states of the same thing. Moreover, we could argue that once we have individuated a temporally extended entity in order to identify its temporal parts, we do not need to give a precise definition of this relation since we already possess a diachronic identity criterion. In conclusion, Lewin's definition of genidentity presupposes that we already have a method to individuate individuals through time, which is highly problematic. In Lewin's account, genidentity is a primitive notion. Overall, therefore, this approach does not seem appropriate to develop an operative concept of genidentity that could be used in physics and biology.

16.2.2 Reichenbach's Conception of Genidentity

Reichenbach ([1956] 1971) offered his own conception of genidentity. It is this conception that will be applied here to biology and physics. (We will focus on the later work of Reichenbach on genidentity; on his earlier conception, see Padovani 2013.) Reichenbach distinguishes *genidentity*, which is applied to *physical* entities, from *logical* identity:

> The *physical identity* of a thing, also called *genidentity*, must be distinguished from logical identity. An event is logically identical to itself; but when we say that different events are states of the same thing, we employ a relation of genidentity holding between these events. A physical thing is thus a series of events; any two events belonging to this series are genidentical. (Reichenbach [1956] 1971, 38)

In this quote, Reichenbach points exactly to the main difficulty inherent to the notion of genidentity. As Quine (1966, 145) later argued, the diachronic identity of a thing cannot be captured by logical identity alone, since such identity is indexed by space-time locations; therefore, something else must be provided.[4] Reichenbach believes that "speaking of things and speaking of events represent merely different modes of speech" ([1956] 1971, 224), and he proposes to define the diachronic identity of a thing as a relation among events. At a minimum, two genidentical events should be related by a *"worldline"* that lies in the light cone of the earliest event (in other words, two events cannot be genidentical if they can be related only by a signal going faster than the speed of light). Being related by a worldline is necessary for genidentity, but it is not sufficient, since there is a certain degree of arbitrariness in how we define worldlines for successive events (it is especially true in the case of fields, for which the direction of the striation is arbitrary). This arbitrariness is one of the reasons that convinced Einstein that the concept of substance was inadequate for modern physics (Reichenbach 1958, 270–271).

In addition to being related by a worldline, then, two genidentical events must be related by a *causal* relation.[5] But not any causal relation will do the trick. Only causal chains that could be said to relate events characterizing states of the *same* thing should be considered. Since there is more than one way to talk about the same thing through time, Reichenbach argues that there will be more than one way to define a genuine genidentical relation. That genidentity could be defined in multiple ways and, therefore, that the notion of an individual is relative to a certain theoretical point of view should not be surprising. In Reichenbach's (and, even more, in our) approach to the relation between metaphysics and science, metaphysical concepts, such as, for example, the concept of an individual, help us organize the scientific discourse, but *they are certainly not a foundation to it*. Except in cases where "individual" is a theoretical term defined explicitly in a theory, the concept of an individual could be abstracted or projected on the scientific discourse. It does not bind it, nor are we forced to include it in our ontology. However, ontological commitment can act as a constraint. Indeed, it could be argued that, in Reichenbach's work at least, the concepts of genidentity and causality are closely connected, to the point that the latter cannot be defined without the former. When Reichenbach discusses the constraints on a causal chain signal in special relativity, he conveys the idea that *something* is propagating through space and time. In consequence, a principle, namely genidentity, is required to individuate this *something* through space-time, in order to discuss the causal relation involved.

The closest way to define a genidentity relation that could reflect how we talk about ideal macroscopic objects, like billiard balls (if these objects were unaltered by any interaction), is what Reichenbach called *material genidentity*. The aim of this definition is to get to the diachronic identity of beings evolving temporally. This relation can be defined by three characteristics: (1) continuity of change, (2) spatial exclusion, and (3) distinguishability of states that differ only in the permutation of two objects ([1956] 1971, 225). The first characteristic expresses the fact that you must be able to follow the evolution of the individual through space and time. Continuity of change is the easiest way to achieve this goal, but this condition could be relaxed in certain cases. The second characteristic specifies that different individuals cannot occupy the same position simultaneously. This guarantees that an interaction between individuals will not make the identity ambiguous. The last characteristic is less intuitive but is justified, we think, by the need to avoid the identity ambiguity when the determination of which events are simultaneous could differ depending on the chosen reference frame. Of course, if no two individuals share all the same properties, they cannot be wrongly identified. This is not in general the case in physics.[6] These characteristics are necessary but not sufficient to guarantee individuality, since, as Reichenbach admits, there are obvious counterexamples ([1956] 1971, 225).

In practice, it would be difficult to find any useful application of the notion of material genidentity in current physics. Nevertheless, it can work as an idealized (paradigmatic) case, illustrating the necessary conditions for a more complex and more useful conception of genidentity. In this second conception, we are interested in the evolution through space and time of a unique individual potentially *interacting with other individuals at a distance*. To capture such a process, Reichenbach proposes the concept of *functional genidentity*, which refers to the retention of the same "function" through time, without retention of matter. For instance, if we claim that an organism is still the "same" after absorbing some nutriments, we are considering the identity through time of an individual that interacts with other individual entities, and, this is not something that material genidentity can explain (because material genidentity presupposes spatial exclusion) (Reichenbach [1956] 1971, 227); we need to take into account the "functional" identity of this organism, and not its "material" identity. Reichenbach remains imprecise about "functional genidentity," but he does give a nice example: transversal water waves. In a transversal wave, water moves only vertically. The continuity of change does not involve transportation of matter. However, there is a continuous momentum and energy propagation perpendicular to this vertical move. Spatial exclusion is violated, since waves can superpose and there could be permutation symmetry since a wave can be identical to another if they share the same dynamical properties. Nevertheless, in many circumstances, you can track a wave in space-time by defining a *function* of the wave parameters that evolves continuously.

Which function should be used to track the "real" individuals is not an appropriate question in this context. Of course, as a minimal requirement, the function should denote a causal process. But among all possible causal processes, which one should be understood as relating states of the same thing is a question without a general answer. Someone could claim that the wave equation describes the collective behavior of water molecules, the legitimate individuals or, even further, that the wave is denoting the change in space-time points' properties, which would be the "real" individuals. Since the connection between a scientific theory and its ontology is not straightforward, all these positions can be argued (Braillard et al. 2011). In section 16.3, we return to the question of whether the existence of a particular function is enough to claim that a wave is an individual.

Overall, Reichenbach has made an important contribution to the definition of genidentity by making several key distinctions:

1. Genidentity is about *physical* identity, which implies more than just *logical* identity.
2. Genidentity corresponds not to any *physical* identity (defined as a worldline of continuous states), but to *causal* physical identity (defined as a worldline of *causally connected* continuous states).
3. Genidentity can be divided into *material* genidentity (defined as retention of matter, distinguishability, and spatial exclusion with regard to other entities) and *functional* genidentity (defined as retention of the same "function" through time, without retention of matter).

After Reichenbach, very few scientists and philosophers have used the notion of genidentity. (There are some exceptions, including Boniolo and Carrara 2004, but because of space limitation we will not discuss them here.)

16.3 Genidentity in Physics

In this section, we show how the notion of genidentity can shed light on the problem of individuation in classical physics. Importantly, we will not adopt here a normative approach to the notion of an individual. In particular, we will not claim that only something like *material genidentity* (defined as retention of matter, distinguishability, and spatial exclusion with regard to other entities) captures the real concept of an individual (see table 16.1), although it is undeniable that this relation is very close to the concept we use in ordinary life when talking about the identity of material things. Instead, we will start with material genidentity, and then relax its requirements in order to generate a more applicable concept of genidentity, namely a *functional* one. Armed with this functional genidentity relation, we will, in section 16.4, apply it to a much-debated problem, the problem of biological individuality.

This move from material genidentity toward functional genidentity will be structured by three paradigmatic cases of growing complexity, each step building on the previous one: (1) the free individual (the form of functional genidentity that is closest to material genidentity), (2) the changing individual without interaction, and finally (3) the fully interacting individual (see table 16.2). Once the three types are described, we can combine them to represent complex situations. All these cases will be illustrated

TABLE 16.1 Necessary Conditions for Material Genidentity

MATERIAL GENIDENTITY		
1. Continuity of change	2. Spatial exclusion	3. Distinguishability

TO BE CONTINUED

TABLE 16.2 Paradigmatic Cases of Functional Individuals

Free functional individuals	Relaxed characteristics 2 and 3 of material individuals	Ex.: free solitons
Changing functional individuals	Constant or continuously changing rate of the function defining free individuals	Ex.: dissipating waves
Interacting functional individuals	Changing/free individuals + interaction events (with or without loss of identity)	Ex.: interacting solitons

in wave theory. The choice of wave theory is not insignificant. Contrary to particle physics, undulatory mechanics has traditionally been hostile to the concept of an individual. So offering a definition of an individual in this context would demonstrate the fruitfulness of our approach. Though philosophers of physics have very rarely paid attention to this aspect of wave theory and to the case of solitons in particular, our conviction is that this is a pivotal domain to understand the notion of genidentity and, more generally, to reflect on identity in classical physics. It is worth mentioning that, since the standard procedure to get to a quantum theory is to start from a classical theory, even if the ontology of a particular quantum theory could be different from its associated classical theory, it is never totally independent from it.

Let us start with the first case, the free individual. A perfect billiard ball moving in a straight line in vacuum is a model of a free individual for which the states are *materially genidentical*. If we relax the second characteristic (spatial exclusion) of material genidentity, a soliton, that is, a kind of self-reinforcing solitary wave, becomes as well a model close to material genidentity (Drazin and Johnson 1989). A soliton is a localized traveling wave in nonlinear systems. It does not obey the superposition principle and does not dissipate. In the case of a soliton, no matter is globally transported. As Reichenbach had emphasized for wave theory, the diachronic identity of a soliton can only be represented by *functional* genidentity. For example, the solution of the nonlinear partial differential equations could serve as the identity function since the shape of the wave is unchanging and the movement is continuous. Other functions of the topology or conservation principles could also be used. Our claim here is that a soliton is functionally as close as one could get to a substantial individual for which we would have a material genidentity relation enabling us to follow its states. Trivially, a solitary soliton is absolutely discernible and it can be continuously followed. For us, it is the paradigmatic case of a functional individual. As to the question "Is this particular soliton solution representing a physical individual or not?," it does not have a general answer. All depends on how this particular piece of mathematics is used in the physical model. It is the physical model that defines what kinds of functions are ontologically significant for diachronic identity. It is worth mentioning that solitons are considered as particles or more precisely quasi-particles in many quantum field models. These particles are different from the so-called elementary particles since they do not arise from the quantization of the wavelike excitations of the fields and possess a topological structure (Mantan and Sutcliffe 2004).

Let us now return to the case of the billiard ball in a vacuum, and relax the third characteristic (distinguishability under permutation) in order to allow other free identical billiard balls moving around without interacting one with the other. The equivalent model would be many identically shaped solitons that travel without meeting or interacting one with the other. In this case, our entities are not *absolutely discernible* anymore but are at best *weakly discernible* (for definitions, see Saunders's and Ladyman's chapters in this volume). Should we in this context renounce calling them "individuals"? We do not think so. Only if a change of reference frames implies an identity ambiguity among them should we renounce talk about "individuals." In many cases, the loss of absolute discernibility does not preclude the definition of a robust genidentity relation, since each soliton could be followed precisely in space-time.

Let us now discuss the second case: the continuously changing individual. Here we have in mind cases where an individual is apparently changing during its movement in space-time, for example, a billiard ball continuously changing color, or, in undulatory theory, a continuously changing solitary wave. The first case, the free individual, gives a baseline with which to understand this new case. We have a function that allows us to follow the individual through its continuous evolution, typically with one or many conserved quantities. Now we allow this function to fluctuate. We measure its derivative to quantify this change. Under which rate of change we should still consider the related events as characterizing the *same* individual is a contextual question. Note that we do not ask how much color change, or topology change, is enough to talk of a new individual, but how intense, or discontinuous, the *rate* of change must be to assert we face a new individual or no individual at all. The question of color or topology change presumes that the identity of the billiard ball is essentially defined by the possession of some properties. The second and third questions, in contrast, are in the spirit of the genidentity conception of the individual. No properties or structure define the identity of the billiard ball or of the wave beyond the causal process followed by the genidentity relation. Please take note that in this case, the source of the change (color, shape ...) is not on the same ontological level as the studied individual. The source of change is considered diffused in the environment. There is no clear vertex with the worldline of another individual. This is the main difference with the next and final case.

Our final case is the fully interacting individual. In the genidentity approach, this case is the really challenging one, since no change of essential properties can inform us when an interaction destroys or creates an individual. Only a careful study of the nature of the interaction itself can provide elements for discussing the identity of the individuals involved. To illustrate this point, let us look at the interaction between a green and a red billiard ball. In the initial state, we have two balls coming toward each other. For the sake of the discussion, let us suppose that we do not have access to the details of this particular interaction. In the last stage, we see a green ball and a red ball apparently identical to the incident balls going away from each other with a certain angle compared to the initial trajectories. What can we say about the conserved identity of the individuals involved? If the interaction is a simple repulsion, it seems probable that the final green ball is genidentical to the initial green ball, and the same would be true for the red one. However, if the repulsion is accompanied by an exchange of color, our answer would be different. The nature of the interaction is what guides us in our assessment of the billiard balls' identity. Of course, if the genidentity of billiard balls is of the *material* kind, then each individual is uniquely associated to a nonintersecting space-time trajectory; therefore we simply have to follow their trajectories during the interaction to be able to know which is which. But in a case of *functional* genidentity, where spatial exclusion is not a given, the study of the nature of the interaction cannot be put aside.

This last case is precisely the subject of the study of soliton interactions. Contrary to linear waves, which do not interact when they are superposed, the nonlinear nature of solitons implies complex interactions. Only a systematic study of these interactions allows us to say something about the genidentity of interacting solitons. For example, let us take the class of one-dimension line solitons involved in a nonresonant elastic interaction. In these situations, the number of solitons is conserved, except during the interaction phase. All soliton properties are completely restored after interaction. For example, the direction of propagation and the linear momentum seem unchanged. Important amplitude change can occur during the interaction. For example, in many cases the total amplitude of the superposed solitons is less than the sum of each individual amplitude, but each amplitude is restored after the interaction (see figure 16.2). The interaction often generates a phase shift, a change of position compared to what would be the position of the solitons if the interaction did not

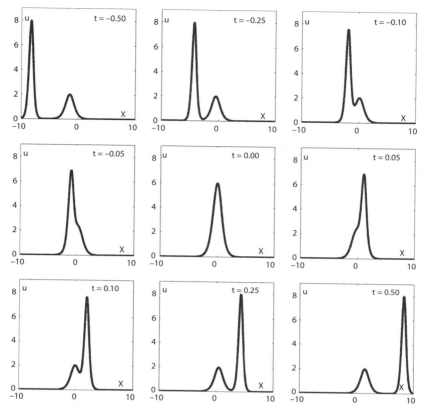

FIGURE 16.2 Temporal evolution of Korteweg–de Vries solitons described by the two-soliton solution $u(x, t) = 12 \times (3 + 4 \cosh(2x - 8t) + \cosh(4x - 64t))/([3 \cosh(x - 28t) + \cosh(3x - 36t)]^2)$ (Soomere 2011, 1583). Note the nonadditivity of amplitudes during interaction

With kind permission from Springer Science and Business Media.

occur (for more details, see Soomere 2011). Depending on the particular type of functional genidentity that is discussed, these characteristics of the interaction will or will not allow us to assert that the solitons emerging from the interaction are the same as the ones entering into it. More precisely, the question that we need to ask is in what way the interaction *perturbs* the genidentity established in the second paradigmatic case. In these examples, even if the interaction is complex, it seems that the worldline of a particular soliton is almost not perturbed except for a phase shift. It seems reasonable to argue that, for most functional genidentity relations, the initial individuals are the ones appearing in the final state. However, this is not the only possible case. In the context of resonant interaction, the fusion of two or more solitons in a new soliton is possible. The time reverse process, fission, is also possible. Obviously, in this case, the initial

TO BE CONTINUED

solitons do not survive. Some cases are more ambiguous and therefore difficult to decide.

This point can be illustrated by an example, that of head-on collisions of two baby skyrmions (BS). Baby skyrmions are topological solitons, in a (2 + 1)-dimensional field theory, which are closely related to the Skyrme model. In a nutshell, a skyrmion is a soliton solution to the Skyrme model; this field model is used, among other things, to represent states of baryons and excited baryons: see (Skyrme 1962). If the parameters are chosen to allow a head-on collision, at the apex of the interaction the two initial baby skyrmions merge to form a ring-like structure. Just after this, two new baby skyrmions emerge, for almost all initial parameters, at 90° from the original incident trajectories. During all the interaction process, radiation is emitted (for more details, see Piette, Schroers, and Zakrewski 1995). Even if the initial BSs are not identical and the final two BSs are also not identical in a similar way, the symmetry of the interaction seems to limit our capacity to argue either way about the survival of the initial BSs. The complexity of the interaction is a strong limitation on the kind of functional genidentity relation that we could reasonably use. This contextuality of identity may seem problematic. If we could accept it for quasi-particles and biological beings, surely the identity criterion of fundamental particles should not be contextually defined? This worry presumes that there is such a thing as a fundamental theory describing fundamental entities. If, as it seems currently probable, we do not have a fundamental theory but only effective ones (see note 2), contextuality is unavoidable. This conclusion does not necessarily lead to arbitrary identities or pseudoidentities. It suggests that identity is relative to a theoretical context, for example, to a certain range of physical parameters shared by most theories, like energy.

Thus, in this section we argued for three points:

1. It is possible in a field theory to define a strong notion of genidentity, at least when worldlines can reasonably be defined. Conservation of matter is not necessary.
2. It is possible to define relaxed genidentity relations that could accommodate changing individuals.
3. Conservation of genidentity in interacting situations depends on the details of the interaction. More precisely, conservation of genidentity depends on how this interaction modifies the rate of change of the individuating function.

Overall, this section has shown that the notion of genidentity suggested here can be applied to some interesting physical cases, and that the confrontation with such cases can in turn help us to refine our notion of genidentity, and to better determine its scope. In the next section, we examine whether the notion of genidentity can be applied to biological cases, and we use what has been shown here about physical cases to shed a new light on the understanding of the identity of biological entities through time.

16.4 Genidentity in Biology

The previous section has presented several applications of the concept of genidentity to physical cases. We would now like to show that the concept of genidentity can also be applied to some biological cases or, more exactly, that it is the concept best suited to understand identity through time in biology.

Though questions pertaining to synchronic identity are important in biology, those pertaining to diachronic identity have long been recognized as central (Sober 2000, 154), in particular, at different temporal scales, in developmental biology and evolutionary biology. How do organisms and species remain the "same" through time, even though they change constantly? How do they "start" and "end" their lives? How can we know that we are talking about the "same" organism or the "same" species at two different moments in time, in particular if massive changes occur between these two moments?

Certainly more than any other philosopher of biology, David Hull (1935–2010) has recognized the crucial importance of the question of diachronic identity in biology. Even more strikingly, Hull explicitly defended the concept of *genidentity* in several texts. For Hull, a fundamental characteristic of living things is that they can undergo massive changes and nonetheless maintain their "identity." It is precisely this idea, Hull claims, that only the concept of genidentity can properly capture:

> Three traditional criteria for individuality in material bodies are retention of substance, retention of structure, and continuous existence through time (genidentity). If organisms are to count as individuals, then the first two criteria are much too restrictive. In point of fact, many organisms totally exchange their substance several times over while they retain their individuality. Others undergo massive metamorphosis as well, changing their structure markedly. If organisms are paradigm individuals, then retention of neither substance nor structure is either necessary or sufficient for continued identity in material bodies. The idea that comes closest to capturing individuality in organisms and possibly individuals as such is genidentity. As its name implies, this criterion allows for change just as long as it is sufficiently continuous. The overall organization of any entity can change but it cannot be disrupted too abruptly. (Hull 1992)[7]

Elsewhere Hull expresses a similar defense of the concept of genidentity (Hull 1986).[8] One might be surprised by the fact that Hull, arguably the most influential of all philosophers of biology in the twentieth century, explicitly used this notion to express one of his most famous theses (about the identity of living things) without having aroused much enthusiasm. Indeed, as we said in the beginning of this chapter, the notion of genidentity is rarely used by philosophers, philosophers of biology included (as an illustration, "genidentity" has only two occurrences, from 1986 to today, in *Biology and Philosophy*, arguably the leading journal in philosophy of biology), and even Hull did not change that situation. Is it because, here again, the notion of genidentity would be too imprecise? We believe, on the contrary, that Hull offers a rather precise and inspiring conception of genidentity, and that the reason that this conception has not been more successful lies elsewhere, namely, in its unusual metaphysical implications. Our strategy here, therefore, will be to reconstruct Hull's conception of genidentity, and then to extend it through an examination of several biological cases and a comparison with the uses of genidentity in physics.

In our view, what is probably Hull's most famous paper (Hull 1978) offers one of the best possible argumentations in favor of biological genidentity. This article is often seen as a defense of the theses that species should be seen evolutionarily as individuals rather than classes and that, consequently, there is no "human nature," no "essence" of humanity, if humans are understood as members of the species *Homo sapiens* (on Hull's and Ghiselin's "individuality thesis," see Haber, this volume). These theses are undeniably important, but we believe that the most crucial component of this paper is the series of diagrams drawn by Hull, and the conception of identity upon which they depend, that is, genidentity. Hull seeks a criterion of identity through time for living things. His thesis is that any organism or any species is a space-time portion, a "branch" on the tree of life, with a beginning and an end, with, between the two, a continuous line of different states. Hull argues that, because living things can undergo massive and unpredictable change, sameness and self-resemblance are inappropriate criteria for biological identity (Hull 1978, 345). He recalls that certain stages in some organisms are so different that biologists had placed them in different species, genera, families, and classes, before they could realize that those stages were in fact the transformations of one and the same organism. For living things, being the same cannot mean looking like oneself, and therefore the only possible criterion for identity in biology is *continuity of change* ("Phenotypic similarity, says Hull, is irrelevant in the individuation of organisms.")

Hull's diagrams offer a description of structural patterns of change in the living world, applicable to both organisms and species, because organisms and species belong to the same ontological category insofar as they both are spatiotemporally localized living things. A first set of structural changes concerns *change* of a living entity, or its *splitting* into two living entities (see figure 16.3 and table 16.3). What is Hull's criterion to postulate the potential emergence of a new entity? His criterion is the degree of *disruption of internal organization*. Of course, this criterion is not always easy to apply, and by definition the observer often faces a continuum of possible situations, but the examples given by Hull help us understand how to apply his view. In case 1a, Hull explains that a living entity can remain the same even if it undergoes a radical change, provided that the continuity between these different states can be established (e.g., a caterpillar becoming a butterfly). Case 1b corresponds to splitting: one individual becomes two individuals, and the initial individual disappears as such. Transverse fission in paramecia is an example. Case 1c corresponds to the appearance

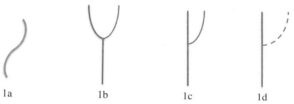

FIGURE 16.3 Ontogenetic change and splitting giving rise to the production of new organisms (in the case of organisms), or phylogenetic change and splitting leading to speciation (in the case of species)
Based on Hull 1978.

TABLE 16.3 Applications of Figure 16.3 to Different Cases of Organisms and Species

	1a	1b	1c	1d
Organism	Metabolic change Ontogenetic change Metamorphosis	Equal splitting (e.g., transverse fission in paramecia)	Budding with important loss and disruption of internal organization (e.g., strobilization in *Scyphozoa*)	Budding with little loss and disruption of internal organization (e.g., budding in *Hydrozoa*)
Species	Change within a species	Splitting	Speciation, one species unchanged, the other diverges	A small, peripheral isolate brings about a genetic revolution

FIGURE 16.4 Total or partial merging between organisms, or between species Based on Hull 1978.

of an individual *on* another individual, with the new individual becoming progressively autonomous. An example is strobilization in certain forms of *Scyphozoa* (sometimes colloquially called "true jellyfish"). In case 1d, a small part of an individual (contrary to case 1c, this is a part of an individual, not a growing individual on an individual) gains independence and becomes itself a new individual. An example is budding in *Hydrozoa* (*Hydrozoa* are Cnidaria that, at least for most of them, have both a polypoid and medusoid stage in their life cycles). Admittedly, it is not trivial to differentiate clearly between 1c and 1d, because it is not easy to make the distinction between a growing individual, and a growing part that will become an individual. But the difference between 1b on the one hand and 1c and 1d on the other is clear: the case is 1b if and only if the initial individual is lost and its internal organization is disrupted. In the two latter cases, the initial individual has certainly lost something (a growing new "individual" in 1c, a "small portion" in 1d), but it is still present and its internal organization has remained roughly the same. Naturally, *material continuity* exists between parents and offspring, but a parent and its offspring are characterized by two different *internal organizations*, and this is precisely the criterion that makes the difference between the continuity of *one* being and the continuity of several beings through *reproduction*.

A second set of structural changes concerns the *merging* of two living entities, or of their parts (see figure 16.4 and table 16.4). Here again, it is the examples analyzed by Hull that help clarify his view. In case 2a, two entities fuse to become one single entity, and they remain one entity for a significant time, so the two initial individuals are lost (that is why fusion in amoebas will often not count as an adequate illustration of 2a, while the fusion of two germ cells will do). In case 2b, a portion of a first individual becomes a portion of a second individual, and the two individuals continue their existence, but both have changed (the first has lost a part, the second has gained a part). Blood transfusion or bacterial conjugation are good examples. In case 2c, a portion

TABLE 16.4 Applications of Figure 16.4 to Different Cases of Organisms and Species

	2a	2b	2c
Organism	Fusion (e.g. fusion of sperm and egg)	Endosymbiosis Blood transfusion Conjugation	Sexual reproduction
Species	Fusion (extremely unlikely)	Introgression	Speciation by polyploidy

of a first individual and a portion of a second individual merge to form a third (new) individual, while the two initial individuals continue their existence. Sexual reproduction is a good example. If we are interested in species, a good example of 2b is introgression, and a good example of 2c is speciation by polyploidy (a rather common event in plants, for instance).

It should now be apparent that one of the most impressive features of Hull's paper is the richness of its examples. It is indeed these examples that give us the key for understanding Hull's conception of genidentity: because living individuals can change massively and because nothing in them seems to be entirely "fixed" for their entire life, the only way to account for the identity through time of living individuals is to determine to what extent they remain *one* entity, with an internal organization that remains practically the same or changes progressively. Of course, more precise accounts of what "internal organization" is are needed. But, first, Hull does give us hints as to how this internal organization is maintained or disrupted, in the case of organisms and of species, respectively (for example, the measure of the extension of gene exchange in the case of species: Hull 1978, 349). Second, Hull's main idea seems extremely clear: he rejects every conception of identity (of an organism or a species) based on *substance* (i.e., the idea that "something" remains in an individual despite its changes) and *resemblance* (i.e., the idea that X is the same if it looks sufficiently like itself). Importantly, we believe that Hull's view is opposed both to *essentialist substantialism* (the idea that a permanent "core" or "substrate" of X remains through time) and to the kind of *functional substantialism* defended, in particular, by Wiggins, because Hull's view is incompatible with the idea that each individual must be understood in relation to a sortal concept, each "sort" being characterized by a common "law of activity" (Wiggins 2011, 57).

Table 16.5 sums up the different conceptions of identity discussed here, their applications to the living world, and some of their proponents.

TO BE CONTINUED

TABLE 16.5 Different Concepts of Identity, and Their Application to the Living World

CONCEPTIONS	SUBSTANTIALISM	IDENTITY-RESEMBLANCE	GENIDENTITY
"Moto"	Something of X remains through time	X looks sufficiently like itself	X is defined by sufficiently continuous states
Examples in biology	Genetic identity "Essence" of a species	Self-resemblance for an organism Species as classes	Metabolism Metamorphosis Symbiosis
Proponents	Many proponents of genetic identity (e.g., Jacob 1982; to some extent Dawkins 1976) and genetic essentialism (see Kripke 1980)	Pheneticism (for species)	Hull 1978, 1992 Boniolo and Carrara 2004

So Hull is undeniably the philosopher of biology who offered the strongest defense of the notion of genidentity applied to the living world. Yet we think that it is possible to go even further in Hull's direction.

Indeed, one additional argument can strengthen Hull's view decisively. It concerns the integration of external components, in particular symbiotic components, into living things. Symbiosis is entirely missing in Hull's picture, except for the rapid mention of the endosymbiotic origin of certain organelles (Hull 1978, 346). This is not really surprising, as it is only recently that symbioses, long thought to be rather rare events, have been recognized as extremely widespread in nature (e.g., McFall-Ngai 2002). Symbiotic elements, and especially symbiotic bacteria, have been shown to decisively influence the ontogeny and phylogeny of many organisms (McFall-Ngai 2002, Pradeu and Carosella 2006, Dupré and O'Malley 2009, Gilbert and Epel 2009, Bouchard 2010, Pradeu 2011). Mammals, for example, are 90% constituted by bacterial cells and only 10% by eukaryotic cells, and mutualistic bacteria are indispensable for digestion, immunity, metabolism, and development. The indispensability of symbionts is in fact a phenomenon that can be found in virtually all animals (McFall-Ngai et al. 2013) and plants (Oldroyd 2013). We suggest that the ubiquity of symbiosis strengthens the genidentity view, making Hull's diagrams switch from "weird and rare" to "weird and common," at both the organism and the

FIGURE 16.5 Forms of integration of external biological material, at the organism level or the species level. The first case corresponds to Hull's 2a (fusion), the second to Hull's 2b (integration with continuation), while the third case is an inversion of Hull's 1c. The third case (absent in Hull's analysis) can be called "internalization," and is described in this chapter as an extremely frequent (though long overlooked) phenomenon in nature

species levels. Indeed, because symbionts are decisively involved in the ontogeny and phylogeny of most living things, cases 2a (fusion) or 2b (integration with continuation) or, even more frequently, "inverted 1c" (internalization) are extremely widespread in nature (see figure 16.5). In the biological world, the need to integrate foreign (cross-kingdom) living things is the rule, not the exception.

What should one deduce from this importance of symbiosis in nature? In order to understand what a living thing X is (be it an organism or a species),[9] one needs to study X *with its symbionts* (e.g., Moya et al. 2008), that is, the "heterogeneous organism" or the "heterogeneous species" (Pradeu 2012). In particular, the genome of a living thing X by itself is not sufficient to understand what X is and does. This is what can be called "the revolution of the microbiome" (the "microbiome" refers to the collective genomes of all the microorganisms living in our bodies (Turnbaugh et al. 2007)) or, perhaps more adequately, "the revolution of collective genomes," with metagenomics as a particularly useful set of tools (Dupré and O'Malley 2007, 2009): in most cases, living entities are composite entities, expressing genomes coming from different species, even indeed from different kingdoms (Pradeu 2012, Bapteste 2014). In addition, it is indispensable to study the dynamics of the *acquisition* of these symbionts, that is, how an organism acquires symbiotic entities through development (McFall-Ngai et al. 2013), and how a species incorporates symbiotic entities through evolution (e.g., Margulis and Sagan 2002, Bright and Bulgheresi 2010). Humans, for example, acquire different types of bacteria throughout their life, and these bacteria influence their development, metabolism, and health (e.g., Scholtens et al. 2012). As a consequence, an organism cannot be biologically defined on the basis of a single and constant genome, or on the basis of self-resemblance, and therefore it seems that only gen-identity can account for this dynamic biological identity through time, by capturing the complex processes of symbiont acquisition at the organism level. But the genidentity view is equally pertinent at the species level, as illustrated by many examples, including the growing recognition of different important forms of introgression in prokaryotes (in particular gene transfer agents, conjugative elements, outer membrane vesicles, viruses, plasmids, etc. [Bapteste 2014]), and the study of the role of endogenous viruses in the evolution of their hosts (Roossinck 2011, Stoye 2012). In all these cases, only the idea of a continuous change, in the double sense of a continuity of states and a sufficiently progressive dynamics of change, enables us to follow in details what contributes to the construction of a given living thing. Therefore, the genidentity view seems the best suited to account for the identity through time of living things, and the ubiquity of symbiosis shows that Hull was even more right than could have been appreciated at the time he expressed his view.

We hope that we have now convinced the reader that the concept of genidentity is potentially extremely relevant to understand identity through time both in physics and in biology. But at this point we would like to draw attention to the way the physical cases studied in section 16.3 shed light on what has been said in the present section about biological cases. From simply potentially relevant, this analogy makes genidentity actually relevant to biology. A preliminary remark is that the physical cases above showed us that the most significant challenge for anyone trying to apply the genidentity view is to determine what process needs to be followed, and how. It seems clear that the same challenge is met in biology. And, here again, it appears that the biologist should follow a *causally significant process*—for instance (as suggested by Hull) internal organization. At the organism level, internal organization can be measured in terms of intensity of interactions and cohesiveness. A more precise suggestion would be to follow well-specified metabolic interactions, which themselves contribute to the cohesiveness of the organism (see Dupré and O'Malley 2009). Or perhaps one should follow metabolic interactions *and* higher-level interactions that themselves exert control on these metabolic interactions: for instance, it can be argued that the action of the immune system can play a very important role in the defense of the genidentity view, because the immune system would exert a control on metabolic interactions, and detect and react to any rapid change in the organism (Pradeu 2012, 248–249). At the species level, internal organization can be measured in terms of intensity of gene exchange (again, following Hull), but other forms of transmissions of biological material could be envisioned (membranes, epigenetic marks, etc.) In all these cases, what matters is the causal process that biologists study: how organisms are maintained through time via physiological and metabolic processes, or how species are maintained through time via the interbreeding of organisms.

But the physical cases can shed light on the biological cases examined here in a much more specific way. It seems in fact possible to draw an analogy between the two categories of cases. In drawing this comparison, it is clear that the least useful case taken from physics is the first one, that of the individual soliton (the free individual). The idea of something that would have no interaction with its environment makes no sense in biology, and even the idea of something that would interact with its environment without being itself modified whatsoever by this interaction seems difficult to conceive (though perhaps Dawkins's definition of a selfish gene as an "immortal coil," or "any proportion of chromosomal material that potentially lasts for enough generations to serve as a unit of natural selection," comes close to the idea [Dawkins 1976, 28]). But the second physical case is much more relevant. In the second case, we were interested in a continuously changing wave, and we saw that the key question was, what *rate* of change makes us consider that X remains the same individual through time, and what *rate* of change makes us consider that we do not face the same X? This shift from the question of the *degree of change* to the question of the *rate of change* is exactly what illuminates Hull's diagrams. It is, in particular, at the very basis of diagram 1a: X can change, and it can even change massively, as long as the change is progressive enough to make us certain that it is indeed the same individual. But the same principle also lies at the heart of all Hull's diagrams: in all cases, it is clear that the rate of "internal organization" disruption will be critical to determine whether we are in the presence of *one* or *two* individuals—or perhaps of no individual at all. Finally, the third physical case is also very pertinent to shed light on the biological cases. In one of the cases of the third category, two individuals interact, and it is only the nature of their interaction that enables us to determine whether the two individuals have changed, and whether we still have two individuals anyway—for instance, solitons can fuse, as we saw. It is exactly this idea that lies of the heart of all the cases of biological fusion that we have examined. In order to determine the difference between 2a (two individuals fuse into one new individual) and 2b (one part of an individual A becomes part of individual B), the only resource we have is to determine *whether or not the genidentical processes that we were following have been disturbed*. In 2b, we say that the two individuals A and B are still the "same" individuals A and B (even though A has lost a part and B has gained this same part) because the *rate of change* in the internal organization of A has not been significantly disrupted (and the same is true of B). It is also the nature of the interaction that will tell us if we are in a case like 2a (fusion of two individuals; the two "former" individuals no longer exist) or "inverted 1c" (one individual is integrated into an individual; the "host" individual still exists, having been modified but not destroyed by this integration) (see figure 16.5). In the case "inverted 1c," the rate of change has not been massively disturbed by the interaction (though, of course, the change itself can be very significant, as when symbionts can induce a specific morphogenesis in the host), while in case 2a, we talk about a true fusion because a new genidentical process starts, one that will be characterized by its own features, and in particular by its own rate of change, and this new process is not the prolongation of former genidentical processes, as these former processes have been completely disturbed, and hence terminated, because of the interaction (as in two germ cells that fuse and form a zygote).

To conclude this section, we can say that what we have learned by analyzing physical cases of genidentity illuminates in a decisive way the examples taken from biology, because they help us see that, even though biology is often perceived as less "abstract," more "substantial," than physics, it is in fact the same structural characteristics that, in biology and in physics, make it possible to follow entities through time: (1) it is key to determine what exactly is the *causal interaction* to be followed; (2) what matters is less the degree of change than the *rate* of change; (3) the most interesting problem concerns the *interaction* of two genidentically defined individuals, and the way this interaction can disturb their genidentity. Therefore, the concept of genidentity is pertinent not only insofar as it can be applied in interesting ways to both physical and biological cases, but also because it offers a unique opportunity to foster a dialogue between physics and biology.

16.5 Genidentity and Process Ontology

An ontology is not just a proposition about the kind of entities that exist in the world, but also a heuristic tool to model the world. As we shall see now, the genidentity view orients us toward an ontology centered on the idea of *change* and *processes*, as opposed to an ontology centered on the idea of *invariant entities*.

During the scientific revolution, natural philosophers made an important move. As decisively emphasized by Wigner (1967, 3), they "devised an artifice which permits the complicated nature of the world to be blamed on something called accidental and thus permits [them] to abstract a domain in which simple laws can be found. The complications are initial conditions; the domain of regularities, laws of nature." It is difficult to overestimate the impact of this heuristic strategy. It allowed physics to become what it is today. One of the reasons it has worked so well is that, in many physical cases, it was possible to fairly well identify the relevant initial conditions, in order to make the discovery of the related law of nature. (For example, think about how Galileo and Newton managed to put aside almost all physical properties to identify the few relevant ones in gravitation phenomena.)

To explain this remarkable success, it is tempting to reify the distinction between laws and initial conditions, in other words between nomic regularities and accidents. For example, we could argue that the laws are independent of the content of the universe, and that, therefore, they are ontologically distinct from accidents that depend on this content. It is what Helen Beebee (2000) called a "governing conception" of laws. If we adopt such a conception, laws are founded on metaphysical invariant features (universals, dispositions . . .). As brilliantly exposed by Aristotle (*Physics* 3–5), in an ontological framework focused on what does not change, the main ontological problem becomes the nature of change. Without rejecting the distinction, the metaphysical alternative is to defend a nongoverning conception of laws. In order to make the distinction between laws and accidents not arbitrary, we have to use another kind of invariance to justify the status of laws. For example, in the Ramsay-Mills-Lewis approach, laws are propositions that occupy a special position in the theory structure (axioms, theorems . . .). What is important to note is that in all these ontological approaches, the focus is on what *does not change*. Change itself is derivative. Change is understood as what happens when invariance fails. It is not an autonomous metaphysical category.

It is today obvious that the strict distinction between laws and initial conditions is often not very useful in biology. Moreover, in many cases it seems to be an obstacle to the development of biological theories and models. If biology is a historical science, most, if not all, of its propositions are contingent. Should we put all biological knowledge in the column of accidents, therefore in the category of what cannot be adequately explained or modeled?[10] This seems inappropriate. In the last century, Whitehead (1978) reminded us that one could, and still can, make a metaphysical choice here. Indeed, to follow the Plato-Aristotle trend is not the only possibility. We could put at the center of our ontology the notion of *change* itself. Consequently, what would be derivative is not change, but the apparent absence of change, in other words regularities and diachronic identity.

This ontological shift of perspective is rarely taken seriously, and this suspicion is understandable given that building an ontology centered on the idea of change and processes is extremely difficult. Past attempts have often been deemed obscure (for example Bergson 1938, Whitehead 1978). Recent moves toward an ontology of processes have been made, especially with regard to the living world (Dupré and O'Malley 2007, 2009, Dupré 2012, Bapteste and Dupré 2013), but a detailed description of how such an ontology should be built is still lacking. It is our claim that individuals defined solely by genidentity relation could be a first step toward getting over this problem. Functional genidentity relations, precisely grounded in science (which, as we saw, are not to be understood in mereological terms), could play the central role in a process ontology. In this context, the notion of individual becomes derivative. Needless to say, a lot of work needs to be done before we could claim to have a contender to neo-Aristotelian approaches, such as Lowe's (2006) or Bird's (2007). As an example, the ontological dependency between processes and events will have to be clarified in a more satisfying way (see Steward 2013, though a process ontology should not be attributed to her). Are processes derivative compared to successions of events, or is the relation between processes and events more equal? Despite the fact that much work is still to be done, we have reasons to be cautiously optimistic. New metaphysical approaches, inspired by current science, for example (Maudlin 2007), are giving us tools to go beyond traditional ontologies. Soon it will not be unreasonable to sustain that processes are ontologically prior, and individuals should be conceived of as specific temporary coalescences of processes. It is scientifically identified processes that will tell us where the individual lies, and what its boundaries are, and not the other way around. We should not start with individuals; we should not build our ontology on the basis of preconceived or phenomenologically determined individuals, because in most cases our intuitive carving of the world into individuals can prove to be misleading in science (in biology, see Hull 1992 and Pradeu 2012; in physics, see Ladyman and Ross 2007 and French and Krause 2006). (Incidentally, we do not claim that *descriptive metaphysics*—i.e., describing the cognitive structures by which we understand the world—is always wrong. After all, at least pragmatically, our everyday ontology is very efficient. But we definitely sustain that such "anthropocentric" metaphysics tends to exclude innovative and important ontological propositions and revisions, which are often based on our best current science.)

16.6 Conclusion

In this chapter, we have defended a genidentity view, that is, the idea that the identity through time of an entity X supervenes on a continuous connection of states. The reason that has led us to this view is partly negative: conceptions of identity based on substance or self-resemblance, which might seem adequate for applications to everyday macroscopic objects, are highly problematic when confronted with cases taken from current physics and biology. In several of the examples studied above, from waves to organisms, nothing really "remains" through time, which makes it impossible to use a substantialist conception of identity; moreover, the overall aspect of the entity under consideration can change so much that it seems impossible to use a conception of identity based on resemblance. For anyone seeking a science-based worldview, therefore, it appears that genidentity is the best-suited view of identity through time. Naturally, the difficulty when applying the genidentity view to specific cases is to determine what exactly needs to be followed and how. Yet we have offered above several examples in which, we believe, it proved possible to address this problem and, therefore, to apply the genidentity view in a fruitful way. We have also shown that there are interesting parallels between the applications of genidentity to physical and to biological cases. Finally, we have tried to show that genidentity leads us to suggest an ontology of processes and change rather than an ontology of laws and substances or essences. In this metaphysical conception, processes, we submit, are ontologically prior, and they make possible the delineation of individuals through time, in physics, in biology, and perhaps beyond.

Acknowledgments

For comments on this chapter, we would like to thank Giovanni Boniolo, Adam Ferner, Steven French, Peter Godfrey-Smith, Matt Haber, Philippe Huneman, and Charles Wolfe.

Notes

1. Some metaphysicians (e.g., Wiggins 2001) put into question the distinction between synchronic and diachronic identity. Yet this distinction is often useful in a scientific context, both in physics (for example, one could argue that quantum particles are individuals even though they do not perdure) and in biology (for example in the discussion about what it means to be part of a given organism at one moment or at two different moments: e.g., Sober 2000, 154).

2. Maybe not for long, since many physicists and philosophers consider fundamental theories, like quantum field theory, as effective theories, therefore not describing a fundamental ontology, but only applicable in a certain range of parameters. In this context, levels of description and parts-whole relations could come back as important concepts in fundamental physics and not only in applied physics.

3. It is not clear whether, for Lewin, these entities are themselves temporally extended or not. Reichenbach (below) is clearer about this point.

4. Already in 1928, Rudolf Carnap had expressed the same view. He defines genidentity as an "association of various 'thing-states' with one object" (Carnap 1967 [1928], 252).

5. The difficulty of defining wordlines and causation in the quantum context is the main reason that genidentity is not used in current physics. However, these obstacles are not present in classical physics and biology, and this is why we think that Reichenbach's view can be useful in these sciences.

6. More generally, we believe we have good reasons to doubt that the third characteristic needs to be as strong as required by Reichenbach. In Simon Saunders's (2003) terminology, Reichenbach requires that events related by material genidentity refer to an individual that is absolutely discernible. Weakly discernible entities, provided they have a certain dynamics, could also be included in the scope of material genidentity. Since our biological cases are absolutely discernible, we will not push this discussion further.

7. Hull uses "substance" to talk about "*material* substance" (i.e., remaining materially the same). This is in sharp contrast with what the metaphysical tradition has called a "substance": for Leibniz and Wiggins, for instance, an individual substance can be maintained through time while changing totally its material constituents (Leibniz, *New Essays*, II, 27). We will show, however, that Hull's critique of substance impacts many forms of substantialism, not just material substantialism.

8. Hull also used the notion of genidentity (1975, 261), but it was in a critique of another paper, and Hull was unsympathetic with this notion if used in the context of ordinary language (in contrast with well-formulated scientific theories).

9. In fact, the same is true at other levels of the biological hierarchy, for instance at the level of a cell in a multicellular organism, but for the sake of clarity we stick here to the two main categories explored by Hull.

10. Even in physics, the scientific status of cosmology (and other "historical" aspects of physics) is also debatable.

References

Aristotle. 1996. *Physics*. Trans. Robin Waterfield. Oxford: Oxford University Press.

Bapteste, Eric. 2014. The Origins of Microbial Adaptations: How Introgressive Descent, Egalitarian Evolutionary Transitions and Expanded Kin Selection Shape the Network of Life. *Frontiers in Microbiology* 5. http://journal.frontiersin.org/Journal/10.3389/fmicb.2014.00083/full.

Bapteste, Eric, and John Dupré. 2013. Towards a Processual Microbial Ontology. *Biology and Philosophy* 28(2): 379–404.

Beebee, Helen. 2000. The Non-governing Conception of Laws of Nature. *Philosophy and Phenomenological Research* 61(3): 571–594.

Bergson, Henri. 1938. *La Pensée et le Mouvant*. Paris: Presses Universitaires de France.

Bird, Alexander. 2007. *Nature's Metaphysics: Laws and Properties*. Oxford: Oxford University Press.

Boniolo, Giovanni, and Massimiliano Carrara. 2004. On Biological Identity. *Biology and Philosophy* 19(3): 443–57.

Bouchard, Frédéric. 2010. Symbiosis, Lateral Function Transfer and the (many) Saplings of Life. *Biology and Philosophy* 25(4): 623–641.

Braillard, Pierre-André, Alexandre Guay, Cyrille Imbert, and Thomas Pradeu. 2011. Une objectivité kaléidoscopique: Construire l'image scientifique du monde. *Philosophie* 110: 46–71.

Bright, Monika, and Silvia Bulgheresi. 2010. A Complex Journey: Transmission of Microbial Symbionts. *Nature Reviews Microbiology* 8(3): 218–230.

Carnap, Rudolf. [1928] 1967. *The Logical Structure of the World and Pseudoproblems in Philosophy*. Berkeley: University of California Press.

Dawkins, Richard. 1976. *The Selfish Gene*. New York: Oxford University Press.

Drazin, Philip G., and Robin Stanley Johnson. 1989. *Solitons: An Introduction*. Cambridge: Cambridge University Press.

Dupré, John. 2012. *Processes of Life: Essays in the Philosophy of Biology*. Oxford: Oxford University Press.

Dupré, John, and Maureen O'Malley. 2007. Metagenomics and Biological Ontology. *Studies in the History and Philosophy of Science C: Biological and Biomedical Sciences* 38: 834–846.

Dupré, John, and Maureen A. O'Malley. 2009. Varieties of Living Things: Life at the Intersection of Lineage and Metabolism. *Philosophy and Theory in Biology* 1. http://www.philosophyandtheoryinbiology.org/.

French, Steven. 2011. Shifting to Structures in Physics and Biology: A Prophylactic for Promiscuous Realism. *Studies in History and Philosophy of Biological and Biomedical Sciences* 42(2): 164–173.

French, Steven, and Décio Krause. 2006. *Identity in Physics: A Historical, Philosophical and Formal Analysis*. Oxford: Oxford University Press.

Gilbert, Scott, and David Epel. 2009. *Ecological Developmental Biology: Integrating Epigenetics, Medicine and Evolution*. Sunderland: Sinauer Associates.

Hull, David L. 1975. Central Subjects and Historical Narratives. *History and Theory* 14(3): 253–274.

Hull, David L. 1978. A Matter of Individuality. *Philosophy of Science* 45(3): 335–360.

Hull, David L. 1986. Conceptual Evolution and the Eye of the Octopus. In *Logic, Methodology and Philosophy of Science*, ed. R. B. Marcus, G. J. W. Dorn, and P. Weingartner, 643–665. Amsterdam: North-Holland.

Hull, David. 1992. Individual. In *Keywords in Evolutionary Biology*, ed. Evelyn Fox Keller and Elisabeth A. Lloyd, 181–87. Cambridge, MA: Harvard University Press.

Jacob, François. 1982. *The Possible and the Actual*. New York: Pantheon.

Kripke, Saul A. 1980. *Naming and Necessity*. Cambridge, MA: Harvard University Press.

Leibniz, Gottfried Wilhelm, Freiherr von. [1765] 1916. *New Essays Concerning Human Understanding*. Chicago: Open Court.

Lewin, Kurt. 1922. *Der Begriff der Genese in Physik, Biologie und Entwicklungsgeschichte: Eine Untersuchung zur vergleichenden Wissenschaftslehre*. Berlin: Springer-Verlag.

Locke, John. [1694] 1975. A*n Essay Concerning Human Understanding*. 2nd ed. Oxford: Oxford University Press and Clarendon Press.

Lowe, Jonathan E. 2006. *The Four-Category Ontology: A Metaphysical Foundation for Natural Science*. Oxford: Oxford University Press.

Manton, Nicholas, and Paul Sutcliffe. 2004. *Topological Solitons*. Cambridge: Cambridge University Press.

Margulis, Lynn, and Dorion Sagan. 2002. *Acquiring Genomes: A Theory of the Origins of Species*. New York: Basic Books.

Maudlin, Tim. 2007. *The Metaphysics within Physics*. New York: Oxford University Press.

McFall-Ngai, Margaret. 2002. Unseen Forces: The Influence of Bacteria on Animal Development. *Developmental Biology* 242(1): 1–14.

McFall-Ngai, Margaret, et al. 2013. Animals in a Bacterial World: A New Imperative for the Life Sciences. *Proceedings of the National Academy of Sciences* 110(9): 3229–3236.

Moya, Andrés, et al. 2008. Learning How to Live Together: Genomic Insights into Prokaryote-Animal Symbioses. *Nature Reviews Genetics* 9: 218–229.

Oldroyd, Giles E. 2013. Speak, Friend, and Enter: Signaling Systems That Promote Beneficial Symbiotic Associations in Plants. *Nature Reviews Microbiology* 11(4): 252–263.

Padovani, Flavia. 2013. Genidentity and Topology of Time: Kurt Lewin and Hans Reichenbach. In *The Berlin Group and the Philosophy of Logical Empiricism*, ed. N. Milkov and V. Peckhaus, 97–122. New York: Springer.

Piette, Bernard M. A. G., B. J. Schroers, and W. J. Zakrewski. Dynamics of Baby Skyrmions. *Nuclear Physics B* 439: 205–235.

Pradeu, Thomas. 2011. A Mixed Self: The Role of Symbiosis in Development. *Biological Theory* 6(1): 80–88.

Pradeu, Thomas. 2012. *The Limits of the Self: Immunology and Biological Identity*. New York: Oxford University Press.

Pradeu, Thomas, and Edgardo D. Carosella. 2006. The Self Model and the Conception of Biological Identity in Immunology. *Biology and Philosophy* 21(2): 235–252.

Quine, Willard Van Orman. 1966. *The Ways of Paradox and Other Essays*. New York: Random House.

Reichenbach, Hans. [1956] 1971. *The Direction of Time*. Berkeley: University of California Press.

Reichenbach, Hans. 1958. *The Philosophy of Space and Time*. New York: Dover Publications.

Roossinck, Marilyn J. 2011. The Good Viruses: Viral Mutualistic Symbioses. *Nature Reviews Microbiology* 9(2): 99–108.

Saunders, Simon. 2003. Physics and Leibniz's Principles. In *Symmetries in Physics: Philosophical Reflections*, ed. Katherine Brading and Elena Castellani, 289–307. Cambridge: Cambridge University Press.

Scholtens, Petra A., et al. 2012. The Early Settlers: Intestinal Microbiology in Early Life. *Annual Review of Food Science and Technology* 3: 425–447.

Skyrme, Tony Hilton Royle. 1962. A Unified Field Theory of Mesons and Baryons. *Nuclear Physics* 31: 556–569.

Smith, Barry, and Kevin Mulligan. 1982. Pieces of a Theory. In *Parts and Moments. Studies in Logic and Formal Ontology*, ed. Barry Smith, 15–109. Munich: Philosophia Verlag.

Sober, Elliott. 2000. *Philosophy of Biology*. Boulder, CO: Westview Press.

Soomere, Tarmo. 2011. Solitons Interactions. In *Mathematics of Complexity and Dynamical Systems*, ed. Robert A. Meyers, 1576–1600. New York: Springer.

Steward, Helen. 2013. Processes, Continuants, and Individuals. *Mind* 122(487): 781–812.

Stoye, Jonathan P. 2012. Studies of Endogenous Retroviruses Reveal a Continuing Evolutionary Saga. *Nature Reviews Microbiology* 10(6): 395–406.

Turnbaugh, Peter J., et al. 2007. The Human Microbiome Project: Exploring the Microbial Part of Ourselves in a Changing World. *Nature* 449(7164): 804–810.

Whitehead, Alfred North. 1978. *Process and Reality*. New York: Free Press.

Wiggins, David. 1968. On Being in the Same Place at the Same Time. *Philosophical Review* 77(1): 90–95.

Wiggins, David. 2001. *Sameness and Substance Renewed*. Cambridge: Cambridge University Press.

Wigner, Eugene P. 1967. *Symmetries and Reflections*. Woodbridge: Ox Bow Press.

CHAPTER 17 | Experimental Realization of Individuality

RUEY-LIN CHEN

17.1 Introduction

In this chapter, I address two main questions: (1) Is there a conception of individuality *in and across* experimental sciences? and (2) Under what conditions can scientists be said to realize the individuality of an object? I will answer the first question by showing that there is a common conception of individuality that is already at work in physical and biological experiments. Some experiments can produce or create entities and properties predicted by theories to be the strongest evidence for ontological commitments to the entities and properties. I will answer the second question by extracting three conditions under which entities produced by those experiments can indeed be treated as individuals. The three conditions are the separation of produced entities from their environments, the manipulation of the entities, and the maintenance of the entities' structural unity in the process of being separated and manipulated.

Scientists have used "experimental realization" to refer to the production of phenomena, properties, and entities in laboratories (see section 17.3). I follow this usage. In this chapter, I will demonstrate that experimental realization is the most powerful means by which a scientist can confirm her *ontological commitment*, which is traditionally regarded as a core element of general metaphysics. What are ontological commitments? Usually, philosophers or metaphysicians approach the question of what there is in the world by logico-linguistic analysis or conceptual speculation. They construct metaphysical theories that posit a set of fundamental and general categories of beings, and thus philosophers have the responsibility to demonstrate that the theories containing the posited categories can provide the best account for all existents. In doing so, they make ontological commitments to those categories (MacDonald 2005). Scientists as well as philosophers make ontological commitments. The difference between philosophers' and scientists' commitments is that scientists may commit to particular entities such as the atoms, electrons, or genes supposed by their theories. Philosophers would say that scientists ontologically commit to some *theoretical entities.*

A theoretical commitment is no more than a belief to be tested; it is not entitled to be a piece of evidence. However, scientists can provide empirical evidence for their commitments by means of observations, measurements, investigations, and experiments. A theoretical commitment is said to be confirmed only if the entity and its properties predicted or posited by a theory are observed via the methods mentioned earlier. However, a scientist may provide evidence by *producing* or *creating* theoretical entities and their *real* properties in the laboratory. In such a case, we say that the scientist *experimentally realizes* her theoretical commitments. Of most concern in this chapter is the question: Are the entities produced by scientists in some particular experiments individuals? Not all kinds of experimental entities are individuals, but *individuality* is a primary property of many entity kinds. How do scientists powerfully confirm their commitments to the individuality of those entity kinds? They can make such a confirmation via the experimental realization of individuality.

Ian Hacking (1983) proposed an experimental argument that using a theoretical entity as a tool to investigate other phenomena provides the strongest evidence for the existence of the entity. Applying Hacking's argument to individuality, I assert that experimental realizations provide the strongest evidence for the theoretical commitments that scientists have to *particular individuals* in the ontological structure of the world. Why the strongest? Theories may be mistaken and therefore need to be tested. Experiments for testing theories may be erroneous, but the experiments that produce something seem to be always correct in the sense that they produce something! This is because one may misidentify or mischaracterize something when it is produced in an experiment, but one cannot reject the fact that something is indeed produced. For example, chemist Joseph Priestley in the 18th century did produce some kind of gas (oxygen) in his experiment, even if he misinterpreted it as "dephlogisticated air." If one wants to investigate whether something is an individual, one must recognize the existence of that entity. In other words, the existence of something is a precedent condition for its individuality. In consequence, the experimental realization of individuality not only is the production of something, but also requires more conditions than does Hacking's experimental argument for the existence of entities.

In what follows, I first review a few traditional approaches to the problem of individuality (section 17.2). By examining the creation of Bose-Einstein condensates in experimental physics and the modification of genes in genetic engineering—in the third and fourth sections respectively—I demonstrate that there are three necessary conditions for the experimental realization of individuality: *manipulation, separation,* and *maintenance of structural unity*. These realization conditions imply a conception of *experimental individuality* and suggest the three attendant criteria: *manipulability, separability*, and *maintainability of structural unity*. In the fifth section, I respond to possible questions by discussing the methodological and ontological meanings of experimental realization. In the concluding remarks, I compare these conditions with the criteria from the metaphysical and theoretical approaches to show that scientific practices cast a new light on the problem of individuality. However, I do not attempt to argue that the conception of experimental individuality can or should replace other theoretical conceptions. Instead, I argue that this conception is proper to scientific practices and that the experimental realization of individuality can provide the strongest evidence for theoretical commitments to individuality.

17.2 Individuality in the Speculative and Theoretical Approaches

Individual is seen as a fundamental category of the *furniture* of the world. To answer the question of what counts as an individual, metaphysicians often propose definitions and attendant criteria of individuality based on conceptual analyses of the ways we use "individual" with respect to things in everyday life. On the one hand, they analyze the meanings of "individual" or "individuate" in logical usages, and on the other hand, they speculate on the characteristics of an ordinary object—say, a dog or a stone—as an individual (Strawson 1959, Wiggins 2001, Lowe 1989, 1998, 2009). David Wiggins's work is a good example. Starting with the verb form of "individuation," he points out that the *Oxford English Dictionary* defines "individuate" in terms of "single out" or "pick out." He interprets "to single x out" as "to isolate x in experience; to determine or fix upon x in particular by drawing its spatio-temporal boundaries and distinguish it in its environment from other things of like and unlike kinds" (Wiggins 2001, 6). Summing up metaphysicians' proposals by the analytical-speculative approach (Wiggins 2001, Loux 2002, Lowe 2009, Chauvier 2008, and Chauvier's chapter in this volume), I list a number of key characteristics, criteria, or conditions of individuals as follows. (Here "M" in the ordinal marks represents "metaphysics.")

(M1) Particularity: *x* is an individual if *x* is particular; or, if one can use a demonstrative to designate *x*, that is, one can say *this x*.

(M2) Distinguishability (or discernibility): *x* is an individual if *x* can be distinguished from its environment and other objects in the environment.

(M3) Countability: *x* is an individual if one can count *x*.

(M4) Delineability: *x* is an individual if one can draw *x's* spatiotemporal boundaries.

(M5) Unity: *x* is an individual if *x* is a unitary whole that consists of its parts.

(M6) Persistence: *x* is an individual if *x* can persist in its unity while undergoing changes through time.

Each condition may be necessary but not sufficient for individuality. What criteria are the most crucial and fundamental? Usually, metaphysicians pay more attention to distinguishability and persistence than to the others, because the two characteristics closely relate to the problem of identity. "No entity without identity" reveals the relevance and importance of identity. If *individual a* is indistinguishable from *individual b*, then by intuition we would say that *a* is identical to *b*. How do we know that *a* is indistinguishable from *b*? A natural way is to examine whether every property of *a* is a property of *b* at the same time and vice versa. From this, we obtain the Principle of the Identity of Indiscernibles (hereafter, PII)—a subprinciple Leibniz's law, as it is called. (The other two subprinciples of Leibniz's law are the Principle of Indiscernibility of Identicals and the Substitution Principle [MacDonald 2005].) In addition, an identical individual should be able to persist and remain fundamentally *itself* while undergoing changes through time. Thus, persistence seems to be a condition of identity. Because the natural sciences seem to tell us that all ordinary objects consist of scientific objects (biological or physical—for instance, organisms, cells, molecules, atoms, electrons, and so on), we may wonder whether those criteria for individuality and identity can be extended to scientific objects. Is there a unitary concept of individuality? To answer this question, one must investigate the individuality conception of scientific objects.

Philosophers of different sciences, however, may have different positions on the relationship between individuality and identity. Philosophers of physics tend to think that the individuation conditions are equal to or imply the identity condition. So they usually attack the problem of individuality by discussing the PII as a starting point (e.g., French 1989, van Fraassen 1989). By contrast, philosophers of biology may see that identity implies individuality in thinking about the concept of biological identity. They might argue that biological identity includes the dimensions of individuality and uniqueness (e.g., Pradeu 2012). Given the divergence, in order to grasp the criteria for individuals in the physical and the living worlds, we should consider the conceptions of individuality in the philosophy of physics separately from those in the philosophy of biology.

Philosophers of physics have often examined three conceptions of individuality by referring to physical theories, for example, classical mechanics, the theory of relativity, and quantum mechanics (French 1989, Ladyman and Ross 2007). The three conceptions are as follows:

> (P1) *Bundle Individuality* means that an individual is nothing more than a "bundle," a "cluster," or a "collection" of properties. This conception is usually associated with PII.
>
> (P2) *Space-Time Individuality* means that an individual can be determined by a spatiotemporal trajectory.
>
> (P3) *Transcendental Individuality* means that all properties of an individual attach to a *bare particular*, which is "over" or "above" all its properties.

In general, empirical scientists do not commit to transcendental individuality because it is inaccessible via experience. The conceptions of (P1) and (P2) plus the assumption of impenetrability function well in classical mechanics. However, the advent of quantum mechanics issued a severe challenge to them. French and Readhead (1988) argued that in general Space-Time Individuality is not available in quantum mechanics; and PII in its weak version (individuals include properties of spatial location) is indeed violated by subatomic particles such as bosons, fermions, and others, because all individuals of each kind of particles are indistinguishable but not identical. Still, these authors suggested that particles can be treated as individuals, and if so, their individuality must be conferred by Transcendental Individuality. Ladyman and Ross (2007) argued that the nonindividual interpretation (ontic structural realism) of particles in quantum mechanics provided the best account. Whose argument is correct? Can logical and conceptual inferences based on the theoretical principles of quantum mechanics provide a crucial and conclusive verdict? I have no answer.

From the viewpoint of biologists and philosophers of biology, individuals in the living world are quite different from those in the physical world. Some organisms are paradigmatic cases about which metaphysicians engage in speculation. Unfortunately, the great kingdom of organisms provides a great many anomalies for some metaphysical criteria. It is not easy to distinguish an individual of corals, slime molds, many fungi, many plants, and the like, from others, to draw its spatial boundaries, and to count it and other counterparts with the unaided eye. Many philosophers of biology tend to think that the problem of biological individuality is best approached through evolutionary theory (Hull 1980, 1992, Millstein 2009 Folse and Roughgarden 2010). Some philosophers argue that theories from more narrowly circumscribed biological disciplines (e.g., physiology, immunology, or epigenetics) also provide important insights (Boniolo and Testa 2012, Pradeu 2012).

To sum up, there are three conceptions that are frequently used to define biological individuality (Pradeu 2012).

(B1) *Phenomenal Individuality* means that a living individual can be determined by observations. This conception is more compatible with the metaphysical criteria than others.

(B2) *Physiological Individuality* means that physiological theories provide good criteria to define a biological individual. For example, the immunity of an entity well defines it as an individual.

(B3) *Evolutionary Individuality* means that the theory of evolution by natural selection suggests the best criterion: a biological individual is any entity upon which natural selection acts. Thus, all biological species are individuals if natural selection acts upon them.

The foregoing discussion shows that the intuitively strong PII may conflict with some conclusions from quantum mechanics; and the view of biological species as individuals based on evolutionary theory may reject the metaphysical view that treats species as natural kinds. Given the conflict between some metaphysical conceptions or criteria supported by our strong intuition and some scientific conclusions or criteria from the received theories supported by much empirical evidence, how should we decide? Naturalism-oriented philosophers of science insist that philosophical accounts of criteria of individuality are, on the one hand, to be developed from the conceptions of individuality that they take to be implicit in the best current theories in science, or these accounts are, on the other hand, to be tested by the best scientific theories. Thus philosophers of physics cannot help search for criteria of physical individuals by reference to the theory of relativity, field theory, and quantum mechanics (Ladyman and Ross 2007). Similarly, philosophers of biology think that defining a biological individual by intuition or common sense is too vague to form a precise criterion; philosophers should approach the problem via the best biological theories (Hull 1992, Pradeu 2012). Whatever notions they hold, we can see that metaphysicians and philosophers of science tend to propose conceptions of individuality via observations or theories—metaphysical, physical, or biological.

Are theoretical conceptions of individuality enough for or proper to scientific experimentation? No. Several preliminary reasons endorse this claim. First, quite a few authors (Kuhn 1977, Hacking 1983, 1988, Galison 1998) argue that experimental sciences are independent of theoretical sciences to some extent. If this is the case, it is difficult for us to simply apply theoretical conceptions of individuality to experimental sciences. Second, theories, both philosophical and scientific, can be wrong. Indeed by nature they must remain open to testing, revision, and rejection. Given the fallible nature of theories, we might want other ways of generating criteria of individuality *for the practice of experimental sciences.* If experimental sciences are relatively stable in contrast with theoretical sciences (Hacking 1988), a natural place to seek these criteria is in the conception of individuality already at work in experimental sciences. Criteria developed in this way would have the advantage of being robust to changes in theory, as well as being closely tied to what scientists actually do. The conception and criteria of individuality implied by the use of certain experimental techniques provides us with an alternative way to ground philosophical accounts of individuality for the practice of experimental sciences (Braillard et al. 2011). In particular, we should pay attention to the experiments that can *create or produce* something. Third, metaphysical as well as scientific theories are thoughts in minds, no matter what their representations are.[1] One can easily see that all criteria developed from theories are observatory or epistemic. Experimentation by nature is performance or doing. Pure observatory or epistemic criteria may be not capable of grasping a conception of individuality from performance. As mentioned earlier, I propose three criteria (*manipulability, separability,* and *maintainability of structural unity*) to capture the conception of individuality that is already at work in scientists' actual experiments. These criteria are extracted from performable conditions: manipulation, separation, and maintenance of structural unity. They are more proper to experimental sciences than are pure epistemic criteria.

Still, one may wonder whether or not a particular experiment depends upon a specific theory. This is a theory-dependency-of-experiments issue. In general, we can divide an experiment into three parts: the design, the performance (or the "material realization" in Radder's phrase), and the interpretation of experimental results (Radder 1995, Chen 2007). Even if the design of an experiment and the interpretation of its results may depend upon a theory, the performance of the experiment is still (causally) independent of the theory. This is because the successful performance of experiments relies upon experimenters' techniques, skills, and practices rather than upon the virtues of theories and designs. A reliable theory and a complete experimental design may promote the probability of an experiment's success, but they are not able to determine whether the experiment would be completed or performed successfully. In other words, the causes of experiments being performed are experimenters' actions rather than theories. Similarly, validly experimental results cannot be theoretically interpreted without being produced beforehand. Even if an experimental result has been accurately predicted before the experiment's performance, this does not prevent us from exploring the conception of experimental individuality independent of theories. The reason for this assertion is that the conception can be drawn from the performance of related experiments. This chapter aims to argue that scientists can identify results as individuals from experimental production without appealing to theoretical interpretations.

In order to search for a conception of individuality proper to experimental sciences, in the next section, I begin to examine the first case: the creation of Bose-Einstein condensates in physics.

17.3 Bose-Einstein Condensates as Individuals in Physical Laboratories

In 1925, Satyendra Nath Bose and Albert Einstein predicted a new state of matter that was later called Bose-Einstein condensation. They formulated the Bose-Einstein statistics to describe the behavior of bosons such as photons, helium4, rubidium87, solidum23, and so on, in an ultracold condition. Later physicists identified that state as the fifth phase of matter (the other four phases are solid, fluid, gas, and plasma). They predicted that dilute gases of bosons would enter into that phase when the gases were cooled to temperatures very near absolute zero. In such a condition of extremely low temperature, where the sum of the energy of all atoms in the cooled gas is approximately zero, quantum effects would become salient. The atoms of the cooled gases were predicted to enter the same and the lowest energy level (physicists call it the degenerate state) and become "cloudy" with the wave property. Very surprisingly, in the theory, all atoms in the phase of Bose-Einstein condensation might be indistinguishable from one another and behave collectively and uniformly.

Seventy years later, American scientists Eric Cornell and Carl Wieman, at the Joint Institute of Laboratory Astrophysics in the University of Colorado and the National Institute of Standards and Technology, used sophisticated new instruments and techniques to create Bose-Einstein condensation in which there were about 2,000 indistinguishable rubidium atoms that existed for approximately 15 seconds (Anderson et al. 1995).[2] What they had done was to cool the atoms in the gas to a temperature of less than 170 nanokelvins (170×10^{-9} K) for rubidium. To enter into the phase, different critical temperatures are required for different kinds of atoms. For example, the critical temperature for a gas of sodium atoms is 2 μK (2×10^{-6} K).

To create Bose-Einstein condensation in laboratories, scientists used three key techniques: laser cooling, optical and magnetic trapping, and evaporative cooling. The laser cooling technique creates a setting in which the atoms absorb photons out of laser light that is cast in the direction opposite to the atoms' motion, thereby slowing down the atoms and cooling them. Laser light is used not only to cool the atoms but also to *trap* them, keeping them away from the room-temperature walls of the glass box that contains a tiny amount of the experimental gas. Six vertically intersected laser beams are arranged to keep the slowing atoms in a restricted space as they have seemingly been *fixed*. These trapped atoms are at an extraordinarily low temperature but not low enough to form Bose-Einstein condensation. The second stage of the cooling process consists in using the techniques of magnetic trapping and evaporative cooling. At this stage, scientists hold the atoms in a magnetic field by carefully controlling the shape and strength of the field, in which the atoms move around much like balls rolling inside a deep bowl. The most energetic atoms escape from the magnetic bowl and carry away more than their share of the energy, leaving the remaining atoms colder. When the energy is reduced to the level below the critical temperature, the Bose-Einstein condensation in gases comes into being. Hence, the astonishing phenomenon predicted in 1925 was seen in a laboratory!

What are the philosophical implications of this experiment? One result is an empirical confirmation that bosons—theoretical entities—do violate the PII. Bosons are indistinguishable from one another, but they are not identical—a claim enforced by Cornell and Wieman's statement in the following:

> The atoms "Bose-condense" into the lowest possible energy state, and the wave packets coalesce into a single, macroscopic packet. The atoms undergo a quantum identity crisis: we can no longer distinguish one atom from another. (Cornell and Wieman 1998, 41)

Why can we no longer distinguish one atom from another? The physicists think that even if the size of each wave packet grows with a collection of atoms becoming colder, it is principally possible to tell atoms apart as long as each wave packet is spatially *separated* from the others. Recall that French and Redhead had argued that PII is violated by bosons, fermions, and others by logical and theoretical inference in 1988. Distinguishability or discernibility, as one criterion for individuality, is thus materially falsified by the experiment of 1995, because in Bose-Einstein condensation we have different individuals (atoms) that are indistinguishable. Still, one may raise a question of whether the atoms in the condensation remain as different individuals in cases where they lose their *individual identities*. Nonetheless, physicists generally think that the counts of the atoms in a Bose condensate are computable, as they report how many atoms have been condensed.[3]

The feature that I want to draw attention to, however, is that the experiment created a new individual. Physicists call it the *Bose-Einstein condensate* or in brief the *Bose condensate*. Cornell and Wieman described their experimental result as follows:

> In June 1995 our research group ... succeeded in creating *a minuscule but marvelous droplet*. By cooling 2,000 rubidium atoms to a temperature less than 100 billionths of a degree above absolute zero (100 billionths of a degree kelvin), we *caused the atoms to lose for a full 10 seconds their individual identities* and behave as though they were *a single "superatom."* (Cornell and Wieman 1998, 40; my emphasis)

It is noteworthy that, moreover, the created *superatom* is observable. Cornell and Wieman stated, "Quantum-mechanical waves extend across the sample of condensate and can be observed with the naked eye. The sub-microscopic thus becomes macroscopic" (Cornell and Wieman 1998, 40). As a result, the experiment really turned the submicroscopic entities into a visible and macroscopic *individual*.

Cornell and Wieman's statements hint at a conception of individuality. It is associated with experimental realization by the physicists' themselves: "Our short-lived, gelid sample was the *experimental realization* of a theoretical construct" (Cornell and Wieman 1998, 40; my emphasis). A metaphysical implication shown by the entire process from Bose and Einstein's prediction to Cornell and Wieman's experiment is that scientists' ontological commitments to entities or properties based on theories could be realized by experimenting. Following the scientists' usage, I call this process *experimental realization*.

Can the criteria for an ordinary individual also be applied to the marvelous entity, inasmuch as this short-lived *individual* is observable like any ordinary object in everyday life? The condensate is indeed particular, countable, and unitary, and it persists for a short but perceivable time. The delineability might not work for the individual, because we cannot draw a clear boundary on any atomic cloud. However, we should be careful about the distinguishability. Although every part (atom) in the whole is indistinguishable at the submicroscopic level, the condensate in itself seems to be distinguishable from its environment at the macroscopic level. But those metaphysical criteria are not sufficient for the Bose-Einstein condensate, because, even if observable, it is a product of physical laboratories. The individual's existence, life, and properties are created by experimental mechanisms and conditions. It is important to know what conditions create its individuality.

Since the first Bose-Einstein condensate was created in 1995—and up till 2005—at least 40 teams and laboratories worldwide have created the marvelous individuals by using different boson gases (Hacking 2006). The central and other newly related techniques for producing the Bose condensates have been so well developed that physicists can easily create the entity and use it to investigate other phenomena of nature, such as cold collision, optical lattice, artificial black holes, slow light, and so on. If one wants to use a scientific-philosophical concept to cover the instrumental use of the Bose condensate, the best one would be "manipulation." The created individual must be manipulated; otherwise one cannot demonstrate that it is really an individual. Thus the *manipulation* is a necessary condition for the emergence of the experimental entity's individuality. The condition is not sufficient because one can manipulate entities that are not individuals, for example, the magnetic field. Scientists can manipulate two magnetic fields but cannot spatially separate them. By contrast, scientists can manipulate and spatially separate two Bose condensates (or atomic clouds). Similarly, one may simply separate oil from water without manipulating either of them. Thus *separation* is taken as another necessary condition for the experimental realization of individuality. In the case of Bose condensate, scientists can investigate the interaction between two individual condensates by separating them spatially.

> Using a single-loop wire and a magnetic-bias field, the two clouds, each containing more than 100,000 atoms, are spatially *separated* above and below the wire center of the double-well MOT [magnetic-optical trap]. The cloud interdistance can be controlled by independently varying the wire current and external bias field. (Lin et al. 2008, 6104; my emphasis)

However, the condensate has a short life. When the temperature around the condensate goes up to the critical point, the degenerate state that constitutes the internal unity of parts (atoms) within the individual is broken down, and the structured entity disappears. This implies that the third necessary condition is *maintenance of structural unity*. By "structural unity," I mean that the components of an individual are structured into a whole in some specific manner. The three necessary conditions jointly constitute a sufficient condition for the individuation in experimental contexts.

A logical relationship between the three realization conditions can be observed. The maintenance of structural unity is necessary for the realization of both manipulation and separation of individuals, because one can manipulate an individual and separate it from others only if one can maintain its structural unity throughout the process. In contrast, one could only show the maintenance of the individual's structural unity without breaking it down while manipulating and separating the individual. That is, this relationship between them is biconditional: (Mx & Sx) ↔ MSUx. (Here "M" represents manipulation, "S" separation, and "MSU" maintenance of structural unity.) Consequently, it can be said that a scientist has realized the experimental individuality of an entity if and only if she can separate the entity, manipulate it, and show that its structural unity was not lost during the periods of manipulation and separation. To sum up, the logical formula among experimental individuation and its realization conditions is EIx ↔ ([Mx & Sx] ↔ MSUx). This relationship will be also shown in the next case: genetic engineering.

17.4 Genes as Individuals in Biotechnology

Experimental manipulation of genes provides another philosophically interesting case study. The development of genetic engineering has important implications for this question. Genetic modification provides the strongest evidence for the ontological commitment to genes as individuals. The ability to locate, isolate, target, insert, and knock out genes implies a conception of individuality on the part of the genetic research community. What conditions does the conception contain?

When we ask whether *a gene* is an individual, what we are asking is whether a token of a gene type (or a genotype) in an individual organism is an individual just as we ask whether a token of an atom type is an individual. This implies that a clear concept of the gene must be established to define and identify a gene type. Is there such a concept?

Historians and philosophers of biology well know that the gene concept has been evolving since it first appeared (Carlson 1991, Maienchein 1992, Kitcher 1992, Portin 1993, Beurton, Falk, and Rheinberger 2000). By the second half of the nineteenth century, Mendel and the rediscoverers speculated that there were particulate "elements" within cells that were responsible for unit characters. In 1909, Willhelm Johannsen coined the term "gene" and used it to represent a unit of heredity. From the 1910s to the 1930s, Thomas Hunt Morgan and his team developed a theory of the gene, defined a gene as a unit functioning for heredity, recombination, and mutation, and located many *genes* on the four chromosomes of *Drosophila* (Darden 1991, Waters 2004). Although Morgan's team provided a large amount of statistical data from hybridization experiments, the referent of "gene" was not materialized and remained contested (Carlson 1991, Maienchein 1992). The discovery of the double helical structure of DNA in 1953 did not solve *the gene problem*, but instead cast doubt on earlier gene conceptions. Philosophers of biology have argued that the referent of the classical "gene" concept cannot correspond to segments of DNA (Kitcher 1982, 1984, Rosenberg 1985), but to date there has not been a very successful positive account of just what a gene is. Even if contemporary molecular biologists define the gene concept as "a segment of DNA for producing a polypeptide," there are no clear boundaries that can be delineated for every *gene*. As a result, Kitcher issued a seemingly desperate lament: "A gene is anything a competent biologist chooses to call a gene" (Kitcher 1992, 131).

When lamenting about the "gene" concept, Kitcher sensitively noted the following:

> Contemporary molecular biology does not worry much about these questions. Sequencing of individual genes goes on without too much fuss about boundaries. The successful sequencer offers the world a list of bases, a list that covers some region of interest. Theoretical discussions center on transcription of DNA, replication of DNA, repair of DNA, without worrying unduly about segmenting the nucleic acids into genes. Indeed, it is hard to see what would be lost by dropping talk of genes from molecular biology and simply discussing the properties of various interesting regions of nucleic acid. (Kitcher 1992, 130)

What molecular biologists are sequencing is really the sequence of nitrogenous bases containing a variety of gene types, whereas biotechnological scientists can separate a sample of a gene type and confirm the existence of the gene. They continue to use the term "gene" although they clearly know that there are no satisfactory definitions for it.

> We often use the term "gene" as being synonymous with "open reading frame" (ORF), i.e. the region between the start and stop codons (although even that definition is still vague as to whether we should or should not include the stop codon itself). In bacteria, this takes place in an uninterrupted sequence. In eukaryotes, the presence of introns ... makes this definition more difficult; the region of the chromosome that contains the information for a specific polypeptide may be many times longer than the actual coding sequence. Basically, it is not possible to produce an entirely satisfactory definition. However, this is rarely a serious problem. (Dale and Schantz 2002, 14)

Indeed, genes can be identified and individuated in experiments by using the technology of recombinant DNA and transgenesis.

The technology of recombinant DNA was invented in the early 1970s and transgenic technology was developed from it (Morange 1998, chap. 16; Thieman and Palladino 2004, chap. 3; Watson and Berry 2004, chap. 4). Hamilton Smith discovered the first restriction enzyme that can cut DNA at recognition sites (i.e., a specific sequence of bases) in 1970. This discovery offers a powerful instrument for scientists to obtain a specific segment of DNA that may have a complete function to express. One year later, Paul Berg used the restriction enzyme *Eco*RI—which was isolated from the *Escherichia coli* (hereafter, *E. coli*)—to cut DNA from *E. coli* and DNA from a primate virus SV40 (simian virus 40). He then added the segments of *E. coli* and SV40 in a tube containing enzyme DNA ligase and created a hybrid DNA. In 1973 and 1974, Stanley Cohen and Herbert Boyer conducted the first complete experiment of recombinant DNA, in which they transferred two different DNA segments that can resist ampicillin and tetracycline respectively into *E.coli* and then realized the transformation of bacterial cells. Both DNA segments are called "antibiotic resistance gene." Shortly after, using the technology of recombinant DNA developed in the experiments with microbes, a team of molecular biologists succeeded in creating the first transgenic animal in 1981. They produced a group of transgenic mice by microinjecting a plasmid carrying the HSV (herpes simplex virus) thymidine kinase gene into the pronuclei of mouse eggs. The experimental result showed that some of the transgenic mice expressed HSV thymidine kinase in their livers and kidneys at different levels (Brinster et al. 1981).

Why did scientists view either the "antibiotic resistance gene" in the bacteria cells or the "HSV thymidine kinase gene" in the experimental mice as a gene? The usage of the term "gene" shows that a conception of individuality is implied in biotechnological scientists' performance of experiments. This conception is used in identifying and individuating a gene. One wonders why scientists use such terminology without theoretical clarification, given the concept of the gene has been proposed by many theories once again. *Experimental techniques make breakthroughs.* Techniques help realize the ontological commitment to this or that gene. It is biotechnology that *isolated* an individual gene, whereas theories had not provided a satisfactory definition for the general concept of the gene. In what follows, I demonstrate these claims by considering the development of bacteria transformation experiments.

The transformation of bacteria was first discovered by Frederick Griffith in 1928. He conducted an experiment on two strains of *Pneumococcus* bacteria, known as the smooth (S) strain and the rough (R) strain for the formation of colonies with smooth or rough surfaces, respectively. It was known that the S-strain was virulent enough to kill mice, and the R-strain avirulent. When Griffith heated S-strain cells, killing them, and injected them into mice, the mice lived. When he injected both heat-killed S-strain and live R-strain cells together into mice, the mice frequently died. The experiment suggested that there must be something in the S-strain heat-killed cells that could convert the R-strain avirulent cells to the lethal form (Chen [2013] offers a brief discussion of Griffith's experiment and discovery). Is this something that caused the bacteria transformation an individual? In 1944, Oswald Avery and his team determined that this something was DNA. In 1953, James Watson and Francis Crick discovered the double helix structure of DNA. The complex and multilevel mechanisms of the DNA replication, the RNA transcription, and the protein synthesis were gradually revealed by molecular biologists.

Although molecular biology offers a great many new terms, the language of "gene" has not been eliminated. As we have shown, no theories of genes can give a satisfactory answer to the question of what *the* gene is, but biotechnological scientists may identify what *a* gene is. This is because if scientists can replicate the transformation of bacteria by manipulating something (a segment of DNA) in laboratories, they can demonstrate that the something that is responsible for the transformation exists. Furthermore, the fact that they manipulate the something in individual organisms in a particular experiment shows that the something is indeed an individual. This first complete recombinant DNA experiment endorses this claim.

In 1974, Cohen and Boyer used plasmids, which are extrachromosomal pieces of DNA in some bacteria and which are small and circular, as vectors that can transfer a foreign segment of DNA into a bacterial cell. They cut out a (supposed) antibiotic resistance *gene* from other bacteria with the restriction enzyme *Eco*RI, linked it with a plasmid by DNA ligase, and transferred the plasmid into *E. coli* that could not resist antibiotics. As a result, the modified *E.coli* appeared to be able to resist antibiotics—They contained the antibiotic resistance gene! In the process of the experiment, the antibiotic gene was separated from its original bacteria and then was manipulated (i.e., linked and transferred). Its structural unity was not broken, thereby allowing it to be expressed in the other kind of bacteria during the process. Scientists thus identify it as *a* gene—an individual.

By the end of the 20th century, recombinant DNA and transgenic technology were so mature that many kinds of microbes, plants, and animals were transplanted with foreign (or exogenous) genes, what scientists call transgenes (Glick and Pasternak 1998, chaps. 18 and 19). To successfully perform such an experiment of transgenesis, scientists must be able to separate and manipulate a gene, insert it into a new and heterogeneous genome, preserve the individual gene and its chemical structure from breaking up in the process, and make it function and express itself as well in the other kind of living things as it does in its original organism. The entire process of these experiments *individuates* the gene. In other words, biotechnological scientists can isolate a token of a gene type, a genuine individual, from a model organism by implementing sets of techniques without any clear theoretical delineation of the gene type. What are those techniques?

Three general methods used for DNA recombinant and transgenic experiments can be extracted from biotechnology textbooks. Each method includes a variety of specific techniques (Watson et al. 1998, Glick and Pasternak 1998, Dale and Schatz 2002, Bains 2004, Thieman and Palladino 2004).

(1) *Cutting of DNA* refers to the method that uses restriction enzymes to cleave specific segments from the recognition sites of a long DNA chain.
(2) *Cloning of DNA* refers to the methods for making copies of DNA fragments. The aim of DNA cloning is to copy a segment of interest or a gene from an organism and paste it on another organism. There are two main methods used for cloning DNA: producing DNA libraries and polymerase chain reaction (PCR).
(3) *Identifying a gene:* There are two effects for different organisms. First, *transformation of bacterial cells* refers to the phenomenon in which a bacterium expresses a novel feature by absorbing an external gene. The transformation is used to identify a specific gene by observing the expression of the gene in the bacterial cells. Second, *knockout of genes in eukaryotic cells* usually refers to the effect in which an intrinsic (endogenous) gene in the genome of organism is disrupted by a foreign (exogenous) gene that is inserted into the genome. Recombinant DNA techniques that cause such an effect can thus be called *the knockout method*.[4] The method is widely used to find out *what a gene does* by knocking the gene out and seeing what effect it has.

Why do these methods and techniques help individuate a gene? Why do the experiments via these methods realize the conditions of *experimental individuality*? Clearly, the method of cutting help realize the condition of separation because it separates a gene from a long chain of DNA; the methods for cloning DNA embody the condition of manipulation because many specific techniques are used to manipulate DNA segments as genes; and the transformation and knockout effects indicate the maintenance of structural unity because they demonstrate that a transgene can keep its integrity and can be expressed in a heterogeneous genome. The following paragraph from a biotechnology textbook summarizes three essential tools that hint at the three realization conditions of individuality:

> Three essential tools form the basis for studying the function of mammalian genes. First, we must be able to *isolate* a gene by cloning. Second, we must be able to *manipulate* the sequence of a gene in the test tube. Third, we must be able to *return* an altered gene to determine how it functions. (Watson et al. 1998, 235; my emphasis)

A scientist is able to isolate and clone a gene only if she is able to separate (cut) it from a DNA chain. In cloning a gene, the scientist is manipulating it. She is able to return an altered gene only if she is able to maintain its structural unity.

Recombinant DNA and transgenic experiments in biology harbor a conception of individuality. Important features of this conception are shared with the conception implicit in the language describing the experimental production of the Bose-Einstein condensate. The two cases, respectively from physics and biology, illustrate that *manipulability* is an important criterion for individuality in experimental contexts. More precisely, in a Hacking-style argument, if one can control and use an object to investigate other phenomena, the fact that this object is an individual is thereby indicated. In being manipulated, the individuality of the object can be shown more obviously if it is separable from other objects. Finally, if the structural unity of the object is not destroyed when it is separated and manipulated, its individuality is warranted in the strongest degree. Nonetheless, there is a slight difference between the two cases: Physical scientists *create* a Bose condensate, but biologists do not create a gene. However, creation is not a criterion of individuality.

17.5 Experimental Realization: Methodological or Metaphysic?

So far one may question whether or not I presuppose the reality of the individuals produced by experimental manipulation. One may argue that manipulating a group of theoretical entities (for instance, using electrons as a tool to investigate the other natural phenomena) always depends on experimental apparatuses. One may ask: How can you distinguish the manipulated objects (say, electrons) from the mere phenomena produced by instruments? How can you warrant the reality of the manipulated individual if it is nothing more than an instrumental effect?

To these interrelated questions, I have two replies: One is methodological and weaker and the other is ontological and stronger.

The methodological answer emphasizes that the term "experimental realization" is used to describe the process by which scientists produce some effects predicted by their theoretical commitment. It may not involve the ontological problems of whether or not the effects are caused by real entities and of whether or not the entities are real. Indeed, scientists view them and talk about them as if they were real. They do believe that the imagined entities in theories have really been created in their experiments. But this does not imply that we have committed to the same ontological position as many scientists have. "Experimental realization" is simply used to represent the scientists' performance of experiments as contrasted with their *ontological commitments* in theories. Moreover, I want to emphasize that I am not discussing the experimental realization of theoretical entities but rather the realization of individuality, by which one can extract a conception of individuality that is at work in the experimental sciences. I find three realization conditions that allow scientists to describe, identify, and count experimental effects as individuals: for instance, a condensate or a gene in the two paradigmatic cases. "Whether the condensates and the genes are real entities" is a question beyond the level of methodology. However, if the existence of entities manipulated in experiments is a precedent condition for the experimental realization of individuality, then the methodological answer must presuppose an ontological position.

Ontologically, one may argue that Hacking's experimental argument with mere manipulation cannot warrant the reality of the manipulated entities *as a kind*. There are two reasons for this argument: First, the effect produced by a particular experiment in a particular situation is particular. Even if the effect can be replicated, it is unclear whether the replicas and the original belong to the same kind if scientists use different instruments. There is an inferential gap from a particular object to a general kind. Second, the manipulation of experimental entities depends upon experimental apparatuses. They may be parts of or changes of instruments. For instance, the flame on a stick is a change of the stick. In other words, the manipulation of entities cannot serve as the evidence strong enough to warrant that the manipulated entities are real rather than mere instrumental effects, because manipulation does not require separating the entities from the experimental instruments. If the manipulated thing is always attached to the instruments as colors are on ordinary objects, then the thing is a mere phenomenon or epiphenomenon rather than a real entity. But these two reasons do not imply that the experimental realization cannot warrant the reality of the manipulated *individuals*.

If a scientist can realize the individuality of an object in a particular experiment, then she has provided the strongest evidence, I think, to warrant the reality of the object. She does create, produce, or confirm an individual. To realize the individuality, one must produce something and then realize the three conditions. First, one must be able to separate an object in order to manipulate it. Separating an object from others implies separating it from the experimental instruments that may have helped produce it, as we have seen in the two cases of Bose-Einstein condensation and transgenesis. Despite being separated from objects, the instruments help produce and keep certain conditions through which the components are structured into a unity, as we have seen in the experiments of the Bose condensate. In the case of the transgenetic experiments, the instruments are used to separate and manipulate genes without breaking their chemical structures. Thus the maintenance of the object's structural unity throughout the entire process of separating and manipulating demonstrate that there is an internal structure within the object and that its persistence is maintained both by its own structure and by the external instruments.

As the fact that industrial products depend on the production of industrial machines does not mean that they are parts or changes of machines, so the instrument-dependency of entities in the experimental process does not reject reality. In other words, the instrument-dependency is merely the dependency of productive processes; it does not imply nonexistence. An artificial entity or individual is still a real entity or individual. In this sense, the experimental realization of individuality is an *ontological (or ontic) realization of individuals*. The produced individuals are indeed instrumental effects but are more than mere phenomena or epiphenomena—they are also real entities.

Still, one may question whether or not I presuppose that the world is composed of individuals without kinds. No. As yet I commit neither to the realist nor to the nominalist view on the ontological structure of the world. But I do believe that individuals exist; otherwise it would be pointless for us to search for criteria of individuality. Moreover, both metaphysical positions, the realist and the nominalist, permit the existence of individuals. Therefore, I affirm that *individuals* produced by experiments are real. Are the theoretical entities produced by experiments real *kinds*? The question is beyond the scope of this chapter.

17.6 Concluding Remarks

Clearly in the case of Bose-Einstein condensates, scientists first create a condensate and manipulate it by using magnetic traps to separate it from other free atoms or other condensates while maintaining the condensates' structural unity in terms of quantum state homogeneity. In the case of genetic modification, scientists produce an organism with a novel phenotype by isolating an exogenous gene to replace an endogenous gene without destroying the structural unity of the exogenous gene as evidenced by its successful function within the context of the new genome. These case studies demonstrate how the three realization conditions characterize a unique conception of individuality grounded in scientific experimentation.

The realization of Bose-Einstein condensates cannot solve the problem of whether bosons are individuals per se. So the conception of experimental individuality is not intended to supersede those from physical or metaphysical theories. Similarly, the realization of the gene modification cannot provide a full answer to the question of biological individuality. The conception of evolutionary or physiological individuality is still necessitated. However, these experimental realizations show that a new conception of individuality is necessary for an experimental science, whether it is physical, chemical, or biological. One can observe that there are quite a few cases for such experimental realizations of the type—for instance, the production of synthetic elements in laboratories, the production of new chemical structures such as plastics and polymers, the production of inverted populations in laser science, the production of transformed cells, and so forth.[5]

The conception of experimental individuality does not reject or supersede all metaphysical criteria listed in section 17.2. It does, however, refute the distinguishability at the submicroscopic level and re-establish it at the macroscopic level. It also supersedes delineability by separability, improving the unity to fit with the experimental context, and restricts the persistence under special experimental conditions. In sum, together, the three properties—manipulability, separability, and maintainability of structural unity—indicate that the individuality of experimental entities is conferred by the experimental setups, techniques, and processes. They are experimentally realized by satisfying the three conditions: manipulation, separation, and maintenance of structural unity.

Acknowledgments

I thank the anonymous referee, Christina Conroy, and Melinda B. Fagan so much for their long and highly valuable reviews. These reviews offered great help for me to revise this chapter. I thank Thomas Pradeu so much for his valuable comments and suggestions, which helped me make the final revision. Many friends read the draft of this chapter or attended my presentation at the Paris conference and gave me valuable comments and suggestions. They are Jean-Sébastien Bolduc, Szu-Ting Chen, Hsi-Heng Cheng, Ellen Clarke, Steven French, Alexandre Guay, Dian Jung Han, Tetsuji Iseda, Chin Li, Alan Love, James Myers, John Pemberton, Jack Powers, Thomas Pradeu, and C. Kenneth Waters. I thank them. Finally, I thank Ian Hacking and his work for inspiring me to write this chapter. This research is funded by Taiwan's NSC (NSC 101-2410-H-194-064-MY3).

Notes

1. There are debates between the syntactical view and the semantic or model view of theories, but those debates have little significance for the distinction between theories and experiments. It is thus unnecessary for us to enter into those debates.

2. Cornell and Wieman published their research in the journal *Science* in July 1995. Three months later, Wolfgang Ketterle and his team in the Department of Physics at the Massachusetts Institute of Technology realized the Bose-Einstein condensation in a gas of sodium atoms and reported their experiment in *Physics Review Letters*. Ketterle's experiment got 5×10^5 atoms to enter into the degenerate state (Davis et al. 1995).

3. Physicists can compute the counts of atoms in a Bose condensate by estimating the total mass of the atoms, but they cannot count them when the Bose-Einstein condensation appears. This is because they cannot distinguish between them. In other words, the atoms in a Bose condensate are not countable. I thank Cristina Conroy for her clarification about this point.

4. The method has two steps: First, the experimenter constructs a target vector that carries the foreign gene with a flanking DNA sequence that is homologous to the region of the targeted gene; second, the flanking DNA sequence is inserted into the region to disrupt the targeted gene (knock it out). The homologous sequence is selected to allow the transgene to be freely expressed without disrupting other genes.

5. I thank John Pemberton for providing me these good instances.

References

Anderson, Mark H., et al. 1995. Observation of Bose-Einstein Condensation in a Diluted Atomic Vapor. *Science* 269 (July): 198–201.

Bains, William. 2004. *Biotechnology from A to Z*. 3rd ed. Oxford: Oxford University Press.

Beurton, Peter J., Raphael Falk, and Hans-Jörg Rheinberger. 2000. *The Concept of the Gene in Development and Evolution: Historical and Epistemological Perspectives*. Cambridge: Cambridge University Press.

Boniolo, Giovanni, and Giuseppe Testa. 2012. The Identity of Living Beings, Epigenetics, and the Modesty of Philosophy. *Erkenntnis* 76: 279–298.

Braillard, Pierre-Alain, Alexandre Guay, Cyrille Imbert, and Thomas Pradeu. 2011. Une Objectivité Kaléidoscopique: Construire L'image Scientifique du Monde. *Philosophie* 110: 46–71.

Brinster, Ralph L., Howard Y. Chen, Myrna Trumbauer, Allen W. Senear, Raphael Warren, and Richard D. Palmiter. 1981. Somatic Expression of Herpes Thymidine Kinase in Mice Following Injection of a Fusion Gene into Eggs. *Cell* 27(1): 223–231.

Carlson, Elof A. 1991. Defining the Gene: An Evolving Concept. *American Journal of Human Genetics* 49: 475–487.

Chauvier, Stéphane. 2008. Particulier, Individus et Individuation. In *L'individu: Perspectives Contemporaines*, ed. Pascal Ludwig and Thomas Pradeu, 11–35. Paris: Vrin.

Chen, Ruey-Lin. 2007. The Structure of Experimentation and the Replication Degree: Reconsidering the Replication of Hertz's Cathode Ray Experiment. In *Naturalized Epistemology and Philosophy of Science*, ed. Chienkuo Michael Mi and Ruey-Lin Chen, 129–149. Amsterdam: Rodopi.

Chen, Ruey-Lin. 2013. Experimental Discovery, Data Models and Mechanisms in biology: An Example from Mendel's Work. In *Mechanism and Causality in Biology and Economics*, ed. Hsiang-Ke Chao, Szu-Ting Chen, and Roberta Millstein, 101–122. Dordrecht: Springer.

Cornell, Eric, and Carl Wieman. 1998. The Bose-Einstein Condensate. *Scientific American*, March, 40–45.

Cornell, Eric, and Carl Wieman. 2001. Bose-Einstein Condensation in a Dilute Gas: The First 70 Years and Some Recent Experiments. Nobel Lecture, December 12, 2001.

Dale, Jeremy, Malcolm von Schantz, and Nick Plant. 2002. *From Genes to Genomes: Concepts and Applications of DNA Technology.* West Sussex: John Wiley and Sons.

Darden, Lindley. 1991. *Theory Change in Science: Strategies from Mendelian Genetics.* Oxford: Oxford University Press.

Davis, Kendall B., et al. 1995. Bose-Einstein Condensation in a Gas of Sodium Atoms. *Physics Review Letter* 75(22): 3969–3974.

Folse, Henri, and Joan Roughgarden. 2010. What Is an Individual Organism? A Multilevel Selection Perspective. *Quarterly Review of Biology* 85(4): 447–472.

French, Steven. 1989. Identity and Individuality in Classical and Quantum Physics. *Australasian Journal of Philosophy* 67(4): 432–446.

French, Steven, and Michael Redhead. 1988. Quantum Physics and the Identity of Indiscernibles. *British Journal for the Philosophy of Science* 39(2): 233–246.

Galison, Peter. 1998. *Image and Logic.* Chicago: University of Chicago Press.

Glick, Bernard, and Jack Pasternak. 1998. *Molecular Biotechnology: Principles and Applications of Recombinant DNA.* 2nd ed. Washington, DC: American Society for Microbiology.

Hacking, Ian. 1983. *Representing and Intervening.* Cambridge: Cambridge University Press.

Hacking, Ian. 1988. On the Stability of the Laboratory Science. *Journal of Philosophy* 85(10): 507–514.

Hacking, Ian. 2006. Another New World Is Being Constructed Right Now: The Ultracold. Max-Plank-Institute for the History of Science, Preprint 316.

Hull, David. 1980. Individuality and Selection. *Annual Review of Ecology and Systematics* 11: 311–332.

Hull, David. 1992. Individual. In *Keywords in Evolutionary Biology,* ed. Evelyn Fox Keller and Elisabeth A. Lloyd, 180–187. Cambridge, MA: Harvard University Press.

Kitcher, Philip. 1982. Genes. *British Journal for Philosophy of Science* 33(4): 337–359.

Kitcher, Philip. 1984. 1953 and All That: A Tale of Two Sciences. *Philosophical Review* 93: 335–373.

Kitcher, Philip. 1992. Gene: Current Usages. In *Keywords in Evolutionary Biology,* ed. Evelyn Fox Keller and Elisabeth A. Lloyd, 128–131. Cambridge, MA: Harvard University Press.

Kuhn, Thomas. 1977. Mathematical vs. Experimental Traditions in the Development of Physical Science. In *The Essential Tension,* 31–65. Chicago: University of Chicago Press.

Ladyman, James, and Don Ross. 2007. *Every Thing Must Go: Metaphysics Naturalized.* Oxford: Oxford University Press.

Lin, Cheng-Tsao, et al. 2008. A Controllable Double-Well Magneto-Optical Trap for Rb and Cs Atoms. *Optics Express* 16(9): 6104–6111.

Loux, Michael J. 2002. *Metaphysics: A Contemporary Introduction.* 2nd ed. London: Routledge.

Lowe, E. Jonathan. 1989. *Kinds of Being*. Oxford: Basil Blackwell.
Lowe, E. Jonathan. 1998. *The Possibility of Metaphysics: Substance, Identity, and Time*. Oxford: Oxford University Press.
Lowe, E. Jonathan. 2009. *More Kinds of Being*. Oxford: Wiley-Blackwell.
MacDonald, Cynthia. 2005. *Varies of Things: Foundations of Contemporary Metaphysics*. Oxford: Blackwell.
Maienchein, Jane. 1992. Gene: Historical Perspectives. In *Keywords in Evolutionary Biology*, ed. Evelyn Fox Keller and Elisabeth A. Lloyd, 122–127. Cambridge, MA: Harvard University Press.
Morange, Michel. 1998. *A History of Molecular Biology*. Trans. Matthew Cobb. Cambridge, MA: Harvard University Press.
Millstein, Roberta. 2009. Populations as Individuals. *Biological Theories* 4(3): 267–273.
Portin, Petter. 1993. The Concept of the Gene: Short History and Present Status. *Quarterly Review of Biology* 68(2): 173–223.
Pradeu, Thomas. 2012. *The Limits of the Self: Immunology and Biological Identity*. New York: Oxford University Press.
Radder, Hans. 1995. Experimenting in the Natural Sciences: A Philosophical Approach. In *Scientific Practice: Theories and Stories of Doing Physics*, ed. Jed Z. Buchwald, 56–86. Chicago: University of Chicago Press.
Rosenberg, Alexander. 1985. *The Structure of Biological Science*. Cambridge: Cambridge University Press.
Strawson, Peter F. 1959. *Individuals: An Essay in Descriptive Metaphysics*. London: Methuen.
Thieman, William J., and Michael A. Palladino. 2004. *Introduction to Biotechnology*. San Francisco: Pearson Education.
van Fraassen, Bas C. 1991. *Quantum Mechanics: An Empiricist View*. Oxford: Oxford University Press.
Waters, C. Kenneth. 2004. What Was Classical Genetics? *Studies in History and Philosophy of Science* 35: 783–809.
Watson, James, and Alfred A. Berry. 2004. *DNA: The Secret of Life*. New York: Alfred A. Knopf.
Watson, James, J. Witkowski, M. Gilman, and M. Zoller. 1998. *Recombinant DNA*. 2nd ed. New York: Scientific American Books.
Wiggins, David. 2001. *Sameness and Substance Renewed*. Cambridge: Cambridge University Press.

CHAPTER 18 | Eliminating Objects Across the Sciences

STEVEN FRENCH

18.1 Introduction

The notion of "object" occupies a central place in our metaphysical schemas, and for good reason: we not only seem to bump up against them every day in our "macro-interactions" (for want of a better word) but they have also long functioned as a convenient ontological hook on which to hang various metaphysical frameworks. However, as soon as we try to export such frameworks from the context of the everyday to that of modern science, fundamental problems arise. Here I want to outline those problems in the physical and biological contexts and use them to motivate an eliminativist stance toward objects in both domains.[1]

Now, such a stance tends to cause alarm in general, but even more so when adopted toward biological entities! However, I shall try to assuage concerns by drawing on certain metaphysical resources that, I claim, enable us to continue to talk about putative objects, while maintaining a "sparse" ontology that does not actually include them.

Thus, I shall begin by reviewing the motivation for eliminativism in physics and then consider the quite different motivations in biology. The former has been drawn upon as part of the elaboration of the "ontic" form of structural realism, and it may be that something similar is appropriate for biology as well. I say "may be" because other ontologies that are not object oriented are also feasible, such as process-based approaches, for example, and I certainly do not want to be imperialist in intention. Nevertheless, I do think that structuralist accounts have a lot to offer in the philosophy of biology (see French 2011a and 2012).

I shall then introduce two sets of metaphysical resources that can be deployed to help ease the ontological pain that eliminativism seems to generate. The first relates to monistic tendencies in general and has been given the attractive moniker of "blobjectivism." Significantly it introduces a contextual account of truth in order to accommodate our statements about entities that, strictly, do not exist. The second lies at the other end of this particular metaphysical spectrum and can be characterized as a form of nihilism. This retains our standard account of truth but insists that the relevant truthmakers of the relevant statements are not the entities apparently referred to by the statements, but more fundamental elements of our ontology. I shall try to make these resources palatable in an effort to show that we don't need an object-oriented ontology in biology, thus opening the door to a range of alternatives—including structuralism, of course!

This appropriation of metaphysical resources should be set in the context of a general stance on the relationship between metaphysics and science presented elsewhere (French 2014, French and McKenzie 2012). This is an issue that has been discussed extensively in the literature, of course, with a number of authors excoriating metaphysics for its lack of engagement with the relevant physics (see Ladyman and Ross 2007). While I share these concerns, I also think that metaphysics—even disengaged from physics or science in general—can serve as a kind of "toolbox" from which we can select certain devices, strategies, and so forth, which can then be used to help articulate various positions, including apparently radical ontological ones such as eliminativism. Of course there is a further issue of whether a strategy or device that appears to work in one context, that of physics, for example, is also appropriate in another, that of biology. It seems to me that this can only be decided on a case-by-case basis, and what I hope to show here, at least in part, is that certain eliminativist strategies that can be applied to physical objects can work with biological ones as well. Let me be clear: insofar as I am advocating a particular metaphysics here, it is only a broad form of object eliminativism, and the point of the toolbox analogy is that there is a variety of metaphysical tools that one can pull out to support such a framework.

18.2 The Problem of Quantum Individuality

This problem has been extensively explored and discussed over the past several years (see, e.g., French and Krause 2006 but also Saunders 2006, Bigaj and Ladyman 2010), but in brief, the situation can be summarized as follows: let us begin with the apparently uncontentious assumption that to be an object is to possess an "individuality profile" in the sense that there is a fact of the matter as to whether a given object is an individual or not (Brading and Skiles 2012). If the object concerned is deemed to be an individual, then we may retain our standard formal (and specifically set-theoretic) framework and, further, "ground" the individuality in any of a number of metaphysical features: primitive thisness, haecceity (the two are not necessarily the same), substance, properties plus the Principle of the Identity of Indiscernibles, tropes, and so on. Metaphysics gives us a wide range of resources that we can draw upon to articulate the sense of individuality here. Of course, one might insist that all such resources are inadequate in some way and that there is no such grounding to be had and that individuality should be taken to be primitive (Morganti 2013). Alternatively, if it is concluded that the object concerned (or class thereof) cannot be regarded as an individual, then we may need to introduce some nonstandard formal framework, such as quasi set theory (French and Krause 2006) and articulate the sense of "nonindividuality" in some way, perhaps via a denial of self-identity (French and Krause 2006). Now, it has been claimed that the "individuality profile" of fundamental objects in physics, such as elementary particles, is in fact underdetermined by the relevant theory, namely quantum mechanics (French and Krause 2006; see also French 2011b); that is, the physics alone does not determine whether such objects should be regarded as individuals or not.

This has been presented as a problem for the standard, "object-oriented" realist, in the context of a broadly naturalistic stance (van Fraassen 1991) since some of the core metaphysical content of their realism is now left unspecified. In essence, the objection goes, what right do they have to refer to objects as part of their realist commitments, if they cannot even tell us whether the objects are individuals or not?

There have been various responses to this underdetermination, but I shall focus on the following three, which I think are some of the more interesting and can be given a broadly realist construal.[2] I shall suggest, however, that simply dropping objects from our metaphysical pantheon entirely is the better alternative in this context.

The first response is to slim down the individuality profile by eschewing many of the aforementioned metaphysical resources and adopt a "thin" conception of objecthood. Thus, following the Quinean precept of "to be is to be a variable" in an appropriately formalized and regimented form of the theory, Saunders has suggested that we do not need to draw on any substantive grounding for the individuality of the objects that are denoted in such a theory. And in the case of quantum physics we can articulate a "weak" form of discernibility by drawing on the entangled relations manifested in this domain. Thus in the case of fermions in a singlet state, say, although we cannot determine, even in principle, which has spin up and which has spin down, we can say that each has opposite spin to the other and that is enough to discern them (Muller and Saunders 2008; for concerns see Hawley 2009). Whether this account can be extended to bosons is contentious (see Muller and Seevinck 2009, Bigaj and Ladyman 2010), but note that this does not resolve the above underdetermination; rather, it simply provides a further grounding for the individuality of the objects, albeit one that is thinner and less metaphysically substantial than, say, primitive thisness, or substance. And it does so within a formal framework that begs the question against the alternative nonindividuals account (French and Krause 2006, 172). More importantly, by drawing on certain quantum relations to ground the individuality of these "thin" objects, this form of weak discernibility might be regarded as broadly structural in nature and thus takes us to the same destination as a fully eliminativist approach (French and Krause 2006, Hawley 2009).[3]

An alternative response is detach this concept of an "individuality profile" from the notion of objecthood entirely. Thus Brading and Skiles reject the above assumption and insist that one can retain such a notion without the usually attendant individuality profile (2012). In essence they argue for a law-constituted view of objects according to which to be an object is to satisfy a certain system of physical laws. This is a view that they trace back to Newton, for example, and it is certainly one that again meshes nicely with forms of structuralism (see French 2014). However, one might want to press a little on what it is to "satisfy" a law. One option would be appeal to the notion of "governance," in the sense that laws are standardly taken to "govern" the entities that fall under them. Thus to be an object would be to be governed by the relevant laws. However, this notion of governance is problematic: it is difficult to accommodate within dispositionalist accounts of laws (Mumford 2004), and Humean views eschew it entirely. Alternatively, one might try to cash out the "satisfaction" here by appealing to some notion of dependence (Lowe 2009), so that objects become dependent on the relevant laws. This is a much-discussed notion in metaphysics and again has been deployed in the structuralist cause; indeed, it can be argued that all there is to a physical object—its very constitution—is dependent on the relevant laws to such an extent that the object can be said to be eliminated in favor of those laws (French 2010 and 2014). Again, this option sets us off toward the same destination as eliminativism.

A third, more outré, alternative is to appeal to another kind of metaphysical resource, in the form of recent accounts of indeterminacy. Thus one might suggest that what the above underdetermination reveals is that the individuality profile of quantum objects is indeterminate. A framework for such suggestions has recently been developed according to which although there is only one reality, there are multiple actualities, each representing a precisification of that which is indeterminate (Barnes and Williams 2011, Williams 2008). Thus, the indeterminacy regarding an object's individuality profile can be understood as there being one actuality in which the object is an individual and another actuality in which it is a nonindividual. Now, as I said, this is certainly "out there," metaphysically speaking. But it might allow the realist to retain a notion of object while insisting that it is indeterminate in one crucial respect, namely that of the individuality profile. Given arguments to the effect that other features of putative objects are indeterminate in the quantum context (Darby 2010, French and Krause 2003) this may not seem such a dramatic move.[4] However, leaving aside any concerns about the very idea of there being one reality but multiple actualities, one might still ask: what grounds the objecthood of the entities in this case? One possibility would be to maintain that no such grounding is required and that objecthood is fundamentally primitive. In that case one might wonder why one does not simply adopt the similar move with regard to individuality and have done with it. Alternatively, one could appeal to the relevant laws (what else could one appeal to, indeed?) but that would take us back to option 2 above.

I could continue, but all such options come with pros and cons. One can avoid having to deal with the latter and having to choose any one of the alternatives by simply eliminating objects as a relevant metaphysical feature to begin with. That brings along with it further worries, but as we'll see, those can be assuaged.

18.3 The Problem of Biological Individuality

Of course we do not have a similarly motivated underdetermination in the biological domain. Nevertheless, well-known concerns about objecthood arise here as well.

Let us begin with what Dupré and O'Malley identify as the implicit assumptions of biological ontology: first, that biology is organized in terms of the "pivotal unit" of the organism; and second, that such organisms constitute biological entities in a hierarchical manner (Dupré and O'Malley 2007).[5] The crucial question, then, is how we delineate the notion of an individual organism, and a useful way to start might be as follows: a biological individual (1) has 3D spatial boundaries, (2) bears properties, and (3) is a causal agent (see Wilson 2013), to which many would be inclined to add (4) is countable and (5) is genetically homogenous (which amounts to what Dupré insists is the further presumption of genomic essentialism).

However, this kind of characterization runs into problems accommodating some now-famous counterexamples, such as the "humungous fungus," where a single contiguous fungus organism can cover several square kilometers, or clonal Aspen colonies, again covering a significant area. Such cases generate well-known concerns with countability in particular and undermine the above characterization of biological individuality.[6]

Further problems arise when we turn to examples of symbiotes, such as that of a coral reef, which consists not just of the polyp plus calcite deposits but also zooanthellae algae that are required for photosynthesis. Another example is that of the Hawaiian bobtail squid, whose bioluminescence (evolved, presumably, as a defense mechanism against predators who hunt by observing shadows and decreases in overhead lighting levels) is due to bacteria that are incorporated into the squid shortly after hatching, most of which are then vented at the break of day, when the squid is hidden and inactive, with the remainder forming the basis of the subsequent colony. The presence of the bacteria confers an evolutionary advantage on the squid and thus renders the squid the individual that it is, from the evolutionary perspective, but these bacteria are, of course, not genetically the same as the squid, nor do they remain spatially contiguous with it.

The problems these examples generate for standard conceptions of biological individuals and objecthood in general are well documented (see, e.g., the symposium "Heterogeneous Individuals," held at PSA 2010, Montreal). Various responses have been countenanced, a number of which I will survey here. None have been embraced "across the board," as it were, and indeed, many commentators have concluded that the best we can do is to acknowledge a plurality of approaches.

So Pradeu offers an immunological approach to individuation, on the grounds that it is the immune system that controls variation, and hence such an account can underpin a notion of biological individuality that will be appropriate from an evolutionary point of view (Pradeu 2012). This yields a conception of the organism as a set of interconnected heterogeneous constituents, interacting with immune receptors, where such heterogeneous organisms express the highest level of individuality by virtue of the immune system acting to eliminate lower-level individuals as variants. One worry with this approach is that it might let in too much: given the existence of virophages and recent suggestions that viruses might be capable of constructing what appear to be immunological type responses, the worry arises that viruses also have to be regarded as heterogeneous organisms (O'Malley 2014; see also Clarke 2010). Of course, this may be seen as an advantage insofar as it extends the immunological account into the microbiological realm (Pradeu, personal communication). The risk, of course, is that this extension may undermine the explanatory power of the account itself (O'Malley 2014).

Alternatively, one might adopt a "tripartite account," according to which an organism is (*a*) a living agent (*b*) that belongs to a reproductive lineage, some of whose members have the potential to possess an intergenerational life cycle, and (c) has minimal functional autonomy (Wilson 2013; see also Wolfe 2010). This is underpinned by the assumption that organisms and the lineages they form have stable spatial and temporal boundaries. However, the example of symbiosis, again, suggests that lineages/individuals should be regarded as fluid and ephemeral, raising concerns for this sort of account (see, for example, Bouchard 2010). Furthermore, the recent increased focus on the microbial world, touched on above and to be returned to below, has encouraged the replacement of the old "tree" of life picture, where the "tree" is composed of such lineages, with a "web or network of life" and hence the idea of stable and well-defined lineages is further eroded. (Nevertheless, as we shall shortly see, lineages may play a crucial role in underpinning a form of biological eliminativism, where this fluidity now works to the latter's advantage.)

Perhaps, then, we should look elsewhere. Godfrey-Smith's approach presents two kinds of biological objects: Darwinian individuals, which are members of a collection in which there is *variation, heredity*, and *differences in reproductive success*; and organisms in general, which are systems comprised of diverse parts that work together to maintain the system's structure (for a useful summary see Godfrey-Smith 2012). Thus, some Darwinian individuals are organisms, some Darwinian individuals are not organisms, such as viruses and genes, and some organisms are not Darwinian individuals, such as symbiotic collectives like the bobtail squid case. And there are two particularly nice features of this view: the first is that the borders between these distinctions are themselves fluid: just as certain metabolic collaborations become Darwinian individuals, so certain of the latter "reach out" to other individuals to form new organisms. The second is that when it comes to Darwinian individuals, there is a continuum between those that can be regarded as "paradigm" individuals and those that are "marginal," with the distinction marked by the tendency to form complex adaptations.

Furthermore, Godfrey-Smith's account can accommodate the kinds of counterexamples mentioned above by associating reproductive success with three kinds of reproducers: simple, collective, and "scaffolded." The first contains the machinery of reproduction and includes bacterial cells; the second are entities whose parts have the capacity to reproduce (and in a way that is not due to the coordinated activity of the whole), and these include colonies and symbiotic arrangements; the third covers entities that are reproduced through the reproduction of some larger entity of which they are a part and includes chromosomes, for example.

Nevertheless, problems still arise. As indicated, reproductive success lies at the heart of Godfrey-Smith's account, and one of the features that he identifies is what he calls the degree of narrowing between generations, or the existence of a reproductive "bottleneck." In paradigmatic cases of reproduction there is a high degree of such narrowing, whereas in cases of growth, for example, or cell aggregation, the degree is zero. Crucially, the former allows local mutations to be transmitted. However, Ereshefsky and Pedroso offer biofilms as a counterexample: they have zero degree of narrowing yet, they insist, count as individuals (Ereshefsky and Pedroso 2013). Biofilms are conglomerative communities of microorganisms typically embedded within a "extracellular polymeric substance" that both holds the cells together and acts as a digestive system (Ereshefsky and Pedroso 2013). They form through coaggregation and interact in various ways, including via lateral gene transfer. It is this that allows genes to be distributed within the collective without any "bottleneck." Thus, Ereshefsky and Pedroso argue, biofilms are not "marginal" individuals (Godfrey-Smith may disagree of course!). Indeed, they note, they seem to perform the "crucial functions" associated with biological individuals such as having repeatable life cycles, being "causally integrated wholes whose parts share genes," and interacting with the environment in such a way that these interactions have a uniform effect on their constituents (Ereshefsky and Pedroso 2013).

Perhaps, then, we should try a different tack that also allows for degrees of individuality. Thus Clarke (2012) has recently brought a fresh perspective on the problem and approaches it via the notion of an organism. She suggests that a biological organism is any collection of living parts that possesses individuating mechanisms, where an individuating mechanism "is a mechanism that either limits an object's capacity to undergo within-object selection ... or increases its capacity to participate in a between-object selection process" (Clarke 2013, 427). Thus, an individuating mechanism is any mechanism that increases capacity for between-unit selection, relative to within-unit selection. Such mechanisms act as the causal basis or realizer of the disposition to change at the between-organism level, rather than at the within-organism level (or higher) in response to natural selection. Biological individuals are then defined as "all and only those objects that possess both kinds of individuating mechanism" (Clarke 2013, 22).

By virtue of the fact that the relevant mechanism can fulfill its function to a greater or lesser extent, individuality can be understood as a "property that an object can possess to a greater or lesser degree" (Clarke 2013, 25). This allows pragmatic considerations to play a role in determining how much individuality an object can possess in order to be included in relevant counting decisions, for example (Clarke 2013). Of course, for some this idea of individuality as sitting on a continuum may be somewhat unnerving—it certainly runs counter to the scholastic idea that individuality should be regarded as conceptually distinct from distinguishability and as that which underpins or is associated with the self-identity of the object. In that context, although things may be more or less distinguishable from one another, individuality is a dichotomous notion. Worries might also arise as to whether the notion of "individuating mechanism" above is either too broad or too thin. However, rather than continue to play the game of bringing in counterexamples, I shall suggest below that this and similar approaches amount to disguised versions of the view I favor.

As a bridge to that account, let us consider what lies at the more radical end of the pluralistic spectrum and accept that the category of organisms that are not Darwinian individuals in Godfrey-Smith's sense is larger than many people appreciate (and, indeed, may include people!). This is the core idea behind Dupré's "metagenomic" perspective that urges a shift from individual organismal lineages to the "overall evolutionary process in which diverse and diversifying metagenomes underlie the differentiation of interactions within evolving and diverging ecosystems" (Dupré and O'Malley 2007, 838). Thus,

> To the extent that ... individual autonomy requires just an individual life or life history, then it surely applies much more broadly than is generally intended by biological theorists. Countless non-cellular entities have individual life-histories, which they achieve through contributing to the lives and life-histories of the larger entities in which they collaborate, and this collaboration constitutes their claim to life. But—and this is our central point—no more and no less could be said of the claims to individual life histories of paradigmatic organisms such as animals or plants; unless, that is, we think of these as the collaborative focus of communities of entities from many different reproductive lineages. (Dupré and O'Malley 2009, 15)

This perspective is accommodated by Dupré's "promiscuous realism," which insists that there are many (indeed, "countless") ways of classifying objects in the world, all of which are equally legitimate and objective (Dupre 1993, 18). On this view, "Individual organisms ... are an abstraction from a much more fundamental entity" (Dupré and O'Malley 2007, 842), and objects are no more than "temporarily stable nexuses in the flow of upward and downward causal interaction" (842).

Elsewhere (French 2011a, 2012) I have argued that we can accept those last two suggestions without having to be quite so promiscuous. Here I want to take a slightly different tack and suggest that we can cut through this Gordian knot of competing definitions and counterexamples by adopting an eliminativist stance toward the problem. Now, I appreciate that eliminativism generates hostility wherever it is proposed, but I shall suggest that the sting of the core concerns can in fact be drawn. Furthermore, I think that the kinds of pluralistic approaches outlined above can be understood as forms of eliminativism manqué.[7] Thus, consider Clarke's multiple realizability account of biological individuality: here all the metaphysical work is done by the "individuating mechanisms" or, more generally, by the relevant biological structures (interrelated in various ways and causally informed). In both this case and those of Godfrey-Smith and Dupré, the putative biological objects are fluid, ephemeral, and contextually or pragmatically determined. This opens up the possibility that they should be seen as "heuristic fictions" (Wolfe 2010) or convenient phenomenological "nodes" in the relevant biological structures, possessing sufficient stability at the relevant level to make the evolutionary bookkeeping come out right, but to be dispensed with as metaphysically foundational entities.[8]

The key motivation here is to keep our ontology as minimal as possible while accommodating the central "statements and thought-contents" of common sense and science. Thus we should strive to avoid the kind of inflationary ontology that the above accounts might lead us to (see, e.g., Thomasson 2007) and focus on the fundamental, where it is accepted of course that this should be discipline relative. We can then draw upon and apply various metaphysical tools to accommodate the statements of biology and physics that appear to refer to such objects. These tools allow us to speak of such objects, even truthfully, without having to accept that they exist as fundamental elements of our metaphysical pantheon. The overall attitude here corresponds to what has been called elsewhere a "toolbox" approach to metaphysics (French and McKenzie 2012), according to which even those areas of metaphysics that appear removed from natural science might be drawn upon to help articulate a relevant metaphysics for that science. In a sense, these areas might be likened to "pure" mathematics, which, as is well known, has, throughout the history of science, led to apparently "surprising" applications (French and McKenzie 2012).

Here then, are some metaphysical tools we might use to bolster an eliminativist attitude (see also French 2013, where these are applied in the physics context).

18.4 Some Handy Metaphysical Tools

The first we might call a kind of "Truth Tweaker," since the central idea is to adjust our notion of truth appropriately so as to accommodate an eliminativist stance. So, for example, Horgan and Potrč defend a form of monistic "austere realism" that they call "blobjectivism" (so, the core idea is that there is only one thing at all, namely the world in its entirety, which they call "the blob"), which hinges on a distinction between truth as direct and "indirect" correspondence. The former is just the standard Tarskian notion, whereas the latter takes certain statements to be deemed to be correct according to certain semantic standards, contextually understood. It is via such a move that they can insist on a monistic ontology while allowing for statements about the apparently pluralistic world in which we seem to find ourselves. Thus,

> Numerous statements and thought-contents involving posits of common sense and science are true, even though the correct ontology does not include these posits. ... Truth for such statements and thought contents is indirect correspondence. (Horgan and Potrč 2008, 3)

In particular, the singular, quantificational, and predicative constituents of these thoughts or sentences need not refer to objects, properties, or relations. Hence, on this account we can take the statement "There are tables" to be true while accepting that there are no tables, qua elements of our fundamental metaphysics. The former statement is true, under the contextually operative standards governing common usage but "There are no tables, qua elements of our fundamental metaphysics" is also true, under the much less common semantic standards applicable to "direct correspondence."

Now, there are all kinds of concerns that have been expressed about Horgan and Potrč's "blobjectivism" and about this form of "existence monism" in general (for a contrast, see Schaffer's "priority monism," which insists not that there exists only one thing but rather that the whole has ontological priority over its constituent parts; Schaffer 2010). But here I am primarily concerned with the truth-tweaking tool in the toolbox and what it might do for us.[9]

Alternatively, we might look to a "Set of Simples" to do the job for us. Thus we can retain our standard conception of truth as direct correspondence only but reconsider what it is that makes statements true. The well-known Quinean view is that the ontological commitments of a theory are whatever is referred to by the variables of an appropriately (and typically logical) regimentation of that theory. The alternative "truthmaker" account takes the ontological commitments of a theory to lie with those things that have to exist in order to make the relevant sentences of the theory true. Typically, the relevant truthmakers are more or less read off from the appropriate statements, so that the truthmaker for "x exists" is simply (and always) taken to be x. However, we might find ourselves in situations in which the truthmaker for "x exists" may not be x; they might, for example, be something more fundamental that metaphysicians refer to as the set of "simples" (simple in the sense of being the most fundamental; see Cameron 2008).

On this view, "There are tables" is again taken to be true, and this time in the standard direct correspondence sense, but we can deny any ontological commitment to tables as elements of our fundamental metaphysics since what makes that statement true are the relevant "simples," "arranged table-wise," as some put it. This retains the literal (noncontextual) truth of sentences and captures the intuition that complex objects such as tables exist, in a derivative sense perhaps, but don't exist qua fundamental elements of our metaphysical pantheon. What really exist, then, are the "simples," and although some may feel that this proposal is, well, too simple, I think that if we understand these in terms of that which is ontologically fundamental, this account yields another tool that we may use to support the elimination of objects in both physics and biology.

18.5 Tooling Up Physics

So, let us consider the ontological status of particles in physics. Of course, we already know that in the context of quantum field theory, the notion of particle as a fundamental object is problematic anyway. So let's consider particles as phenomenological or secondary objects in general, or as treated by quantum statistics (French and Krause 2006). Then it seems we must face the metaphysically dilemmas outlined above—in particular, do we regard these objects as individuals or nonindividuals? Should we opt for weak discernibility or some kind of metaphysical indeterminacy? Or should we eliminate them entirely as fundamental objects? I would suggest the last represents the easiest option and one that brings a certain minimalism into our metaphysics of physics. But to still the howls of outrage we can then apply one or other of the above tools, so that we, and physicists, can still talk of particles and in particular utter true statements about them, while denying they exist, fundamentally.

So we could tweak the truth and accept that "There are particles" is (contextually) true in the indirect correspondence sense; but insist that there are no particles qua objects. According to Horgan and Potrč, there is just the "blob" that they then identify with the (as in, the one and only) quantum field (an identification that is problematic to be sure). Of course, we don't have to accept the details of their blobjectivism. Neither do we have to accept their monism in general,[10] although I think it can be made acceptable in the context of modern physics (French, forthcoming). The point is just to deploy the tool they have given us in order to underpin our eliminativism: it is only in the appropriate context that statements apparently referring to such objects can be regarded as true, but in the fundamental context, where truth as direct correspondence applies, they come out false.

Or we could use the second tool outlined above and insist that "There are particles" remains (literally) true, in the direct correspondence sense, but maintain that, again, there are no particles qua objects since the relevant truthmakers are some set of metaphysical simples, arranged "particle-like." In other words, particles, like tables, are derivative entities.

Obviously two concerns then arise; the first is what is to count as the relevant "simples"? And the second is what is meant by being "arranged particle-like"? Elsewhere I've argued that both can be dealt with by adopting a form of ontic structural realism where the simples turn out to be not quite so simple, since the relevant structural elements that make true statements about particles involve the relevant laws and symmetries, and the latter help to articulate the relevant sense of "arrangement" (French 2014). Again, I shan't go into details here, since my aim is just to wield the tools a little and show how they might be used.[11]

Now, how might this work in the biological context?

18.6 Tooling Up Biology

Consider, as the obvious counterpart to elementary particles, genes, and take the following statement:

> Gene duplication is an important source of phenotypic change and adaptive evolution. We leverage a haploid hydatidiform mole to identify highly identical sequences missing from the reference genome, confirming that the cortical development gene Slit-Robo Rho GTPase-activating protein 2 (SRGAP2) duplicated three times exclusively in humans. (Dennis et al. 2012)

Again, we can take this as true (or approximately so) in the relevant biological context, in the sense of Horgan and Potrč's notion of indirect correspondence, which does not commit us to the existence of genes qua fundamental biological entities. Indeed we know full well that the notion is problematic anyway, as Keller (2000) and others have insisted. What statements would be taken to be true in the direct correspondence sense would depend on how monistically we could understand biology. There is interesting work to be done here, but we might, for example, take as the biological equivalent of Horgan and Potrč's "blob" the entire biochemical structure of the world! Here, of course, forms of holism might be deployed in support of monism, just as they are in the physics context (Schaffer 2010). This is contentious (see Weber and Esfeld 2003), but I think monism of whatever form can be made less so in general by importing structuralist elements (French, forthcoming). Certainly, biological structuralism[12] (see French 2011a, 2012) might sit well with a monistic ontology, but the former can also be supported by the alternative simples tool.

Here we would take the truthmaker of the above statement to be, not genes understood as fundamental biological objects, but the relevant biochemical structure, for example. This would make the statement true and allow us to eliminate genes via an understanding of how such structures "arrange" to yield the phenomenological features denoted by "genes." And it should now be clear how the kinds of moves made by advocates of the various forms of pluralism sketched above can be appropriated by the eliminativist: on Clarke's, Dupré's, and Godfrey-Smith's accounts, biological individuals are, in various respects and degrees, contextual entities, possessed of a certain "fluidity" that, according to the eliminativist, arises within the relevant context,[13] or when the appropriate underlying "arrangement" is in place. Thus, as already indicated, we can adopt Clarke's notion of "individuating mechanism," and, from the perspective of the monistic approach, understand it as giving rise to the appearance of individuality within the relevant context (where the semantic standards of truth as indirect correspondence apply); or, from the "simples" perspective, take this mechanism as yielding the appropriate arrangement from the fundamental biological structure that, again, gives the appearance of objects with biological individuality.[14]

Thus, consider our friend the bobtail squid again, and the following statement:

> The life history of the bobtail squid ... shows a very short life span, reaching sexual maturity at 2 months and dieing [*sic*] anywhere between 3 and 10 months. It is a semelparous species, reproducing once in its lifetime. It has been suggested that E. scolopes has a high level of neural complexity, on par with more behaviorally advanced cephalopod molluscs.[15]

Again, this can still be taken as true, even if we adopt an eliminativist stance toward squids as biological objects, where it may be understood to be true in the sense of indirect correspondence and the relevant biological "blob," as it were, is the relevant ecosystem. Again, ecological holism is a well-known set of positions, and the twist given to it here would be to put it to the service of a form of eliminativist monism. Or the statement can be taken as true in the standard, literal sense but with the truthmaker understood to be not the squid, as an object, but, again, the relevant bio-eco-structure, say. (Remember: we are not denying the existence of the bobtail squid as that which is observed swimming about etc. but as that which should be regarded as a biological "object," where the latter raises the sorts of issues canvassed above regarding its individuality.)

In both cases, as already suggested, the kinds of features already presented in the context of discussions of biological individuality can then be appealed to in order to resolve concerns with these forms of eliminativism.[16]

So, in the case of Horgan and Potrč's monism, an obvious worry has to do with how we recover the appearances of a diverse and differentiated world from the underlying monistic "blob." We cannot appeal to the usual story about parts and composition since for them, there are no parts to be composed—there is just the one thing, the blob. Thus they appeal to "manners of instantiation," in the sense that the appearances are "instantiated" but in different manners. To say the least, this is hardly illuminating! Even more acutely, it leads to the accusation that the ontological minimalism of existence monism is bought at the expense of ideological inflation, in the sense that a new and mysterious relationship has to be brought into play (Schaffer, forthcoming). However, we might flesh out the metaphysical bare bones of this relationship by appealing to metabolic collaborations, individuating mechanisms and the like as the biological counterparts to these "manners of instantiation." Likewise, if we opt for the simple truthmakers approach, we can understand metaphysicians' talk of "arrangements of simples" in terms of the appropriate biological mechanisms (embodied perhaps).

Alternatively, and recalling what was said above, we might draw on the concept of a "lineage" again, but now in the context of underpinning two forms of eliminativism.[17]. We recall that a lineage can be defined as a "continuous line of descent; a series of organisms, populations, cells, or genes connected by ancestor/descendent relationships."[18] Thus Mishler (1999) invokes lineages in order to eliminate species, which he sees as having generated problematic issues involved in equating and comparing quite different biological groupings, including the sorts of countability concerns noted above. Instead, he urges, biologists should focus on the lineages that underlie and, crucially, contain these groupings. In this context (the elimination of species) the lineages will be at organism and population level. But one can go further and take organisms to be constituted by cells, for example, themselves regarded as lineages of their components and so on.

There are two features of this proposal to immediately note:[19] the first is that if we think of lineages in terms of patterns—of descent, transmission, persistence, and so on.—and also take them to be uniquely biological, then we may have the basis for a biological pattern-based form of structural realism, in accordance with Ladyman and Ross's pattern-based ontology in general (Ladyman and Ross 2007). The second has to do with the issues whether it's lineages "all the way down/up," as it were,[20] and where they might be said to stop. Haber suggests that the possibility of a problematic regress might be blocked by specifying minimal and maximal lineages. Thus he recommends the adoption of a "levels of lineage" approach according to which biological objects are effectively blurred not only spatiotemporally but across levels of the biological hierarchy as well (Haber 2013). These levels are constituted by lineage-generating entities (where the criteria for what can generate a lineage are contentious and reflect the kinds of concerns raised above regarding biological objects), and no single level of lineage should be taken to be privileged when it comes to phylogeny, for example. Furthermore, parts of the biological hierarchy are constituted by, and constitute, other parts of the hierarchy, and according to Haber, "Interpreting these relations is central to characterizing the entities, processes, mechanisms, and structures of biology in a levels of lineage perspective" (2013, 611). Furthermore, as already indicated, this is extended to biological organisms, iteratively, but with any regress halted via the imposition of maximal and minimal levels (the former might be constituted by maximal clades, for example, although what fulfills that role for the latter remains an issue, but might include self-replicating RNA, "chemotons" or reproducers in general [Haber 2013, 616]). Crucially, given their multilevel, overlapping, and interwoven nature, lineages are multiply decomposable which may then throw light on the so-called species problem and further illuminate species pluralism (Haber 2013, 617–618).

There is much more to be said on this topic of course. But in this context it leads to the suggestion that we might take lineages as our biological simples, on the pluralist truthmaker account or the totality of lineages as our biological blob, according to the existence monist.[21] With regard to the former, as Haber notes, lineages may not exhaust the set of simples, and empirical considerations may further intrude. But as truthmakers, lineages clearly offer a productive approach to grounding the truth of typical biological statements about species, organisms, and so on, in a way that avoids well-known discordance. With regard to the monist stance, application of the Truth Tweaker tool and its inherent contextualism would shed light on how the relevant processes can be understood as yielding the sorts of individuating mechanisms needed to provide the context necessary for us to speak truly about putative biological objects.[22]

What I am proposing, then, is the application of items from the metaphysical toolbox—contextual truth and truthmakers, although there are more such tools, of course—to features of biological "reality"—mechanisms, lineages, and so on—in order to underpin and effect the elimination of biological objects. In effect this is to respond to Sider's challenge (to the monist in particular, although it can be generalized) to provide a "detailed grounding story" that will enable us to relate the everyday or more generally, observable "appearances" with the "fundamental facts" (Sider 2008). This story will have a scientific component and a metaphysical component, where the former provides a scientific explanation in terms of the "scientifically ultimate facts" and the latter grounds such facts in metaphysically ultimate features. For the monist there is only one such feature, namely the whole, and individuating mechanisms and the like take us from this feature to the appearances. For the advocate of simples as truthmakers, the metaphysically ultimate features could be taken to correspond to the relevant biological structures, and Clarke's mechanisms will then likewise underpin the grounding story of how these structures come to be arranged so as to yield familiar objects.

This then bears on the relationship between science and metaphysics in general, an issue that I touched on at the beginning of this chapter and that has been largely discussed in the context of physics (see, e.g., Callender 2011, French and McKenzie 2012) but which could profitably be developed further in the biological context. An immediate concern is that once metaphysical talk of "manners of instantiation" and "simples arranged however-wise" is fleshed out in terms of, or underpinned by, individuating mechanisms and so forth, we can drop the former talk entirely. Indeed, following that line, the "detailed grounding story" could be told entirely in scientific terms, with no need for the metaphysical component at all, since all the relevant work would be done by the detailed scientific mechanisms, features, and so forth. Here we face the possibility that this "fleshing out" of the tools of metaphysics suggests a kind of meta-eliminativism of metaphysics itself![23] Now this might be a result that physicists and biologists themselves would welcome, but Haber has suggested that we should entertain a more modest, "scaled down" metaphysics according to which the only metaphysical commitments we should accept are those entailed by our theories. Of course, the worry here is whether such commitments can really be called "metaphysical." Alternatively, one might take "simples" or "blob" to be "umbrella" terms that enable us to relate what they denote in physics to what they denote in biology or vice versa (Ladyman and Ross 2007), but this of course retains only a diminished role for metaphysics.

However, these are further issues to be discussed elsewhere. The crucial point is that by adopting such metaphysical devices we lose nothing in terms of our ability to accommodate the statements of science, but we gain a minimalist ontology that allows us to sidestep the problems of individuality and in terms of which a metaphysics of science can be constructed.

18.7 Conclusion

The notion of "object" is a tool fashioned out of everyday materials that is not fit for purpose in today's science. In physics, the flashes on scintillation counters, tracks in bubble chambers, and (computer generated) outputs of colliders all suggest an object-oriented ontology that is typically articulated, naively it has to be said, by drawing on those everyday materials. But when this object-oriented metaphysics is imported into the quantum domain, we run up against the metaphysical underdetermination noted above, where physics cannot help us in deciding whether particle as objects should be regarded as individuals or not. Likewise, in biology "biologists implicitly import concepts of the individual into their models and discussions without general consensus about which concept should be used. In the absence of such consensus, theoretical debates about fitness, adaptation, sociality, the evolution of sex and more are hampered because biologists are unaware that they are talking about slightly different things" (Clarke 2013, 414).[24] Of course, one might argue that this lack of consensus is only temporary and that the imposition of a particular account of biological individuality will provide a resolution.[25] I am less sanguine that philosophy can "police" science in this way and suspect that biologists, as philosophically reflective as they are (at least as compared to most physicists), are metaphysically opportunistic and will seize whatever account they see as best fitting the circumstances at hand.

Eliminating objects entirely allows us to sidestep all of these issues and concerns and avoid the "hampering" that Clarke mentions right from the start. If both biologists and philosophers of biology simply refrain from importing, implicitly or otherwise, concepts of the individual into their models and discussions, they would not need to reach consensus to begin with and could instead focus on the crucial features of concern. Of course, they and we, more generally, may then differ as to what we include as the fundamentals of our metaphysics, whether physical or biological, or even whether there should be any fundamentals at all. For myself, I think that in the case of physics, an appropriate reading off from the most fundamental theories supports a structuralist ontology. I also think something like this can be extended to biology, albeit with significant differences (see French 2011a, 2012). But perhaps an alternative ontology can be articulated in this domain, such as some form of process metaphysics (although personally I have my doubts!).[26] The point here is not to defend either of these alternatives but to indicate how supposedly "pure" metaphysics might provide us with the tools to draw the sting from eliminativism by recovering and allowing us to retain the relevant statements of science and common sense (whether they apparently refer to elementary particles or squids).

The final take-home message, then, is a general one: although much of contemporary metaphysics remains out of touch with modern physics (Ladyman and Ross 2007, Callender 2011), it can still provide us with a selection of useful tools and devices that we can deploy for our own purposes (French and McKenzie 2012).

Acknowledgments

I would like to thank various participants at the "Individuals Across the Sciences" conference in Paris, May 2012, for useful comments and discussion, but in particular, John Dupré, Alexandre Guay, Matt Haber, and

Matteo Morganti. I would also like to thank Alexandre and Thomas Pradeu for their kind invitation to participate in the conference and contribute to this collection of essays.

Notes

1. It has been pointed out that whereas in (the philosophy of) physics a distinction is typically drawn between the notions of "object" and "individual," with the focus on the former, that is not the case in biology, where the focus lies with the latter. However, I suspect that this conceptual distinction only became manifest in the physics context with the development of the logic and metaphysics of "nonindividual" objects, and, as far as I know, there has been no such similar development in the philosophy of biology. For the purposes of this chapter, I shall take the notion of biological "object" to be synonymous with that of a biological individual.

2. There is a broader response, which I shall not discuss here, as I have considered and dismissed it elsewhere (French 2010, 2014). This is to insist that such underdetermination is innocuous since only a naive naturalist would ever expect metaphysics to be capable of being read off our physics. Now I am of course a student of the post-Redhead "school of thought" that long maintained that one only gets out as much metaphysics from a given theory as is put in, as it were. However, the underdetermination here is much more fundamental than that between properties as instantiated universals and as tropes, for example. It concerns the very basic features of our ontology—whether objects should be counted as individuals or not—and consequently the very basic formal mechanisms for representing them—set theory and standard classical logic or quasi-set theory and "Schrödinger" logic. for example (French and Krause 2006). My argument—articulated in detail elsewhere (e.g., French 2014)—hinges on the insistence that if we cannot determine (on the basis of the physics) whether our objects are individuals or not, then we are not entitled to include them in our fundamental ontology. Some, perhaps many, would disagree, but that's a matter for another discussion.

3. Perhaps for this reason, certain of the advocates of this approach have taken to refer to the objects that are weakly discerned as "relationals" rather than individuals; I am grateful to one of the referees for reminding me of this.

4. Such other features are not adequately captured via the multiple actualities account, as Darby (2010) notes, but here I am only concerned with the putative indeterminacy of the "individuality profile."

5. Thus the category of biological individuals is larger than that of organisms.

6. Of course, we might just drop the above characterization or, more radically perhaps, count "counterintuitively." The former move simply reinforces the point I'm trying to emphasize here, with the possibility of a plethora of context-dependent individuality criteria. The latter is more problematic, as it's not clear what this could mean. Perhaps there is a possible analogy with the situation in physics, where problems having to do with individuality have led to the suggestion that when it comes to particles in an aggregate, we can only count in terms of assigning a "quasi-cardinal"

number, but not an ordinal (French and Krause 2006). However, such technical moves remain to be fully developed in the biological context.

7. Haber (2013) considers an eliminativist approach toward organisms, although he takes the conflicting accounts of individuality in this context to be the expected outcome of evolution. I am grateful to him for drawing my attention to this work.

8. It has been noted, by a number of commentators, that structural realism is motivated by a concern to respond to the implications of radical theory change for scientific realism by insisting that the relevant physical structure—as represented by mathematical equations—can be identified as remaining through such changes. This is doubly inaccurate as a characterization. First of all, historically, structural realism was equally motivated by a concern to capture the ontological implications of (primarily but not exclusively) physics (see French 2014). Of course, one would hope that these two motivations would work in concert and push in the same direction, as it were, but it has already been acknowledged that they may pull apart in certain situations, raising the concern as to which should be taken as primary (see French 2006). Second, as has been repeatedly emphasized, the use of mathematical equations is not the only way in which the relevant structure may be presented (see French 2011a and 2014).

9. Just to bang home that point, it may be that a form of priority monism might also be regarded as an appropriate ontological framework here. Having already examined that option myself, I'm prepared to be persuaded, but what I want to emphasize here is the *metaphysical strategy* deployed as a tool, rather than the ontological underpinning per se.

10. Despite this insistence, one of the referees has asked why I endorse Horgan and Potrc's existence monism rather than Schaffer's priority monism (see n. 6 above). Actually in forthcoming work (French, forthcoming) I consider both and indicate how, by introducing relevant considerations from physics, they yield interesting structuralist variants.

11. A referee has commented that this view suffers from problems in defining structure and in articulating the relevant dependence and priority relations at work here. I think we simply disagree as to the nature of these problems and their solutions, which I have presented in a series of papers and in French 2014 (as to dependence and priority relations, there is of course the line that these have no place within eliminativist structuralism, as there is only structure, so nothing that is dependent upon it! But again, I have addressed this and related issues in French 2014).

12. Here I take no stance on reductionism. Of course some might argue that biological or biochemical structures ultimately reduce (ontologically) to physical structures, but although I am not averse to taking that line, it needs more consideration than I can give in this chapter.

13. Note, again, this does not imply a form of antireductionism, as the contexts referred to arise within the biological domain.

14. Again my thanks to Matt Haber for suggesting that a notion of "embodied mechanism" might be appropriate in this context.

15. http://www.bio.davidson.edu/people/midorcas/animalphysiology/websites/2005/plekon/bobtailsquid.htm.

16. A concern has been raised that it is not clear what use these metaphysical strategies are for the practicing biologist. Of course, a blunt response might be that

I am not writing for the biologist but for the philosopher of biology who should be concerned about the nature of the ontology she is advocating! A less offensive response might be to run that line again but in a slightly more nuanced fashion, noting that there is considerable overlap between practicing biologists and philosophers of biology and that the more reflective of the former should surely be interested in what metaphysical frameworks and devices might be available in terms of which they can conceptualize their ontologies.

17. Again I am hugely grateful to Matt Haber for this suggestion and for the points that follow.

18. Understanding Evolution website: http://evolution.berkeley.edu/evolibrary/glossary/glossary.php?start=g&end=m.

19. And once again, I am taking these points from Haber's very helpful comments on an earlier version of this chapter.

20. Here again there is an obvious analogy with similar issues that arise in the context of physics-based structural realism; in particular whether it's a case of "structures all the way down."

21. And of course, the "or" here is understood in the inclusive sense, so that, as Haber notes, focusing in lineages gives a way of combining the "simples" and blobjectivist accounts. Thus, the former would correspond to the minimal lineages and the latter to the maximal.

22. Of course, in both cases much more needs to be said before these considerations can be taken to support a form of biological structuralism, in particular with regard to how lineages can be conceptualized from the ontic structuralism standpoint.

23. And this does not only apply to eliminativism (at the "object" level) of course: if one states that certain biological features are "grounded in" but not eliminated by others, fleshing out the scientific details of that grounding in a way that makes it less a label and more a substantive claim might be argued to undermine the point of introducing the notion in the first place, since all the work doing the "grounding" is performed by those scientific details!

24. Here Clarke is using "things" in the generic sense to refer to different features, notions, etc. and not specifically to "things" in an object-oriented sense.

25. As indeed Clarke does, although as indicated above, I would argue that her view amounts to a form of eliminativism "manqué"!

26. Primarily this is because I am not clear what is meant by the notion of "process" in this context. Of course, similar concerns have been raised about structure but, I would argue, they can be dealt with by articulating not only the presentation and representation of structure but also via an appropriate metaphysics (French 2014). I look forward to when processual ontologies have reached a similar stage of articulation so that an appropriate comparison can be made.

References

Barnes, Elizabeth, and J. Robert G. Williams. 2011. A Theory of Metaphysical Indeterminacy. In *Oxford Studies in Metaphysics* 6, ed. Karen Bennett and Dean W. Zimmerman, 103–148. Oxford: Oxford University Press.

Bigaj, Tomas, and James Ladyman. 2010. The Principle of Identity of Indiscernibles and Quantum Mechanics. *Philosophy of Science* 77: 117–36.

Bouchard, Frédéric. 2010. Symbiosis, Lateral Function Transfer and the (Many) Saplings of Life. *Biology and Philosophy* 24: 623–41.

Brading, Katherine, and Alexander Skiles. 2012. Underdetermination as a Path to Ontic Structural Realism. In *Structural Realism: Structure, Objects, and Causality*, ed. Elaine Landry and Dean Rickles, 99–116. Dordrecht: Springer.

Callender, Craig. 2011. Philosophy of Science and Metaphysics. In *The Continuum Companion to the Philosophy of Science*, ed. Steven French and Juha Saatsi, 33–54. London: Continuum.

Cameron, Ross. 2008. Truthmakers and Ontological Commitment: Or How to Deal with Complex Objects and Mathematical Ontology without Getting into Trouble. *Philosophical Studies* 140:1–18.

Clarke, Ellen. 2010. The Problem of Biological Individuality. *Biological Theory* 5: 312–325.

Clarke, Ellen. 2012. The Organism as a Problem in Biological Ontology. Talk given at the Objects, Kinds and Mechanisms in Biology Workshop, Leeds, April.

Clarke, Ellen. 2013. The Multiple Realizability of Biological Individuals. *Journal of Philosophy* 110: 413–435.

Darby, George. 2010. Quantum Mechanics and Metaphysical Indeterminacy. *Australasian Journal of Philosophy* 88: 227–45.

Dennis, Megan, et. al. 2012. Evolution of Human-Specific Neural SRGAP2 Genes by Incomplete Segmental Duplication. *Cell* 149: 912–922.

Dupré, John. 1993. *The Disorder of Things: Metaphysical Foundations of the Disunity of Science.* Cambridge, MA: Harvard University Press.

Dupré, John, and Maureen O'Malley. 2007. Metagenomics and Biological Ontology. *Studies in the History and Philosophy of Science C: Biological and Biomedical Sciences* 28: 834–846.

Dupré, John, and Maureen O'Malley. 2009. Varieties of Living Things: Life at the Intersection of Lineage and Metabolism. *Philosophy and Theory in Biology* 1: 1–25.

Ereshefsky, Marc, and Makmiller Pedroso. 2013. Biological Individuality: The Case of Biofilms. *Biology and Philosophy* 28: 331–349.

French, Steven. 2006. Structure as a Weapon of the Realist. *Proceedings of the Aristotelian Society* 106: 167–185.

French, Steven. 2010. The Interdependence of Structures, Objects and Dependence. *Synthese* 175: 89–109.

French, Steven. 2011a. Shifting to Structures in Physics and Biology: A Prophylactic for Promiscuous Realism. *Studies in History and Philosophy of Science Part C: Biological and Biomedical Sciences* 42: 164–173.

French, Steven. 2011b. Metaphysical Underdetermination: Why Worry? *Synthese* 180: 205–21.

French, Steven. 2012. The Resilience of Laws and the Ephemerality of Objects: Can a Form of Structuralism Be Extended to Biology? In *Probability, Laws and Structures*, ed. Dennis Dieks, Wenceslao J. González, Stephan Hartmann, Michael Stöltzner, and Marcel Weber, 187–200. Dordrecht: Springer.

French, Steven. 2013. Whither Wave-Function Realism? In *The Wave Function*, ed. David Albert and Alyssa Ney, 76–90. Oxford: Oxford University Press.

French, Steven. 2014. *The Structure of the World: Its Metaphysics and Representation*. Oxford: Oxford University Press.

French, Steven. Forthcoming. Is Monism a Viable Option in the Philosophy of Physics?

French, Steven, and Décio Krause. 2003. Quantum Vagueness. *Erkenntnis* 59: 97–124.

French, Steven, and Décio Krause. 2006. *Identity in Physics: A Historical, Philosophical, and Formal Analysis*. Oxford: Oxford University Press.

French, Steven, and Kerry McKenzie. 2012. Thinking Outside the (Tool)Box: Towards a More Productive Engagement between Metaphysics and Philosophy of Physics. *European Journal of Analytic Philosophy* 8: 42–59.

Godfrey-Smith, Peter 2012. Darwinian Individuals. In *From Groups to Individuals: Perspectives on Biological Associations and Emerging Individuality*, ed. Frédéric Bouchard and Philippe Huneman, 17–36. Cambridge, MA: MIT Press.

Haber, Matthew H. 2013. Colonies Are Individuals: Revisiting the Superorganism Revival. In *From Groups to Individuals: Evolution and Emerging Individuality*, ed. Frédéric Bouchard and Philippe Huneman, 195–217. Cambridge, MA: MIT Press.

Hawley, Katherine. 2009. Identity and Indiscernibility. *Mind* 118: 101–119.

Horgan, Terence, and Matjaž Potrč. 2008. *Austere Realism*. Cambridge, MA: MIT Press.

Keller, Evelyn Fox. 2000. *The Century of the Gene*. Cambridge, MA: Harvard University Press.

Ladyman, James, and Don Ross. 2007. *Every Thing Must Go: Metaphysics Naturalized*. Oxford: Oxford University Press.

Lowe, E. Jonathan. 2009. Ontological Dependence. *Stanford Encyclopedia of Philosophy*, Spring 2010 ed., ed. Edward N. Zalta. http://plato.stanford.edu/archives/spr2010/entries/dependence-ontological/.

Mishler, Brent D. 1999. Getting Rid of Species. In *Species: New Interdisciplinary Essays*, ed. Robert A. Wilson, 141–185. Cambridge, MA: MIT Press.

Morganti, Matteo 2013. *Combining Science and Metaphysics: Contemporary Physics, Conceptual Revision and Common Sense*. Basingstoke: Palgrave Macmillan.

Muller, Fred, and Saunders, Simon. 2008. Discerning Fermions. *British Journal for the Philosophy of Science* 59: 499–548.

Muller, Fred and Michael Seevinck. 2009. Discerning Elementary Particles. *Philosophy of Science* 76: 179–200.

Mumford, Steven 2004. *Laws in Nature*. London: Routledge.

O'Malley Maureen. 2014. Review of Thomas Pradeu (2012), *The Limits of the Self: Immunology and Biological Identity*. *British Journal for the Philosophy of Science* 65: 179–183.

Pradeu, Thomas. 2012. *The Limits of the Self: Immunology and Biological Identity*. Oxford: Oxford University Press.

Saunders, Simon. 2006. Are Quantum Particles Objects? *Analysis* 66: 52–63.

Schaffer, Jonathan. 2010. Monism: The Priority of the Whole. *Philosophical Review* 119: 31–76.

Schaffer, Jonathan. Forthcoming. Why the World Has Parts: Reply to Horgan and Potrč.

Sider, Theodore. 2008. Monism and Statespace Structure. In *Being: Developments in Contemporary Metaphysics*, ed. Robin Le Poidevin, 129–150. Cambridge: Cambridge University Press.

Thomasson, Amy. 2007. *Ordinary Objects*. Oxford: Oxford University Press.

van Fraassen, Bas. 1991. *Quantum Mechanics: An Empiricist View*. Oxford: Oxford University Press.

Weber, Marcel, and Michael Esfeld. 2003. Holism in the Sciences. In *Unity of Knowledge in Transdisciplinary Research for Sustainability*, ed. G. H. Hadorn, 1–16. Oxford: EOLLS Publishers.

Williams, J. Robert G. 2008. Multiple Actualities and Ontically Vague Identity. *Philosophical Quarterly* 58: 134–154.

Wilson, Robert A. 2013. The Biological Notion of Individual. In *Stanford Encyclopedia of Philosophy*, Spring 2014 ed., ed. Edward N. Zalta. http://plato.stanford.edu/entries/biology-individual/.

Wolfe, Charles. 2010. Do Organisms Have an Ontological Status? *History and Philosophy of the Life Sciences* 32: 195–231.

NAME INDEX

Adams, R. M. 9, 196, 213
Albert, D. 189, 237, 267
Arenhart, J. 13, 61, 64, 69, 77, 281, 291
Aristotle 1, 5, 11, 27, 30, 36, 42–43, 85, 88, 97, 99, 341
Armstrong, D. M.44, 110, 113

Bacciagaluppi, G. vii, 233, 235, 246
Bapteste, E. 337–338, 341
Barrett, J. 257, 263, 267
Bell, J. 198, 204, 234
Bernays, P. 67, 189, 275
Bigaj, T. 202, 204, 279, 292, 372–373
Bird, A. vii, 3, 46, 342
Black, M. 47–48, 53–54, 56–58, 70, 167, 189
Bohm, D. 15, 63, 74–75, 226–229, 231, 233–246, 291
Bohr, N. 14, 124, 137–140, 226, 228, 242, 245
Boniolo, G. vii, 317–318, 325, 336, 343
Born, M.70, 226, 287
Bouchard, F. 1, 6, 99, 119, 336, 377
Bourke, A. F. G.149–150, 156–158
Brading, K. 8, 372, 374
Braillard, P.-A. 324, 354
Bratman, M.146–147, 152–153, 158–159
Brogaard, B. 303–304, 307
Brown, H. 228, 233–235, 241, 243
Buss, L. 1, 6, 18, 99, 101, 160, 162
Butterfield, J. 1, 53, 199, 207, 209, 211, 213, 216, 222, 267, 283

Callender, C. 387–388
Carnap, R. 3, 188, 343
Carrara, M. 317–318, 325, 336
Caulton, A. vii, 1, 53, 188, 199, 207, 209, 211, 213, 222, 276, 279, 283
Chauvier, S. 43, 350
Chauvinus, S. 27, 34
Chen, R. L. 354, 361
Clarke, E. vii, 1, 6, 119, 148–149, 160, 367, 376, 378–379, 384, 386, 388, 391
Conroy, C. 16, 250, 257, 266, 268, 367–368

Da Costa, N. C. A. 69, 291
Darby, G. 375, 389
Darwin, C. 87, 99, 310
Dasgupta, S. 222–223
Dawkins, R. 92, 100, 336, 339
De Gosson, M. 236–238
Dewdney, C. 232, 234, 240
Dieks, D. 189, 279
Dorato, M. 1, 292
Dupré, J. vii, 1, 6, 99–100, 111, 124, 140, 336–338, 341, 375, 379–380, 384, 388

Earman, J. 8, 222, 282
Einstein, A. 10, 17, 74, 77, 210–211, 222, 227, 285, 287, 322, 350, 355–357, 364–368

Elby, A. 228, 233–234, 241, 243
Esfeld, M. vii, 208, 239, 383
Everett, H. 16, 187, 224, 250–252, 257, 266, 268

Fagan, M. 127, 135–136, 139, 141, 309, 367
Fraser, D. 202, 287
Frege, G. 51, 76, 175
French, S. 1–3, 8, 10, 17, 68, 77–79, 194, 196, 199–200, 222, 226, 229, 233, 245, 276–277, 291, 317, 320, 342, 343, 351–352, 356, 367, 371–375, 379–380, 382–383, 387–391

Gardner, A. 114–115, 118, 150, 157, 162
Ghirardi, G. 187, 189
Ghiselin, M. 295, 297–299, 302–303, 332
Gilbert, M. 147, 153
Glick, D. 15, 207, 246
Godfrey-Smith, P. 6–7, 13–14, 99–100, 103–110, 113, 115, 118–119, 125, 142, 160, 305, 307, 343, 377–380, 384
Goldstein, S. 189, 231, 241
Goodman, N. 295, 298–299, 300
Gould, S. J. 1, 6–7, 100, 111
Grafen, A. 150, 162
Guay, A. 1, 8–9, 11, 17, 140, 204, 222, 246, 312, 317, 367, 388
Guillemin, V. W. 236, 238

Haber, M. 97, 99, 100, 140, 299, 305, 310, 332, 343, 385–388, 390–391
Hacking, I. 349, 353–354, 357, 364–365, 367
Hamilton, A. 297, 305
Hawley, K. 200, 282, 373–374
Healey, R.222–223
Heisenberg, W. 136, 226, 287
Hennig, W.87–89, 297
Hilbert, D. 67, 165–166, 169, 172–174, 176, 178, 189, 211, 275, 280, 283
Hiley, B. J. 15, 228–229, 234, 236, 238–239, 242–246

Holland, P. R. 232, 240
Horgan, T. 288, 380–384, 390
Huggett, N. 189, 198, 276, 279
Hull, D. L. 1, 5–7, 13–14, 16–17, 92, 103–104, 110–115, 118, 125, 193, 295, 297–299, 302–305, 331–339, 342, 344, 353

Jammer, M. 243–244
Janzen, D. H. 97, 161, 375

Kant, I. 2, 29, 43
Kaye, R. D.235–236
Ketterle, W. 77, 367
Kistler, M. vii, 3
Kitcher, P. 302, 359–360
Krause, D. 1, 8, 13, 61, 64, 68–69, 73, 77–79, 194, 200, 222, 226, 233, 245, 281, 291, 317, 342, 372–375, 382, 389–390
Kyprianidis, A. 232, 240

Ladyman, J. 1–3, 10, 15, 17, 193, 195–196, 199, 201–204, 208, 226, 228–230, 233–234, 243–246, 279, 281, 283, 291, 308, 311, 320, 327, 342, 352–353, 372–373, 385, 387–388
Laplane, L. 7, 140, 142
Leibniz, G. W. 1, 42, 44, 52, 55–58, 65–68, 70, 166, 197, 210, 215, 221, 273, 275–276, 278–280, 283–284, 318–319, 344, 351
Lewin, K. 17, 317, 319, 321–322, 343
Lewis, D. 9, 190, 214, 216, 254, 267–268, 341
Lewontin, R. 86–87, 89, 105–106, 116, 118
Locke, J. 1, 5, 52, 60, 64, 319
Lowe, J. vii, 2, 44, 46–48, 64, 70–71, 74, 78, 193, 196, 213, 220–221, 342, 350, 374

Macdonald, C. vii, 2, 267, 349, 351
MacKenzie, K. vii, 3, 189, 204, 372, 380, 387–388

Maudlin, T. 2–3, 209, 214–216, 243, 268, 342
Maynard-Smith, J. 1, 6, 19, 99, 102, 156, 163
Mayr, E. 145, 163
McFall-Ngai, M. 308–309, 336–337
Michod, R. E. 1, 6, 160–161
Morganti, M. 1, 3, 9, 11, 16, 273, 284, 291–292, 373, 389
Muller, F. 1, 68, 70, 189, 194, 196, 198–199, 276, 279, 373
Mumford, S. 2–3, 374

Norton, J. 140, 198, 222, 276, 279

Ockham, W. 40, 43, 282
Okasha, S. 6, 162, 297, 307, 310
Olson, E. 255–256, 268
O'Malley, M. vii, 1, 6, 99–100, 111, 124, 140, 336–338, 341, 375–376, 379

Parfit, D. 16, 250, 253–256, 259–260, 262, 266–268
Paternotte, C. 147, 155, 160, 162
Pättiniemi, I. 15, 226, 246
Philippidis, C. 235–236
Plato 30, 39, 43–44, 341
Pooley, O. 222, 283
Potrč, M. 288, 380–384, 390
Pradeu, T. 1, 6, 8–9, 11, 17, 42, 99–100, 110, 119, 125, 140–142, 159, 161, 312, 317, 319, 336–338, 342, 352–353, 367, 376, 389
Putnam, H. 15, 175, 190, 241, 288
Pylkkänen, P. 15, 242, 246

Queller, D. C. 156, 161
Quine, W. V. O.7–8, 26, 28, 67, 165, 167, 174–180, 188–190, 193, 197, 275, 278–280, 322, 373, 381

Ramsey, F. P. 66, 189
Redhead, M. 2, 194, 196, 229, 276–277, 356, 389

Reichenbach, H. 8, 17, 317, 322–326, 343–344
Rosenberg, A. 4, 307, 359
Ross, D. 1–3, 15, 17, 195, 203, 226, 228–230, 233–234, 243–246, 281, 291, 308, 342, 352–353, 372, 385, 387–388
Russell, B. 65–66, 75–76, 175, 188, 200

Sarkar, S. 308, 311
Saunders, S. 1, 8, 10–11, 14–15, 68, 70, 87, 136–138, 140, 142, 165, 172, 189–190, 194, 196–199, 201, 204, 216, 219, 222, 230, 253, 257–258, 267–268, 276–277, 327, 344, 372–373
Schaffer, J. 222–223, 381, 383, 385, 390
Schleiden, M. 95–96, 142
Schrödinger, E. 63, 78, 185, 187, 217, 227, 231–232, 236–240, 243, 277, 389
Seevinck, M. P. 1, 196, 199, 276, 279, 373
Sider, T. 254, 267, 301, 386
Simons, P. 43, 299, 301
Sjöqvist, E. 233, 235
Skiles, A. 372, 374
Sober, E. 6, 111, 157, 298, 331, 343
Stachel, J. 8, 207, 209–210, 212–214, 221–222, 283
Sterelny, K. 100, 145
Sternberg, S. 236, 238
Strawson, P. 1, 2, 28, 194, 350
Szathmary, E. 1, 6, 99, 156

Teller, P. 65, 190, 222, 284
Toraldo di Francia, G. 61, 63, 78
Tuomela, R. 147–148, 158, 160

Vaidman, L. 257, 267–268
van Fraassen, B. C. 351, 373
Varzi, A.299–301, 306
von Neumann, J.75–76

Wallace, D. vii, 188, 190, 202, 253, 257, 266–268
Waters, K. 359, 367

Weber, M. 187, 383
Weingard, R. 228, 233–234, 241, 243
Weisberg, M. 1, 311
Werner, S. A. 234, 237
Weyl, H. 226, 240
Whitehead, A. N. 65–66, 341
Wiggins, D. 1–2, 62, 104, 110, 253, 318–319, 335, 343–344, 350

Wilson, D. S. 111, 157, 298
Wilson, J. 1, 6, 141
Wilson, R. 1, 6, 122, 125–126, 140, 300, 375, 377
Wolfe, C. 343, 377, 380

Woodger, J. H. 295, 298, 303, 311

Zurek, W. H. 73, 187

GENERAL INDEX

adaptation 98, 103, 112, 117, 150–152, 161–162, 377, 388
aggregate (*see also* plurality) 4, 34–36, 38, 40–44, 103–107, 116–119, 144, 156, 281, 288, 378, 389
Aharonov-Bohm effect 229, 235–240
alternation of generations 86, 89, 91, 92, 96, 98
autonomy 31–34, 39, 122, 125–126, 134, 160, 229, 242, 274, 334, 377, 379

Bell inequalities 198, 204
biofilm 13, 103–119, 378
biotechnology 359–363
Black symmetrical world 47–48, 53–54, 57–58, 70–71, 167, 189
blobjectivism 372, 380–382, 384
Bose-Einstein condensate 10, 17, 77, 210–211, 350, 355–357, 364–368
Bose-Einstein statistics, *see* Bose-Einstein condensate
boson 72, 77, 169, 189, 197–198, 203, 210–211, 220, 223, 276–277, 279, 280, 287, 352, 355–357, 366, 373
  Higgs boson 189, 287
  Nambu-Goldstone boson 203
bottleneck 90–100, 103, 106–108, 113, 115, 119, 161, 307, 377–378
boundaries 35–36, 43, 107, 122, 124–126, 129, 133, 309–310, 312, 319, 342, 350–352, 359–360, 375, 377
branching 16, 250–253, 257, 259, 261, 265–267
bundle 8, 15, 64–65, 195, 200–201, 230, 284, 291, 352

causal interpretation of the quantum theory 227
cell 90, 104–105, 112, 116, 122–140, 305, 308, 344, 378–379
  cell as an individual 122
  See also "stem cells"
complementarity (applied to stem cells) 136–138
contextuality 69, 138, 227, 282–284, 286, 330
contiguity 125, 155, 161, 375–376
continuant 85–86, 90, 94–97, 100
cooperation 114, 117–118, 144–145, 148, 156
countability 52, 75, 77, 93–94, 119, 279, 351–353, 368, 375–376, 389

decoherence 187–188, 190, 257
degrees (of individuality) 7, 26–27, 30, 36–41, 43, 103, 107, 113

399

development 96, 106, 107, 116–117, 124, 128, 132–139, 299, 305, 331, 337
diachronic/synchronic 7, 8, 10, 62, 203, 250, 252–254, 256, 259, 262, 265–267, 274, 319, 322–323, 326–327, 331, 341, 343
differentiation 37–38, 44, 105, 127–136, 141, 308
discernibility (if different from principle of indiscernibles) 13, 38, 44, 53–58, 73–74, 76–78, 168–171, 189, 274, 301, 320, 326–327, 344, 351, 356, 374, 382
division of labor 98–99, 103, 106–108, 115, 119, 122, 149, 153, 156, 161

entity 12, 28, 46, 49–50, 52, 59, 63, 104, 145, 244, 317, 321, 333–335, 349, 351, 366
essence 29, 31, 52, 59, 223, 332, 336
essentialism 15, 64, 207–209, 212, 214–218, 223, 283, 297–298, 301, 310, 318, 335–336, 375
  structural essentialism 208–209, 212, 215–217
experimental realization 349–350, 357–359, 364–366

fermion 169–170, 185–186, 189, 197–198, 211, 216, 218, 220, 223, 231, 233, 237, 276–277, 352, 356, 373
fitness 6, 86, 89, 91, 97, 106, 113, 116, 118–119, 150–156, 161, 388
fock space 190, 280
fusion 36, 43, 93, 97–99, 329, 334–335, 337, 339–340

gene 30, 92, 113–114, 145, 148, 289, 307, 339, 359–364, 383
general relativity 194, 207, 209
genidentity 11, 17, 317–332, 335–338, 340–344
  functional genidentity 324–330, 342
  material genidentity 323–326, 344

germ-soma separation 106, 134, 149
Gibbs paradox 190
Goldstone's theorem 202
group 41, 105, 112–113, 115, 122, 125, 144–151, 156–162, 278, 295, 306–308
  group identification 157
GRWP theory 187

Haag's theorem 288
haecceity, *see* individual, haecceity
Hamilton-Jacobi theory 227, 231, 236
heredity 13, 86, 91–92, 359, 377
heritability 6, 89, 97, 99
hole argument 209–211, 214–215, 218, 222–223, 283
homogeneity 51, 132, 138, 141, 366, 375
Humean supervenience 214

identity 5, 7–11, 28–35, 46–49, 51–53, 56–59, 61–68, 70–74, 76–79, 104, 110, 157–159, 166–168, 189, 193–205, 220–223, 226–229, 239–240, 244, 250–267, 273–275, 277–292, 297–301, 304, 317–344, 351, 356, 378
  branch-relative identity 16, 250, 256, 258, 260–264, 266
  diachronic identity 7–8, 16, 62, 63, 250, 252–254, 256, 259, 261–262, 265–267, 319, 322–323, 326–327, 331, 341, 343
  genetic identity 149, 155, 158, 162, 336
  personal identity 16, 193, 253, 259, 262, 266–268
  psychological connectedness 250, 254–256, 259, 262, 265
  synchronic identity 10, 62, 194, 250, 261, 319–320, 331
  transitivity of identity 252–253, 255–256, 259–262, 265
  transtemporal identity, *see* diachronic/synchronic
  transworld identity 8, 194, 284

immunology 6, 11, 110, 134, 141, 307, 319, 338, 353, 376
impenetrability 65, 197, 275, 352
indexical 174
indiscernibles (principle), *see* Leibniz's law of identity
individual
  bare particular 189, 196, 352
  biological (definition of) 5, 6, 126, 145, 148–149, 243, 289, 307, 331–340, 353, 376, 384
  Bohmian 15, 63–64, 74–75, 226–229, 231, 233, 235, 238–240, 242–245, 338–340, 342
  bundle theory 8, 15, 64–65, 194–195, 200–201, 230, 284, 291, 352
  colony 36, 111, 133, 149, 295, 306, 361, 375–377
  contextual 137–138, 226–227, 273, 283–286, 288, 327, 330, 372, 380–382, 384, 386
  countability 9, 28, 52, 75, 77, 85, 87, 94–94, 106, 119–221, 277, 279–280, 282–283, 285–286, 351–353, 357, 368, 375–376, 385, 389
  Darwinian 13, 87, 90, 95, 97, 377, 379
  eliminativism 17, 371–372, 374, 377, 379, 382, 384–385, 388, 391
  emergent 188, 190, 195, 202
  evolutionary 6, 97, 125, 134–135, 140–141, 144, 160, 306, 353, 366
  experimental 129, 278, 354, 363, 367
  general definition 5, 27, 30–41, 49–50, 59, 61–62, 124–125, 165, 196, 213, 318–320, 322, 375
  genidentical 318, 321–322, 331–332, 336, 339–340
    internal organisations 333–334
    metabolic interactions 338
  haecceity 9, 64, 189, 195–196, 199, 202, 204, 223, 229–230, 233, 244, 283–284, 373
  identity-resemblance 318, 335–336

immunological 6, 11, 110, 319, 338, 353, 376
indistinguishable 14, 38, 66, 74–75, 77–78, 165, 167–171, 173, 178, 180–183, 189, 209–210, 219, 223, 229, 234, 276, 295, 320, 351–352, 355–357
logico-cognitive 27–28, 61, 71, 348
mechanism 379, 384, 386
multi-level 7, 361, 386
  nomological 110, 335, 374–375
nonindividual 9, 12, 15, 28, 32, 39, 41, 43–44, 46, 49–50, 59, 61–64, 71–75, 77–79, 194–195, 198–199, 208, 214, 223, 226, 233, 240, 244, 277–281, 285–286, 288, 292, 352, 373–374, 382, 389
permutation invariance 8, 15, 67, 73–74, 78, 165, 168–169, 171–180, 182, 184, 188–189, 194, 197, 202, 207, 210–212, 214, 219, 221–223, 229, 277, 283–284, 323–324, 327
persistence 5, 7–8, 13–14, 35, 85–87, 90, 93, 165, 175, 178, 194, 204, 244, 301, 308–309, 351, 357, 365, 367, 385
phenomenal 6, 353
physical 325–331
physiological 6, 87, 99, 122, 125–126, 318, 353, 366
primitive 9, 16, 37, 52, 61, 64, 67–68, 78–79, 166–168, 190, 196, 202, 213, 241, 274, 280, 282–283, 286, 290–291, 300, 322, 372–375
process 9, 13, 17, 91, 104, 110, 114, 116, 119, 122, 124–128, 131, 133–135, 141, 144–145, 151–152, 159, 204, 228–229, 239, 288, 244–245, 309–310, 317–320, 324, 327, 330, 338–343, 358, 362, 371, 378–379, 386, 388
pseudo-individual 46, 59

individual (cont.)
- quantum 63, 72, 76, 165, 194, 200, 226, 228–229, 240, 244, 274–275, 277, 280, 282, 284, 290, 372, 382
- quasi-individual 46, 59
- relational 8, 15, 52, 194–195, 197, 199, 201, 208, 212, 214–217, 219–223, 227, 258, 275–276, 278, 284, 286, 304, 310, 389
- structural 208, 213, 220–221, 224
- subindividual 47, 59
  - symbiotic 85, 87, 108–109, 113, 156, 289, 308, 336–337, 377
  - symmetry-breaking description (in terms of) 168
- theory-based 6–8, 13, 104, 110, 118, 195, 201, 240, 244–245, 274–281, 295, 299, 305–306, 323–327, 349, 352–353, 381
- trajectory 15, 74, 168–170, 173, 176–178, 189–190, 197, 204, 229–231, 233, 235–236, 239–240, 244, 291, 317, 328, 352
- transcendental 30, 43, 64, 65, 196, 230, 352

individuation 2, 6, 10–11, 29–31, 43, 61–66, 75, 129, 193–204, 212–213, 281–282, 309–311, 325, 332, 350–351, 358–359, 376
- vs. individualization 29

information 63, 75, 117, 138, 155, 171, 229, 239, 241–242, 244, 360
inheritance 89, 103, 115–118
integration 106–109, 125, 129, 134, 153, 160–161, 337
interactor 92, 104, 109–118, 160
intuition 11, 13, 43, 61–62, 73, 75, 78, 86, 98, 113, 119, 126, 130, 147–149, 158, 161, 203, 210, 289, 342, 351, 353, 375, 382

jellyfish 95–97, 100, 334
joint action 144, 146–148

law of nature 3, 42, 110, 138, 203–204, 209–210, 241, 251, 302, 319–320, 335, 340–341, 343, 374–375, 383
- governing 341
- Ramsay-Mills-Lewis 341

Leibniz's law of identity 8, 53–58, 65, 67–68, 166–167, 195–197, 199, 202, 204, 230, 275–276, 279–280, 282, 285, 351, 353, 354
life cycle 13, 86–105, 116–117, 152, 161, 334, 377, 378
logic 11, 37, 52, 54, 61–62, 65–71, 77–79, 166, 173–174, 188, 284, 298, 389
Lowenheim-Skolem theorem 190

mass 27–28, 51, 172, 179, 183
mereology 16, 295–296, 298–302, 304, 306, 311–312, 317, 320–321
See also part-whole relationship
metabolism 6, 87, 90, 97, 109–110, 112, 126, 145, 160, 306–308, 333, 336–338, 377, 385
metaphysics 1–4, 7, 11, 13, 16–17, 25–26, 30, 36, 41, 46, 48, 52, 61, 73, 75, 78, 85, 89, 104, 113, 119, 165, 207–209, 212–214, 216, 221, 229, 245, 259, 265–266, 273–275, 278–179, 281, 284, 289, 291–292, 296, 305, 307–308, 310–311, 323, 342, 348, 350, 372–374, 380–382, 387–389, 391
- experimentalmetaphysics 278
- metaphysics as a toolbox 3, 372, 380–381, 386
- metaphysics of science 2–5, 208, 273, 387
- naturalistic metaphysics 16, 202, 204, 273–274, 279, 281–283, 285, 288–292, 296, 311, 353, 373, 389
- ontological commitment 203, 323, 348–349, 357, 359, 361, 364, 381

ontological relativity 15, 174–175
particular 12, 15, 26–31, 34, 41–42, 49–50, 63–64, 74, 173–174, 177, 189, 196, 278, 352
primitive ontology 190, 241, 288
principle of generosity 233–234
principle of parsimony 234
process ontology 318, 320, 340–342
substitution principle in metaphysics 351
universal 27–28, 39, 43–44, 50, 282, 291, 341, 389
metaplectic group 236–238
microbe 104, 141, 309, 337, 361–362, 376–377
miscibility 34–43
multicellularity 9, 93, 96–97, 106, 123–142, 156, 161, 305, 309, 320, 344
mutualism 112, 117, 336

nominalism 282, 291, 299, 301, 303, 366
nonindividual, *see* individual, nonindividual

object 2, 7, 10, 25–26, 49–50, 57, 59, 62–79, 87, 92–94, 97, 127, 165–168, 193–205, 207–208, 212–215, 221–222, 226–231, 239–240, 243, 245, 275–287, 291, 297–301, 317, 319, 323, 342–342, 348, 350–351, 357, 364, 365, 371–391
ontological commitment, *see* metaphysics, ontological commitment
ordinary language 166, 170, 344
organicism 145
organism 6, 33, 40, 85–90, 106, 109, 111–112, 114, 122–129, 139, 145, 148, 150, 295, 303–307, 331–338, 375–379
organism is a society of cells 122
model organism 124, 132–135, 362

part-whole relationship 9, 89, 295–299, 302–306, 310, 312, 320, 343
See also mereology
particle 8–10, 14–15, 44, 46–47, 65, 79, 136–138, 142, 165, 168–173, 176–183, 185–190, 194–195, 197–204, 208, 210, 212–213, 215–224, 226–236, 238–246, 251–252, 258, 261–264, 275–282, 284–288, 292, 317, 319–320, 326–327, 330, 343, 352, 373, 382–383, 387–389
particular, *see* metaphysics particular
permutation 8, 73–74, 78, 165, 168–184, 188–189, 194, 197, 202, 207, 210–212, 214, 221, 223, 229, 277, 283–284, 323–324, 327
persistence 7, 86, 93, 152, 178, 351, 365, 367, 385
photon 189, 197, 239, 355–356
phylogeny 91, 100, 297, 305–309, 333, 336–337, 385
plurality (*See also* aggregate) 50–51
policing 149, 153–154, 161
population 6, 13, 112, 131, 140–142, 145, 156, 304, 307–309, 385
predicate, *see* property
principle of identity of indiscernible, *see* Leibniz's law of identity
principle of individuation 11, 64, 75, 193, 195, 200, 204, 281, 319
property
  demarcating property 179–180, 182–184, 186–188
  emergent property 179, 188, 190
  intrinsic property 14–15, 26, 173, 189, 193, 197, 200–203, 208, 214, 227, 230, 275–276, 280, 282, 285, 290, 297, 304, 310, 363
  monadic property 68, 167, 168, 174, 275–276, 278, 284, 290
  relational property 8, 15, 52, 194–195, 197, 199, 201, 208, 212, 214–217, 219–223, 227, 258, 275–276, 278, 284, 286, 304, 310, 389

General Index | 403

psychology 11, 145–146, 151, 157, 193, 250, 254–256, 259, 262, 265, 268
Putnam's paradox 15, 175, 190

quantum entanglement 185, 189, 204, 216, 220, 223, 259
quantum field 14, 74, 180, 189–190, 202–203, 228, 233, 244–280, 245, 286–288, 290, 327, 343, 382
quantum measurement problem 184, 187–188, 190, 251, 267
quantum mechanics 10, 16, 64, 70, 74–75, 77, 79, 136, 169–171, 173, 175–176, 178–179, 184–185, 187, 190, 197–198, 200–202, 207, 229, 236–238, 250–251, 266, 273–275, 280, 285, 287–288, 290–291, 317, 352–353, 373
  Copenhagen interpretation 69
  Everettian quantum mechanics 16, 187, 224, 250–253, 257, 259, 266, 268
  many-worlds interpretation 257, 267–268
  Relative Facts Interpretation 257–258, 268
quantum nonlocality 228, 239–240, 242
quantum potential 227–229, 231–336, 239–243
quantum statistics 9, 74, 177, 190, 209–211, 277–278, 280, 283–285, 292, 382
quark 183
quasi-set theory 73, 77–79, 389

renormalization group 203
replicator 92, 104, 111–118
reproducer 91, 103–104, 106, 108–109, 115–119, 377, 386
reproduction 86–87, 89–100, 103, 105–106, 111, 116, 119, 125, 149, 161, 305, 307, 334–335, 377

selection, natural 6, 38, 86, 103–119, 126, 135, 141, 149, 151–152, 155, 157, 289, 295, 339, 353, 378
singularity 31, 37
sociology 4
soliton 326–330, 339
species 30, 34, 38, 86–89, 103–119, 122, 156, 193, 289, 295–310, 331–338, 353, 386
  species as individuals 297–299
Standard Model 183, 287
stem cells 122–140, 309
  modeling approach to stem cells 127
"stemness" 139
structuralism 10, 172, 201, 207–208, 212–214, 222, 228–229, 244–245, 278, 281, 283, 320, 371–372, 374, 383, 390–391
substance 11, 27, 49, 85, 196, 319–320, 322, 331, 335, 342–344, 373–374
substantialism 196, 318, 320, 335–336, 344
superorganism 85, 150, 162
symbiosis 87, 97, 108–109–113, 289, 308–309, 335–340, 376–377
symplectic symmetry 229, 236–238, 245

transition
  Egalitarian vs. fraternal 156–158
  major evolutionary 6, 156, 158, 161
transitivity 67, 252–262, 265–266, 300, 302, 304, 306–308, 311–312
transplantation 33, 40, 43, 254, 362
truth tweaker 380, 386

uncertainty principle 10, 132, 136, 238
  for stem cells 132
uniqueness 29–30, 38, 43–44, 77, 165, 169–170, 179, 182, 184–186, 238, 273, 275–276, 282, 291, 324, 328, 352

unity 31, 34, 43, 50–51, 52, 59, 61, 63, 79, 114, 152, 161, 179, 243, 318, 348, 350, 351, 354, 358–359, 362–367
universal, *see* metaphysics, universal
Unruh effect 288

vagueness 43, 62, 310

weak discernibility 8, 48, 53–58, 68, 77, 167–168, 195–202, 204, 219–220, 224, 228, 230, 274, 276, 278–282, 286, 288, 327, 373–374, 382

Zermelo-Fraenkel
  set theory 66–76